Lecture Notes in Computer Science 1726

Edited by G. Goos, J. Hartmanis and J. van Leeuwen

Springer

Berlin
Heidelberg
New York
Barcelona
Hong Kong
London
Milan
Paris
Singapore
Tokyo

Vijay Varadharajan   Yi Mu (Eds.)

Information
and Communication
Security

Second International Conference, ICICS'99
Sydney, Australia, November 9-11, 1999
Proceedings

Springer

Series Editors

Gerhard Goos, Karlsruhe University, Germany
Juris Hartmanis, Cornell University, NY, USA
Jan van Leeuwen, Utrecht University, The Netherlands

Volume Editors

Vijay Varadharajan
Yi Mu
School of Computing and Information Technology
University of Western Sydney NEPEAN
P.O. Box 10, Kingswood, NSW 2747, Australia
E-mail: {vijay/y.mu}@cit.nepean.uws.edu.au

Cataloging-in-Publication data applied for

Die Deutsche Bibliothek - CIP-Einheitsaufnahme

Information and communication security : second international
conference ; proceedings / ICICS '99, Sydney, Australia, November
9 - 11, 1999. Vijay Varadharajan ; Yi Mu (ed.). - Berlin ; Heidelberg ;
New York ; Barcelona ; Hong Kong ; London ; Milan ; Paris ;
Singapore ; Tokyo : Springer, 1999
  (Lecture notes in computer science ; Vol. 1726)
  ISBN 3-540-66682-6

CR Subject Classification (1998): E.3, G.2.1, D.4.6, F.2.1-2, C.2, K.6.5

ISSN 0302-9743
ISBN 3-540-66682-6 Springer-Verlag Berlin Heidelberg New York

Typesetting: Camera-ready by author
SPIN: 10705505   06/3142 – 5 4 3 2 1 0   Printed on acid-free paper

PREFACE

ICICS'99, the Second International Conference on Information and Communication Security, was held in Sydney, Australia, 9-11 November 1999. The conference was sponsored by the Distributed System and Network Security Research Unit, University of Western Sydney, Nepean, the Australian Computer Society, IEEE Computer Chapter (NSW), and Harvey World Travel. I am grateful to all these organizations for their support of the conference.

The conference brought together researchers, designers, implementors and users of information security systems and technologies. A range of aspects was addressed from security theory and modeling to system and protocol designs and implementations to applications and management. The conference consisted of a series of refereed technical papers and invited technical presentations. The program committee invited two distinguished key note speakers. The first keynote speech by Doug McGowan, a Senior Manager from Hewlett-Packard, USA, discussed cryptography in an international setting. Doug described the current status of international cryptography and explored possible future trends and new technologies. The second keynote speech was delivered by Sushil Jajodia of George Mason University, USA. Sushil's talk addressed the protection of critical information systems. He discussed issues and methods for survivability of systems under malicious attacks and proposed a fault-tolerance based approach. The conference also hosted a panel on the currently much debated topic of Internet censorship. The panel addressed the issue of censorship from various viewpoints namely legal, industrial, governmental and technical. It was chaired by Vijay Varadharajan with panel members from the Australian National Office of the Information Economy, Electronic Frontiers Association, the Australian Computer Society, and the Internet Industry Association.

There were 62 technical papers submitted to the conference from an international authorship. The program committee accepted 24 papers and these were presented in eight sessions covering cryptanalysis, language based approach to security, electronic commerce and secret sharing, digital signatures, security protocols, applications, cryptography, and complexity functions. The accepted papers came from a range of countries, including some 7 papers from Australia, 6 from Japan, 3 from the USA, 2 from Sweden, 2 from the UK, 1 from China, France, Spain, and Singapore. I would like to thank the authors of all the papers submitted to the conference, both those whose work is included in the proceedings and those whose work could not be accommodated.

I would like to thank all the people involved in organizing the conference. In particular, I would like to thank the members of the program committee for their effort in reviewing the papers and putting together an excellent program. I would like to thank all the speakers, session chairs, and panelists for their time and effort. Special thanks to members of the organizing committee for their tireless work in organizing and helping with many local details, especially Yi Mu, Irene Ee, Rajan Shankaran, Kenny Nguyen, and Chuan Kun Wu. Finally, I would like to thank all the participants of ICICS'99. I hope that the professional contacts made at this conference, the presentations, and the proceedings have offered you additional insights and ideas that you can apply to your own efforts in information and communication security.

August 1999   Vijay Varadharajan

The Second Conference on Information and Communication Security ICICS'99 (Sydney, Australia)

Sponsored by
Distributed System & Network Security Research Unit, UWSN
Institute of Electrical & Electronics Engineers (IEEE Computer, NSW)
Australian Computer Society
Harvey World Travel (Penrith), Australia

General Chair:

Vijay Varadharajan	*University of Western Sydney*

Program Chairs:

Vijay Varadharajan	*University of Western Sydney*
Josef Pieprzyk	*University of Wollongong*

Program Committee:

Ross Anderson	*Cambridge Uni., UK*
Ed Dawson	*Queensland Uni. of Tech., Australia*
Robert Deng	*KRDL, Singapore*
Yvo Desmedt	*Florida State Uni., USA*
Yong Fei Han	*GemPlus, Singapore*
Kwangjo Kim	*ICU, South Korea*
Wenbo Mao	*HP Labs, UK*
Mitsuru Matsui	*Mitsubishi, Japan*
Cathy Meadows	*NRL, USA*
Chris Mitchell	*London Uni., UK*
Yi Mu	*UWSN, Australia*
Luke O'Connor	*IBM, Switzerland*
Eiji Okamoto	*JAIST, Japan*
Tatsuaki Okamoto	*NTT, Japan*
Josef Pieprzyk	*Wollongong Uni., Australia*
Ravi Sandhu	*George Mason Uni., USA*
Jim Schindler	*HP, USA*
Jennifer Seberry	*Wollongong Uni., Australia*
Vijay Varadharajan	*UWSN, Australia*

CONTENTS

Keynote Speech

Security Protocols

Applications

Cryptography

Complexity and Security Functions

KEYNOTE SPEECH

International Cryptography

Doug McGowan

Hewlett-Packard, USA

Abstract

Cryptography is a fundamental security technology. It is used in applications ranging from authentication and digital signatures to protecting the privacy and integrity of emails and other forms of interactions. Besides a technology, cryptography has become a topic for front page news. Governments around the world are debating and passing laws concerning the export and (sometimes) the use of cryptography. This talk will discuss the current international scene for cryptography technology. We will discuss today's situation with regards to cryptography export as well as future trends. We will also discuss some new technologies that may provide solutions for manufacturers and software developers while complying with international regulations.

Reaction Attacks Against Several Public-Key Cryptosystems

Chris Hall[1], Ian Goldberg[2], and Bruce Schneier[1]

[1] Counterpane Systems
hall,schneier@counterpane.com
101 E. Minnehaha Pkwy
Minneapolis, MN 55419
(612) 832-1098

[2] U.C. at Berkeley
iang@cs.berkeley.edu
Soda Hall
Berkeley, CA 94720-1776

Abstract. We present attacks against the McEliece Public-Key Cryptosystem, the Atjai-Dwork Public-Key Cryptosystem, and variants of those systems. Most of these systems base their security on the apparent intractibility of one or more problems. The attacks we present do not violate the intractibility of the underlying problems, but instead obtain information about the private key or plaintext by watching the reaction of someone decrypting a given ciphertext with the private key. In the case of the McEliece system we must repeat the attack for each ciphertext we wish to decrypt, whereas for the Ajtai-Dwork system we are able to recover the private key.

Keywords: public-key cryptosystems, lattice-based cryptosystems, error-correcting codes, Ajtai-Dwork lattice cryptosystem, McEliece.

1 Introduction

In an attempt to design cryptosystems based upon (believed) intractible problems different from factoring and discrete logarithms, several public-key cryptosystems have been presented, including the two systems in [M78,AD97]. The first system is based upon the intractibility of decoding an arbitrary error-correcting code, and the second upon finding the shortest vector in a lattice. The authors of the former rely upon the fact that the respective problem is known to be **NP**-complete. However, it is not known whether the particular instances used in their cryptosystem are equally difficult to solve. The general shortest vector problem is known to be **NP**-hard, but the lattice-based systems depend on the apparent difficulty of an easier shortest vector lattice problem (the unique shortest vector lattice problem). We refer the reader to [AD97] for more details.

In this paper we present attacks against the McEliece Public-Key Cryptosystem (PKC), a McEliece variant [HR88], the Ajtai-Dwork PKC, and a modified version of the Ajtai-Dwork PKC that appears in [GGH97]. In these attacks an attacker presents the owner of the private key with a ciphertext that may contain one or more errors (that is, the ciphertext may decrypt to a plaintext which fails a simple signature or checksum verification). By watching the reaction of the owner in order to determine whether or not the ciphertext decrypted correctly, the attacker can usually determine information about the plaintext (in the McEliece system) or the private key (in the Ajtai-Dwork systems).

We feel that this is a legitimate class of attacks that one must be careful to guard against when implementing systems that use these ciphers. In the simplest case one may have a tamper-resistance module (such as a smartcard) which contains a copy of the secret key. By feeding erroneous ciphertexts to the card, one may be able to decrypt ciphertexts without the private key or even recover the private key. In other systems one may have to rely upon social engineering to finesse the attacks, but that's a small price if one can recover the private key. The basic principle behind the attack is that someone's reaction to a question often reveals information that they didn't intend to give.

The organization of our paper is as follows. In section 2 we describe the McEliece PKC, as well as a variant of it, and attacks against them. In section 3 we describe the original Ajtai-Dwork PKC as presented in [AD97], a modified version that appears in [GGH97], and attacks against both of these systems. Finally, in section 4 we discuss the properties of these public-key cryptosystems which allowed our attacks to work in order to try and give some design criterion for new public-key cryptosystems so that they will not be vulnerable to the same sort of attack.

2 McEliece

In [M78], McEliece outlined a public-key cryptosystem based upon error correcting codes. A user chooses a $n \times k$ generator matrix G (for a (n, k) error-correcting code which can correct up to t errors), a $k \times k$ non-singular matrix S, a $n \times n$ permutation matrix P, and publishes the matrix $G' = SGP$ as his public key. To encrypt a message a user chooses a random vector Z of weight t and sends:

$$C = MG' + Z.$$

To decrypt the message, the owner of the key computes:

$$CP^{-1} = MSG + ZP^{-1}$$

and corrects the t-bit error ZP^{-1} using the known error-correction algorithm. After that M can be recovered using S^{-1}.

This system, like other systems [HR88,J83,N86], depends on the fact that in the worst case, decoding an arbitrary error-correcting code is NP-complete [BMvT78]. It is hoped that G' represents one of these difficult cases.

Since their introduction, public-key cryptosytems based on error-correcting codes have largely been of theoretical interest. They require enormous keys in order to achieve comparible security to currently implemented public-key cryptosystems based on factoring or the discrete logarithm problem. For example, the purported attack in [KT91] can break a cryptosystem based on a $(1024, 654)$ BCH code in 60 hours. Such a key would require a 654-kilobit generator matrix for part of the public key. This would seem to imply that secure keys would require one or more megabytes of storage for the generator matrix alone. Compare this to a 1024-bit RSA key which needs only a little more than 1Kb of storage and provides excellent security.

Large keys aside, error-correcting code cryptosystems have been touted as a potential saviour of public-key cryptography. These systems are based on the syndrome-decoding problem which was shown to be NP-complete in [BMvT78]. Other public-key cryptosystems, such as RSA, are based on problems such as factoring which are only thought to be difficult. It is possible that someone will discover an efficient factoring algorithm, whereas it would require proving that $P = NP$ to find an efficient general syndrome-decoding algorithm. Note, NP-completeness only implies that the worst case is hard to solve. It may be possible that the instances of problems produced by PKCs such as [M78] are much easier to solve.

Until recently, the two best attacks against McEliece's system appear in [AM87,KT91]. The attack in [AM87] relies on choosing k bits in an n-bit ciphertext that do not contain any errors. Given that the t incorrect bits are unknown, the probability of this event happening is low. However, once it occurs, one can solve for the message M using an algorithm outlined in the paper.

The attack outlined in [KT91] requires $O(n^3)$ operations for an arbitrary (n, k) code. It determines the error vector for the ciphertext C. The basic premise behind the attack is that a polynomial-time algorithm is introduced which will determine the error pattern for a received vector provided that the error pattern has weight at most $[(d-1)/2]$ (where d is the minimum distance of the code). The discovery of this algorithm does not contradict the NP-completeness proof of [BMvT78] because it only corrects error patterns of weight at most $[(d-1)/2]$. If the received vector is further than $[(d-1)/2]$ from every codeword, then this algorithm cannot decode the vector to a codeword. Note, while the basic attack was outlined in [KT91], further information about the attack has failed to appear. It is possible that the authors were not able to make their attack scale as they had wished (when they present it, they had only run it against a toy problem).

An even more recent attack against the McEliece PKC was presented in [B97]. This attack relied about known linear relationships between two different ciphertexts (really their underlying plaintexts) in order to determine the error vectors used in encrypting the plaintexts. The author is quick to point out that the attack does not "break" the cryptosystem in the sense that it does not recover the private key, but it is still an interesting attack to note.

In the next subsections, we present a new attack which determines the error vector using $O(n)$ operations. It is a chosen ciphertext attack and does not require that the attacker see the resulting plaintext. The attacker only needs to be able to determine whether the chosen ciphertext had too many errors to be decoded correctly, or that it decoded to a plaintext other than the plaintext in question. The former can be distinguished in systems which send some sort of message indicating that a retransmission is necessary (such as network protocols). The latter can be distinguished in systems where a randomly chosen ciphertext has only a low probability of decrypting to a meaningful plaintext. In those cases we assume that some sort of retransmission request will also be sent.

2.1 Decoding Algorithms

Our attack rests largely upon one premise:

If a (n, k) error-correcting code is used which can correct t or fewer errors, then a decoder will not attempt to correct a vector which has $t + 1$ or more errors.

There are several decoders for Reed-Solomon and Goppa codes which meet this criterion, including those described in [B73,P75,SKHN76]. A common component of each of these algorithms is that they work to find an error-location polynomial $\sigma(z)$ of degree *at most* t. The decoders then determine the roots of $\sigma(z)$ in order to determine the location of the errors in the received vector (there is a one-to-one correspondance between the roots and the errors). Because $\deg(\sigma(z)) \leq t$, there are at most t roots and hence t errors that can be corrected. This means that these decoding algorithms are not be capabable of correcting $t + 1$ errors.

In principle these error-correcting codes are capable of correcting $t + 1$ errors *in some cases*. However, the conventional error-correction algorithms, based on the Berlekamp-Massey algorithm [B67,B68,M69,P75] and Euclid's Algorithm [SKHN74,SKHN76], are not capabable of correcting $t + 1$ errors, as they are of the above type. When a decoding algorithm is capable of decoding $t + 1$ errors, we note that it is often possible to recover part of the permutation matrix P. This follows by observing which of the permuted error vectors of Hamming-Weight $t + 1$ are correctable and then determining the (possibly) unique permutation matrix P which would produce the permuted error vectors. One simply needs to observe the probability that bit i will be corrected and match it to a bit in the distribution on the unpermuted code.

Note, at least one system explicitly states that vectors with more than t errors should be ignored. For example, see [HW92].

2.2 Removing an Error Vector

Suppose that we have a ciphertext C which we wish to decrypt to its corresponding plaintext M. To find the message, we present an algorithm which allows the attacker to determine the error vector Z used to encrypt M. We do this by making small one-bit modifications to C until we obtain a ciphertext C' which either will not decode at all (it contains $t+1$ or more errors) or decrypts to a different plaintext (contains t or fewer errors, but decodes to the wrong codeword).

Once we obtain such a ciphertext we can flip each bit once in n different trials, and see whether or not the resulting ciphertext decrypts correctly. If it does, then we know that we have found an erroneous bit and we keep track of it. Otherwise we know that the bit is correct. Once all erroneous bits have been found they can be flipped to produce $C - Z$.

We now give the algorithms used in full:

Algorithm A: *Determining C' (a modified ciphertext with $t+1$ errors)*

1. Let $i = 1$.
2. Flip bits 1 through i of C to form C'.
3. Have C' decrypted and determine whether or not it decrypted to M. Based on a comment made in the introduction, this should be possible (we watch the reaction of the key owner).
4. If C' did not decrypt correctly, halt the procedure and continue onto the next algorithm. Otherwise, increment i and goto step 2.

This algorithm will halt within $2t+1$ steps for in the worst case we will "correct" the t incorrect bits and introduce $t+1$ additional errors (we never correct an error that we introduce).

Once this algorithm halts we know that we have a vector C' with exactly $t+1$ errors. If it had fewer, then the decoder would have succeeded in removing the errors and decrypting to M. If it had more ($t+j$ errors for $j > 1$), then at some previous trial we must have had a vector C' with $t+1$ errors (because we know that C has at most t errors and we made one bit modifications to obtain C'). However, that vector should not have decrypted to M because that would violate the known properties of the decoder. Therefore we would have stopped the procedure sooner. So we must have that C' is a message with exactly $t+1$ errors. Note, $t < n/2$ for these error correction codes.

We then continue with the next algorithm:

Algorithm B: *(determining which bits of C' are errors)*

1. Let $i = 1$.
2. Flip bit i of C' to form C''.
3. Have C'' decrypted and determine whether or not it decrypted to M.
4. If C'' decrypted correctly, then bit i is in error. Record i, add 1 to i, and goto step 2 if $i \leq n$. Otherwise halt.

In the worst case we only need to examine the first $n-1$ bits of C'. Hence we will require at most $n-1$ queries.

Once this algorithm halts, we can flip all those bits of C' recorded in step 4 of the above algorithm. This will result in a ciphertext without any errors.

It works because any C'' which decrypts correctly must have at most t errors. However, it has Hamming Distance 1 from C' which we know has $t + 1$ errors, so C'' must have exactly t errors. This means that the bit we flipped was an incorrect bit. If C'' does not decrypt correctly, then it must have $t + 2$ bit errors and hence we introduced another error when we flipped bit i.

The first algorithm will halt within $2t + 1$ steps and the second will require at most $n - 1$ steps. So the total attack takes at most $n + 2t$ queries and comparable time.

2.3 Attacking McEliece

Once we have executed the algorithms described in the previous section, we will obtain a ciphertext $C - Z$ which is error-free and we can then solve for the message M using the method outlined in [AM87]. Let C_c denote the corrected ciphertext vector.

Briefly, the algorithm selects k bits from the vector C_c to form C'_c and the corresponding k columns from the generator matrix G. The resulting matrix G' will be a $k \times k$ matrix and we know that

$$C'_c = M G'$$

so we can solve for M by inverting G'

$$M = C'_c G'^{-1}.$$

The attack as it is presented in [AM87] requires that the k selected bits be error-free in order to recover M. Consequently their attack requires $C(n, k)/C(n-t, k)$ times as much work since that is how many G' they must try on average. Others have improved on this attack, including [LB88,vT90]. However, the attacks are still expensive.

2.4 Attacking Hwang-Rao

The algorithms described in section 2.1 can also be used to attack the system described in [HR88]. The system is very similar to the McEliece cryptosystem in that it adds a random error vector E to a codeword prior to transmission. Again, it is assumed that decoding the resulting codeword is as hard as decoding an aribtrary error-correcting code.

Once we have managed to remove the error-vector from the ciphertext, we can apply an attack suggested by the Hwang and Rao in their paper. There are a couple of different versions of their system described in their paper and corresponding attacks are also described. We refer the reader to [HR88] for details.

3 Ajtai-Dwork

In [AD97] the authors present a PKC based on the assumption that the Unique Shortest Vector Problem[1] (SVP) is computationally infeasible in the worst-case

---

[1] Unique up to a polynomial factor.

complexity. Briefly, a public key in the system consists of $m + n$ n-dimensional vectors $\mathbf{w} = (\mathbf{w}_1, \ldots, \mathbf{w}_n)$ and $\mathbf{v} = (\mathbf{v}_1, \ldots, \mathbf{v}_m)$. How these vectors are chosen is beyond the scope of this paper, but details can be found in [AD97] and [GGH97].

Encryption is simple in this system, but only one bit can be encrypted at a time. However, there are multiple encryptions of both '0' and '1', and under the asumption that SVP is infeasible, it is impossible to distinguish between two ciphertexts without knowing u. To encrypt a '0' a user chooses $b_1, \ldots, b_m \in \{0,1\}$ and computes

$$x = (\sum_{i=1}^{m}(b_i \cdot v_i)) \bmod P(\mathbf{w})$$

where $P(\mathbf{w})$ is the parallelepiped formed by the w_i. To encrypt a '1' a user chooses x uniformly in the parallelepiped $P(\mathbf{w})$.

To decrypt a message, the user computes $\tau = \,<x,u>$. If τ is within $1/n$ of an integer, then the ciphertext is decrypted as '0'. Otherwise the ciphertext is decrypted as '1'. It is interesting to note that an encrypted '1' will decrypt to a '0' with probability roughly $2/n$. However, such a point is beyond the scope of this paper (although it is addressed by the authors in [GGH97] and is the motivation for their modification).

We have not included our chosen-ciphertext attack against Ajtai-Dwork which allows us to recover the private key. It is a slightly more complicated version of the attack described in the next section and we leave it as an exercise to the reader to extend the attack.

3.1 Attacking the GGH Variant

In [GGH97] the authors present a modification of the Ajtai-Dwork PKC. In the original Ajtai-Dwork system as presented in [AD97], a '0' will always decrypt to a '0'. However, there is a $2/n$ probability that a '1' will decrypt to a '0' (where n is the dimension of the associated lattice). The enhanced system eliminates this problem by first specifying a vector v_{i_1} such that $<u, v_{i_1}>$ is an odd integer.

To encrypt '0', one still uniformly selects $b_1, \ldots, b_m \in \{0,1\}$ and reduces the vector $\sum_{i=1}^{m} b_i \cdot v_i$ modulo the parallelepiped $P(\mathbf{w})$. The resulting vector x is the ciphertext. To encrypt '1', one uniformly selects $b_1, \ldots, b_m \in \{0,1\}$ and reduces the vector $\frac{1}{2}v_{i_1} + \sum_{i=1}^{m} b_i \cdot v_i$ modulo the parallelepiped $P(\mathbf{w})$. Again, the resulting vecotr x is the ciphertext.

To decrypt, the user computes $\tau = \,<u,x>$ and decrypts the ciphertext as '0' if τ is within $1/4$ of some integer and decrypts the ciphertext as '1' otherwise. Our attack makes use of an oracle $O(v)$ which computes $< u, v >$ and returns the resulting plaintext. Note, such an oracle exists since we could send a legitimate message to a user which has a MAC computed as if the ciphertext in question decrypted to a '0'. If the ciphertext decrypts correctly, then we know that $O(v) = $'0'. Otherwise we know that $O(v) = $'1'.

Given that we can send a user a vector v such that $v_i = x$ for some non-zero x and $v_j = 0$ for $i \neq j$, we can focus on one component of u at a time. Hence we assume a simplified oracle $O_i(x)$ which computes $u_i \cdot x$ and returns '0' (resp. '1') if the result is (resp. is not) within 1/4 of an integer. Assuming that $|u_i| = d_0.d_1 \ldots d_r$ we can determine $|u_i|$ using the following algorithm:

Algorithm D: *(determining $|u_i|$)*

1. Let $j = 0$.
2. If $O_i(2^{j+1}) = $ '0', then $d_j d_{j+1} \in \{00, 11\}$. Otherwise $d_j d_{j+1} \in \{01, 10\}$.
3. Let $j = j + 1$. If $j \leq r$ goto step 2.
4. Pick $d_0 = 0$ and choose the remaining d_j appropriately. [2]

This algorithm will determine $|u_i|$ within 2^{-r} of the correct value. That's because $1.0 = 0.\bar{1}$ in binary and our test (in step 2) may choose the alternate expansion for the remaining digits of $|u_i|$.

It's important to understand that this attack determines the projection of the hyperplanes onto all of the coordinate axis. For any particular axis it sets all other coordinates to zero and performs a binary search on the boundary midway between the hyperplane through the origin and the next hyperplane along the axis. However, there's nothing that says the attacker must use the same notion of coordinates axis as everyone else. In fact, she could rotate the axis and perform a binary search on the modified axis (that's in fact what we do below in order to determine the sign of the u_i). Therefore one cannot counter this attack simply by refusing to decrypt all ciphertext such that $u_i = 0$ for all but one i.

Once we determine $|u_i|$ for all i it remains to determine the respective signs. Since the private key u is equivalent to the private key $-u$ we assume without loss of generality that $u_1 \geq 0$ (more properly that the first non-zero $u_i > 0$, but we can permute the vectors and bases appropriately so that $u_1 > 0$). Given $|u_j|$ for some $j > i$, we can compute $u_i + u_j$ for both values of the sign for u_j. The result will be different in each case and let k be the first bit that both sums differ in. Then by computing

$$O(0, \ldots, 2^{k-1} \cdot u_i, \ldots, 2^{k-1} \cdot u_j, \ldots)$$

we can determine the sign of u_j. That is, we use the oracle to determine which sum is correct when considering bit k and hence whether $u_j > 0$ or $u_j < 0$.

Using this the oracle $O(v)$ (which was used to construct $O_i(x)$) we are able to compute u. Notice that each invocation of the oracle returns one bit of information. Correspondingly we use each invocation of the oracle to determine one bit of the private key. Hence our attack determines the private key u using the information theoretic minimum number of oracle invocations. Therefore we know that this is the best possible attack against u using an oracle.

We should point out that this is a *non-adaptive* attack. Unfortunately, one must modify the attack to use one adaptation in order to achieve the information theoretically minimal attack. This is due to the fact that one must see

---

[2] If $|u_i| = 1$ for some i, then $u_j = 0$ for all $j \neq i$. This situation occurs with negligible probability, but it is also easy to detect, so we assume that $|u_i| < 1$ for all i and hence $d_0 = 0$.

the $|u_i|$ before one can determine ciphertexts which will give their signs. From a reaction attack standpoint, this is a minute difference since we gauge the reaction to each ciphertext separately. However, it makes a difference from a chosen-ciphertext standpoint because non-adaptive attacks are considered more powerful than adaptive attacks.

4 Future Systems

The success of our attacks rely upon a common weakness in the PKCs we examined: given a ciphertext C, an attacker can produce a second ciphertext C' which has non-neglible probabilities of decoding to the same plaintext and to a different plaintext. For the McEliece system of [M78], whether or not C' decodes correctly gives information For the Ajtai-Dwork and Goldreich-Goldwasser-Halevi systems of [AD97,GGH97], whether or not C' decodes correctly gives information about the private key.

Each of these systems could be considered a closest-point cryptosystem. That is, the ability to decode a ciphertext depends upon the ability to determine the closest "point" to the ciphertext in some linear space. For error-correcting code systems, this equates to the ability to determine the closest codeword in the linear space of codewords. For lattice-based systems, this equates to finding the closest point in a lattice. In all of these systems one could consider the class of ciphertexts corresponding to a particular plaintext to be a sphere surrounding a point in the respective space (where the boundaries of the sphere are determined by an appropriate distance metric).

By examining ciphertexts which are close to each other in the space (but possibly in different classes) we can determine the boundaries of this sphere and hence the center of the sphere. For some of the systems this will only give us the closest point in the respective space. In those systems, an attacker can already produce points in the space at will using the public key. So determining the closest point doesn't give any more information about the private key that the public key does. For other systems, the security of the system relies upon the inability to determine points in the respective space. Hence the ability to determine a closest point leads to the ability to determine the private key.

We feel that any other public-key cryptosystem with these properties will be vulnerable to the same sorts of attacks we present here.

5 Conclusion

We feel that the existance of these attacks effectively limits these ciphers to theoretical considerations only. That is, any implementation of the ciphers will be subject to the attacks we present and hence not safe. We should point out that we have *not* broken these cryptosystems in the sense that we are able to mathematically derive the private key. Instead our attacks rely upon the fact that someone within the system knows the private key (and hence the answer to the intractible problem). Therefore we would encourage further research into the properties of these cryptosystems in order to determine if a modified cryptosystem can be designed with the same theoretical basis and that is also safe against our attacks.

References

[AM87]   C. M. Adams, H. Meijer, "Security-Related Comments Regarding McEliece's Public-Key Cryptosystem," *Advances in Cryptology—Proceedings of CRYPTO '87*, Springer-Verlag, 1988, pp.224–230.

[AD97]   M. Ajtai, C. Dwork, "A Public-Key Cryptosystem with Worst-Case/Average-Case Equivalence," In *29th ACM Symposium on Theory of Computing*, 1997, pp. 284–293.

[B67]   E.R. Berlekamp, "Nonbinary BCH Decoding," paper presented at the 1967 International Symposium on Information Theory, San Remo, Italy.

[B68]   E.R. Berlekamp, *Algebraic Coding Theory*, New York: mcGraw-Hill, 1968.

[B73]   E.R. Berlekamp, "Goppa Codes," *IEEE Transactions on Information Theory*, Vol. IT-19, No. 5, pp. 590–592, September 1973.

[BMvT78]   E.R. Berlekamp, R.J. McEliece, H. van Tilborg, "On the Inherent Intractability of Certain Coding Problems," *IEEE Transactions on Information Theory*, Vol. 24, 1978, pp. 384–386.

[B97]   T. Berson, "Failure of the McEliece Public-Key Cryptosystem Under Message-Resend and Related-Message Attack," *Advances in Cryptology—CRYPTO '97 Proceedings*, Springer-Verlag, 1997, pp. 213–220.

[GGH97]   O. Goldreich, S. Goldwasser, S. Halevi, "Eliminating Errors in the Ajtai-Dwork Cryptosystem," *Advances in Cryptology—CRYPTO '97 Proceedings*, Springer-Verlag, 1997, pp. 105–111.

[HR88]   T. Hwang, T.R.N. Rao, "Secret Error-Correcting Codes (SECC)," *Advances in Cryptology—CRYPTO '88 Proceedings*, Springer-Verlag, 1990, pp. 540–563.

[HW92]   L. Harn, D.C. Wang, "Cryptanalysis and Modification of Digital Signature Scheme Based on Error-Correcting Codes," *Electronics Letters*, v. 28, n. 2, 10 Jan 1992, p. 157–159.

[J83]   J.P. Jordan, "A Variant of a Public-Key Cryptosystem Based on Goppa Codes," *Sigact News*, 1983, pp. 61–66.

[KT91]   V.I. Korzhik, A.I. Turkin, "Cryptanalysis of McEliece's Public Key Cryptosystem," *Advances in Cryptology—EUROCRYPT '91 Proceedings*, Springer–Verlag, 1991, pp. 68–70.

[LB88]   P.J. Lee, E.F. Brickell, "An Observation on the Security of McEliece's Public-Key Cryptosystem," *Advances in Cryptology—EUROCRYPT '88 Proceedings*, Springer–Verlag, 1988, pp. 275–280.

[LDW94]   Y.X. Li, R.H. Deng, X.M. Wang, "On the Equivalence of McEliece's and Niederreiter's Public-Key Cryptosystems," *IEEE Transactions on Information Theory*, Vol. 40, 1994, pp. 271–273.

[M69]   J.L. Massey, "Shift Register Synthesis and BCH Decoding," *IEEE Transactions on Information Theory*, Vol. IT-15, No. 1, pp. 122–127, Jan. 1969.

[M78]   R.J. McEliece, "A Public-Key Cryptosystem Based on Algebraic Coding Theory," *Deep Space Network Progress Report 42-44*, Jet Propulsion Laboratory, California Institute of Technology, 1978, pp. 104–113.

[N86]   H. Niederreiter, "Knapsack-Type Cryptosystems and Algebraic Coding Theory," *Problems of Control and Information Theory*, v. 15, n. 2, 1986, pp. 159–166.

[P75]   N.J. Patterson, "The Algebraic Decoding of Goppa Codes," *IEEE Transactions on Information Theory*, Vol. IT-21, No. 2, pp. 203–207, March 1975.

[SKHN74] Y. Sugiyama, M. Kasahara, S. Hirasawa, T. Namekawa, "A method for solving the key equation for decoding Goppa codes," presented at the IEEE Int. Symp. Information Theory, Notre Dame, Ind., Oct. 27–31, 1974.

[SKHN76] Y. Sagiyama, M. Kasahara, S. Hirasawa, T. Namekawa, "An Erasures-and-Errors Decoding Algorithm for Goppa Codes," *IEEE Transactions on Information Theory*, pp. 238–241, March 1976.

[vT90]   J. van Tilburg, "On the McEliece Cryptosystem," *Advances in Cryptology—CRYPTO '88 Proceedings*, Springer-Verlag, 1990, pp. 119–131.

Cryptanalysis of Some AES Candidate Algorithms

Wenling Wu   Bao Li   Denguo Feng   Sihan Qing

Engineering Research Center for Information Security Technology,
Chinese Academy of Sciences,   Beijing 100080, China
{wwl, lb, fdg, qsh}@ercist.iscas.ac.cn

Abstract: *Decorrelated Fast Cipher(DFC),LOKI97 and MAGENTA are candidates for the Advanced Encryption Standard(AES). In this paper we analyze their security respectively. Since cryptographic properties of the confusion permutation of DFC are weak, its security depends on its subkey addition completely. It seems that LOKI97 is not stronger than LOKI89 and LOKI91. Because of characteristics of the round function, LOKI97 is susceptible to linear cryptanalysis for some keys. MAGENTA is a Feistel cipher with only 6 rounds and its key schedule is very simple. Based on these limitations, we present a chosen plaintext attack against it using 2^{64} chosen plaintexts. In conclusion, we do not think DFC, LOKI97 and MAGENTA are qualified candidates for the AES.*

1   Introduction

For many applications, the Data Encryption Standard algorithm is nearing the end of its lifetime. Its 56-bit key is too small, as shown by a recent distributed key search process [1]. Although triple-DES can solve the key length problem, the DES algorithm was primarily designed for hardware encryption, yet the great majority of its applications today implement it in software, where it is relatively inefficient.

For these reasons, the US National Institute of Standards and Technology (NIST) has issued a call for a successor algorithm, to be called the Advanced Encryption Standard or AES. Decorrelated Fast Cipher (DFC), LOKI97 and MAGENTA are candidates for AES which have been issued by NIST. They have a 128- bit block length and a 256-bit key length (though keys of 128 and 192 bits must also be supported).

This paper is organized as follows: Section 2, 3 and 4 analyze DFC, LOKI97, and MAGENTA respectively. Section 5 contains our conclusions.

2   Decorrelated Fast Cipher

2.1   The structure of DFC

DFC encrypts a 128-bits plaintext block, which is split into two 64-bit halves. In each round the right half and a 128-bit subkey are fed into the function F. The output of F is XORed with the left half, then the halves are swapped. This process is repeated for all but the final round, where the final swap is left out. At the end of the rounds, the halves are concatenated to form the ciphertext block. Decryption follows the same procedure, except that the subkeys are used in reverse order.

2.2 The F-function

The F function (as for "Round Function") is fed with one 128 -bit parameter (subkey), or equivalently two 64-bit parameters: an "a-parameter" and a "b-parameter". It processes a 64-bit input x and outputs a 64-bit string.

$$F_{a/b}(X) = CP(|((ax + b)mod\ 2^{64} + 13)mod\ 2^{64}|_{64})$$

where CP is a permutation over the set of all 64-bit strings.

2.3 The CP permutation

The CP permutation (as for "confusion Permutation") uses a look-up table RT (as for "Round table") which takes a 6-bit integer as input and provides a 32-bit sting output.
Let $y = y_1|y_r$ be the input of CP where y_1 and y_r are two 32-bit strings. We define

$$CP(y) = |((y_r \oplus RT(trunc_6(y_1)))|(y_1 \oplus KC) + KD)mod\ 2^{64}|_{64},$$

where $KC = eb64749a$ and $KD = 86d1bf275b9b241d$.

2.4 Cryptographic properties of CP permutation

Let

$$CP = (f_0, f_1, ..., f_{63}) : GF(2)^{64} \rightarrow GF(2)^{64}$$
$$X \rightarrow Z$$

Then

$$z_0 = f_0(x_0, ..., x_{63})$$
$$z_1 = f_1(x_0, ..., x_{31}, x_{33}, ..., x_{63})$$
$$z_2 = f_2(x_0, ..., x_{31}, x_{34}, ..., x_{63})$$
......
$$z_i = f_i(x_0, ..., x_{31}, x_{31+i+1}, ..., x_{63})$$
......
$$z_{31} = f_{31}(x_0, ..., x_{31}, x_{63})$$
$$z_{32} = f_{32}(x_0, ..., x_{31})$$
$$z_{33} = f_{33}(x_1, ..., x_{31})$$
......
$$z_j = f_j(x_{j-32}, ...x_{31})$$
......
$$z_{62} = f_{62}(x_{30}, x_{31}) = x_{31} \oplus x_{30} \oplus 1$$
$$z_{63} = f_{63}(x_{31}) = x_{31} \oplus 1$$

We have the following result:

Therom 1. 1) CP is not complete,

        2) The nonlinearity of CP is 0,

        3) The differentity of CP is 1.

Proof: 1) and 2) are from $z_{63} = x_{31} \oplus 1$.

    3) Let $e_i = (0, ..., 0, 1, 0, ..., 0)$, $CP(X) \oplus CP(X \oplus e_{32}) = e_1$. Hence, the differentity of CP is 1.

2.5 VDFC-variant of DFC

The structure of VDFC is the same as DFC, its round function is $CP(X, K_i) = CP(X \oplus K_i)$, K_i is the subkey of the ith round. The difference between DFC and VDFC is only subkey addition.

Lemma 1. Let $G(X) = X + KD : GF(2)^{64} \to GF(2)^{64}$, then

        1) $|\{X \in GF(2)^{64} : G(X) \oplus G(X \oplus e_{24}) = e_{24}\}| \geq 5 \times 2^{61}$,

        2) $|\{X \in GF(2)^{64} : G(X) \oplus G(X \oplus e_{56}) = e_{56}\}| \geq 5 \times 2^{61}$.

Proof: Let $KD = 86d1bf275b9b241d$,

$G(X) = A + x_{24}2^{39} + x_{25}2^{38} + (x_{26}+1)2^{37} + x_{27}2^{36} + x_{28}2^{35} + (x_{29}+1)2^{34} + B$,

$G(X \oplus e_{24}) = A + (x^{24} \oplus 1)2^{39} + x_{25}2^{38} + (x_{26}+1)2^{37} + x_{27}2^{36} + x_{28}2^{35} + (x_{29}+1)2^{34} + B$.

When $x_{25} = 0$, we have

$(x_{26}+1)2^{37} + x_{27}2^{36} + x_{28}2^{35} + (x_{29}+1)2^{34} + B < 2^{38} + 2^{36} + 2^{35} + 2^{36} < 2^{39}$.

Thus, $G(X) \oplus G(X \oplus e_{24}) = e_{24}$.

When $x_{26} = x_{27} = 0$, similarly, we have $G(X) \oplus G(X \oplus e_{24}) = e_{24}$.

Therefore, $\{X \in GF(2)^{64}| : G(X) \oplus G(X \oplus e_{24}) = e_{24}\}| \geq 2^{63} + 2^{61} = 5 \times 2^{61}$.

Proof of 2) is similar to 1).

Table 1 gives its differential characteristics and probability. It is clear that for a 8-round VDFC with a fixed key the probability of a differential is larger than $(5/8)^5$, hence VDFC does not resist to differential cryptanalysis.

Round number	Differential characteristics	Probability
1	$(e_{56}, 0) \to (0, e_{56})$	1
2	$(0, e_{56}) \to (e_{56}, e_{24})$	5/8
3	$(e_{56}, e_{24}) \to (e_{24}, 0)$	5/8
4	$(e_{24}, 0) \to (0, e_{24})$	1
5	$(0, e_{24}) \to (e_{24}, e_{56})$	5/8
6	$(e_{24}, e_{56}) \to (e_{56}, 0)$	5/8

Table 1

3 LOKI97

LOKI97 is a new block cipher which has evolved from the earlier LOKI89 and LOKI91. LOKI89 is a 64-bit, 16 round, symmetric block cipher with a 64-bit key schedule. It was originally designed by L. Brown, J. Pieprzyk and J. Seberry in 1990, and later re-designed (and renamed LOKI91), with improved resistance to differential cryptanalysis, by L. Brown, M. Kwan, J. Pieprzyk and J. Seberry. LOKI89 was analyzed by Biham and Shamir [7] who found that whilst reduced round versions of LOKI89 could be susceptible to differential cryptanalysis, the full 16 round version was not. Tokita Sorimachi and Matsui[8] have examined LOKI91 and have concluded that its 12 or more round versions are immune to linear cryptanalysis. The LOKI97 S-boxes are designed to have maximum nonlinearity $N(S) = 2^{n-1} - 2^{\frac{n-1}{2}}$. S_1 may be approximated by a linear approximation with the probability

$$\frac{1}{2} - 2^{-7} \le p \le \frac{1}{2} + 2^{-7}.$$

S_2 may be approximated by a linear approximation with the probability

$$\frac{1}{2} - 2^{-6} \le p \le \frac{1}{2} + 2^{-6}.$$

The following results are given by [3].
The best linear approximation of a single round is

$$\frac{1}{2} - 2^{-11} \le p_{best} \le \frac{1}{2} + 2^{-11}.$$

The best linear approximation for 14 rounds is

$$\frac{1}{2} - 2^{-141} \le p_{best} \le \frac{1}{2} + 2^{-141}.$$

The best linear approximation for full 16 rounds is

$$\frac{1}{2} - 2^{-161} \le p_{best} \le \frac{1}{2} + 2^{-161}.$$

Applying the results above and from [4], it is shown that it needs 2^{232} known-plaintexts to attack LOKI97 and that LOKI97 is immune to linear cryptanalysis. In this paper, we will point out that LOKI97 is not immune to linear cryptanalysis for some keys according to the characteristics of the round function of LOKI97.

3.1 The Structure

LOKI97 is a new 16 round Feistel cipher. It encrypts a 128-bit plaintext block to create a 128-bit ciphertext block. The 128-bit plaintext input value $[L|R]$ is split into two 64-bit words:

$$L_0 = L$$
$$R_0 = R$$

These are then processed through 16 round ($i = 1, ..., 16$) of a balanced Feistel network:

$$R_i = L_{i-1} \oplus F(R_{i-1} + K_{3i-2}, K_{3i-1})$$
$$L_i = R_{i-1} + K_{3i-2} + K_{3i}$$

The resulting 128-bit ciphertext output value is composed of :

$$[R_{16}|L_{16}].$$

The decryption computation involves splitting the ciphertext into two 64-bit words:

$$[R_{16}|L_{16}].$$

and then running the rounds in reverse, ie. use 16 rounds ($i = 16, ..., 1$)

$$L_{i-1} = R_i \oplus F(L_i - K_{3i}, K_{3i-1})$$
$$R_{i-1} = L_i - K_{3i-2} - K_{3i}$$

The decrypted 128-bit plaintext value is given by

$$[L_0|R_0].$$

3.2 Round Function

The round function $F : F_2^{64} \times F_2^{64} \to F_2^{64}$ is specified as follows:
$$F(A, B) = Sb(P(Sa(E(KP(A, B))))), B).$$

$KP(A, B)$ a very simple keyed permutation which splits its 64-bit input A into two 32-bit words, and uses the lower (rightmost)32-bit of input B to decide whether to exchange corresponding pairs of bits in these words (if key bit is 1), or not (if key bit is 0).

E an expansion function which creates a 96 bit output value from the 64 inputs bits as follows:
$$[4 - 0, 63 - 56|58 - 48|52 - 40|42 - 32|34 - 24|28 - 16|18 - 8|12 - 0].$$

S_a, S_b are two columns of S-boxes, composed by concatenating boxes S_1 and S_2, with $S_a = [S_1, S_2, S_1, S_2, S_2, S_1, S_2, S_1]$, and $S_b = [S_2, S_2, S_1, S_1, S_2, S_2, S_1, S_1]$. In S_a the inputs are data+key from the output of E, whilst in S_b the upper bits are pure key bits (from the lower, rightmost 32-bits of B).

P maps input bits $[63 - 0]$ to output bits:

$$[56, 48, 40, 32, 24, 16, 08, 00, 57, 49, 41, 33, 25, 17, 09, 01,$$
$$58, 50, 42, 34, 26, 18, 10, 02, 59, 51, 43, 35, 27, 19, 11, 03,$$
$$60, 52, 44, 36, 28, 20, 12, 04, 61, 53, 45, 37, 29, 21, 13, 05,$$
$$62, 54, 46, 38, 30, 22, 14, 06, 63, 55, 47, 39, 31, 23, 15, 07]$$

ie. input bit 63 goes to output bit 56, input 62 to output 48 etc.

S-boxes   The S-boxes chosen for LOKI97 use cubing in a Galois field $GF(2^n)$ with n odd. The S-box functions are:

$$S_1[X] = ((X \, xor \, 1FFF)^3 \, mod \, 2911) \& FF \quad in \; GF(2^{13})$$
$$S_2[X] = ((X \, xor \, 7FF)^3 \, mod \, AA7) \& FF \quad in \; GF(2^{11})$$

3.3 Linear approximations of LOKI97

Let $S_1(x_{12}, ..., x_2, x_1, x_0) = (f_7, f_6, f_5, f_4, f_3, f_2, f_1, f_0) : F_2^{13} \to F_2^8$. By computation, we can deduce that:

$$f_0 = x_{12} \oplus x_{12}x_{11} \oplus x_{10} \oplus x_9 \oplus x_{12}x_9 \oplus x_9x_8 \oplus x_7 \oplus x_{11}x_7 \oplus x_{10}x_7 \oplus x_6$$
$$\oplus x_{12}x_6 \oplus x_{11}x_6 \oplus x_{10}x_6 \oplus x_9x_6 \oplus x_7x_8 \oplus x_{12}x_7 \oplus x_9x_7 \oplus x_6x_5 \oplus x_{12}x_4$$
$$\oplus x_{10}x_4 \oplus x_9x_4 \oplus x_5x_4 \oplus x_{11}x_3 \oplus x_8x_3 \oplus x_6x_3 \oplus x_5x_3 \oplus x_{12}x_2 \oplus x_{11}x_2 \oplus$$
$$x_9x_2 \oplus x_8x_2 \oplus x_{11}x_3 \oplus x_{10}x_1 \oplus x_9x_1 \oplus x_8x_1 \oplus x_7x_1 \oplus x_6x_1 \oplus x_9x_0 \oplus 1.$$

Fixing $(x_{12}, x_{11}, x_{10}, x_9, x_8) = i$ ($0 \le i \le 31$), we get Boolean functions $g_i(x_7, x_6, x_5, x_4, x_3, x_2, x_1, x_0)$, and the Hamming weight of g_i is as follows:

$$W_H(g_0) = 144, \; W_H(g_1) = 128, \; W_H(g_2) = 128, \; W_H(g_3) = 128,$$
$$W_H(g_4) = 144, \; W_H(g_5) = 128, \; W_H(g_6) = 128, \; W_H(g_7) = 128,$$
$$W_H(g_8) = 128, \; W_H(g_9) = 144, W_H(g_{10}) = 128, W_H(g_{11}) = 128,$$
$$W_H(g_{12}) = 128, W_H(g_{13}) = 112, W_H(g_{14}) = 128, W_H(g_{15}) = 128,$$
$$W_H(g_{16}) = 128, W_H(g_{17}) = 112, W_H(g_{18}) = 128, W_H(g_{19}) = 128,$$
$$W_H(g_{20}) = 128, W_H(g_{21}) = 144, W_H(g_{22}) = 128, W_H(g_{23}) = 128,$$
$$W_H(g_{24}) = 112, W_H(g_{25}) = 128, W_H(g_{26}) = 128, W_H(g_{27}) = 128,$$
$$W_H(g_{28}) = 112, W_H(g_{29}) = 128, W_H(g_{30}) = 128, W_H(g_{31}) = 128,$$

It is shown that g_0, g_4, g_9,g_{13}, g_{17},g_{21}, g_{24}, g_{28} are unbalanced Boolean functions and their nonlinearities are less than 16.

In S_b layer of the round function, the upper bits are pure key bits(from the lower, rightmost 32-bits of B). So for some keys , applying these properties,we obtain the following linear approximation of the round function $F(X, K) = Y$ with probability $p = \frac{1}{2} + \frac{1}{2^4}$, which involves output bits and key:

$$Y[0] = K[h(36 - 32)] \qquad (1)$$

where $K[h(36 - 32)] = h(k_{36}, k_{35}, k_{34}, k_{33}, k_{32})$.

Let (L_{i-1}, R_{i-1}) and (L_i, R_i) be input and output of the ith round respectively.

$$L_i = R_{i-1} + K_i^1 + K_i^3 \qquad (2)$$
$$R_i = L_{i-1} \oplus F(R_{i-1}, K_i^2) \qquad (3)$$

For equation (2), we have the following linear approximation with probability 1:

$$E_i : \ L_i[0] = R_{i-1}[0] \oplus K_i^1[0] \oplus K_i^3[0].$$

For equation (3), by applying linear approximation (1) of the round function, we construct the following linear approximation with probability $\frac{1}{2} + \frac{1}{2^4}$:

$$D_i : \ R_i[0] = L_{i-1}[0] \oplus K_i^2[h(36 - 32)].$$

Let (L_0, R_0) and (L_{16}, L_{16}) be input and output of 16-round LOKI97 respectively. Through path $(E_1, D_2, E_3, D_4, E_5, D_6, E_7, D_8, E_9, D_{10}, E_{11}, D_{12}, E_{13}, D_{14}, E_{15}, D_{16})$, we have the following linear approximation with probability $\frac{1}{2} + 2^{-25}$:

$$R_0[0] \oplus (\bigoplus_{\substack{1 \leq i \leq 16 \\ i\,odd}} (K_i^1[0] \oplus K_i^3[0])) \oplus (\bigoplus_{\substack{1 \leq i \leq 16 \\ i\,even}} (K_i^2[h(36 - 32)])) = R_{16}[0]. \qquad (4)$$

Through path $(D_1, E_2, D_3, E_4, D_5, E_6, D_7, E_8, D_9, E_{10}, D_{11}, E_{12}, D_{13}, E_{14}, D_{15}, E_{16})$, we obtain the following linear approximation with probability $\frac{1}{2} + 2^{-25}$:

$$L_0[0] \oplus (\bigoplus_{\substack{1 \leq i \leq 16 \\ i\,even}} (K_i^1[0] \oplus K_i^3[0])) \oplus (\bigoplus_{\substack{1 \leq i \leq 16 \\ i\,odd}} (K_i^2[h(36 - 32)])) = L_{16}[0]. \qquad (5)$$

Through path $(E_2, D_3, E_4, D_5, E_6, D_7, E_8, D_9, E_{10}, D_{11}, E_{12}, D_{13}, E_{14}, D_{15}, E_{16})$, we obtain the following linear approximation with probability $\frac{1}{2} + 2^{-21}$:

$$L_0[0] \oplus F(R_0 + K_1^1, K_1^2)[0] \oplus k = L_{16}[0] \qquad (6)$$

where $k = (\bigoplus_{\substack{2 \leq i \leq 16 \\ i\,odd}} (K_i^1[0] \oplus K_i^3[0])) \oplus (\bigoplus_{\substack{2 \leq i \leq 16 \\ i\,even}} (K_i^2[h(36 - 32)]))$.

3.4 Linear cryptanalysis of LOKI97

By (4), we can deduce five key bits of K_i^2 (i even, $1 \leq i \leq 16$), namely, $K_i^2[36]$ - $K_i^2[32]$, with success rate 97.7% using about 2^{50} known-plaintexts.

By (5), we can deduce five key bits of K_i^2 (i odd, $1 \leq i \leq 16$), namely, $K_i^2[36]$ - $K_i^2[32]$, with success rate 97.7% using about 2^{50} known-plaintexts.

Now we indicate the process for (4).

Step 1. for any even $j(1 \leq j \leq 16)$,fixing h of $K_i^2[h(36 - 32)](i \neq j)$ in (4).

Step 2. let $K_j^2[h(36 - 32)] = K_j^2[32]$, deducing $k_0 = k' \oplus K_j^2[32]$ by (4).Where

$$k' = (\bigoplus_{\substack{1 \leq i \leq 16 \\ i\,odd}} (K_i^1[0] \oplus K_i^3[0])) \oplus (\bigoplus_{\substack{i \neq j \\ i\,even}} (K_i^2[h(36 - 32)])),$$

$$K_j^2[h(36 - 32)] = K_j^2[32].$$

Step 3. let $K_j^2[h(36-32)] = K_j^2[33]$, deducing $k_1 = k' \oplus K_j^2[33]$ by (4).

Step 4. let $K_j^2[h(36-32)] = K_j^2[32,33]$,deducing $k_2 = k' \oplus K_j^2[32,33]$ by (4).

Step 5. let $K_j^2[h(36-32)] = K_j^2[32,33,34]$, deducing $k_3 = k' \oplus K_j^2[32,33,34]$ by (4).

Step 6. let $K_j^2[h(36-32)] = K_j^2[32,33,34,35]$, deducing $k_4 = k' \oplus K_j^2[32,33,34,35]$ by (4).

Step 7. let $K_j^2[h(36-32)] = K_j^2[32,33,34,35,36]$, deducing $k_5 = k' \oplus K_j^2[32,33,34,35,36]$ by (4).

Step 8. counting $K_j^2[32] = k_0 \oplus k_2$, $K_j^2[33] = k_1 \oplus k_2$, $K_j^2[34] = k_3 \oplus k_2$, $K_j^2[35] = k_3 \oplus k_4$, $K_j^2[36] = k_5 \oplus k_4$.

It follows from the above discussion that 92 key bits of LOKI97 can be deduced with success rate 97.7 % using about 2^{50} known-plaintexts and applying algorithm 1 of Ref.[6] . In the next part, we will attack LOKI97 using some 128-bit keys and applying algorithm 2 of Ref.[6].

Let $K = (K_1, K_2)$, by applying key schedule, we have the following equation:

$$\begin{cases} K_1^1 = K_1 \oplus g(K_2, f(K_2, K_1), f(K_1, K_2)) \\ K_1^2 = K_2 \oplus g(f(K_2, K_1), f(K_1, K_2), K_1^1) \end{cases} \quad (7)$$

Suppose that it is easy to find solutions K_1 and K_2 of equation (7) with known K_1^1 and K_1^2.

$F(R_0 + K_1^1, K_1^2)[0]$ is independent of $K_1^2[37]$ - $K_1^2[63]$, and $K_1^2[32]$ - $K_1^2[36]$ are known, therefore (K_1^1, K_1^2) may have 2^{96} candidates. Firstly, by applying (6) and algorithm 2 of Ref.[6], we deduce K_1^1 and rightmost 32-bits of K_1^2 with success rate 0.967 using about 2^{45} known-plaintexts. Then, we solve equation (7) for every K_1^2 (about 2^{27}) and finally check the result.

4   MAGENTA

4.1   Description of MAGENTA

MAGENTA is a Feistel cipher with 6 or 8 rounds which encryption function can be described as follows:

$$Enc_K(M) = \begin{cases} F_{K_1}(F_{K_1}(F_{K_2}(F_{K_2}(F_{K_1}(F_{K_1}(M)))))) \\ F_{K_1}(F_{K_2}(F_{K_3}(F_{K_3}(F_{K_2}(F_{K_1}(M)))))) \\ F_{K_1}(F_{K_2}(F_{K_3}(F_{K_4}(F_{K_4}(F_{K_3}(F_{K_2}(F_{K_1}(M)))))))) \end{cases}$$

$$K = \begin{cases} (K_1, K_2) \\ (K_1, K_2, K_3) \\ (K_1, K_2, K_3, K_4) \end{cases}$$

where $F_Y(X) = ((x_8, ..., x_{15}), (x_0, ..., x_7) \oplus E(x_8, ..., x_{15}, y_0, ..., y_7))$.

4.2 Differential cryptanalysis of MAGENTA

Now we give an algorithm which can be used to attack MAGENTA having a 128-bit key $K = (K_1, K_2)$ with 2^{64} chosen-plaintexts.

Algorithm 1.

For all 2^{64} possible values K_1,

Step 1. choose plaintext/ciphertext pair (L_0, R_0), (L_6, R_6) and (L_0^*, R_0^*), (L_6^*, R_6^*) satisfying $R_0 \oplus E(L_0 \oplus E(R_0, K_1), K_1) = R_0^* \oplus E(L_0^* \oplus E(R_0^*, K_1), K_1)$.

Step 2. compute $L_2 = L_0 \oplus E(R_0, K_1)$, $L_2^* = L_0^* \oplus E(R_0^*, K_1)$, $L_4 = L_6 \oplus E(R_6 \oplus E(L_6, K_1), K_1)$, $L_4^* = L_6^* \oplus E(R_6^* \oplus E(L_6^*, K_1), K_1)$.

Step 3. if $L_2 \oplus L_2^* = L_4 \oplus L_4^*$, then output K_1.

In order to decide whether output K_1 is the true key, we choose another plaintext/ciphertext pair to test. Hence it needs about 2^{64} chosen-plaintexts to deduce subkey K_1. The hinge of this algorithm is the selection of chosen-plaintext. The approach is as follows.

Step 1. For any given X_1, $X_2 \in F_2^{64}$, compute $E(X_1, K_1) \oplus E(X_2, K_1) = \alpha (\alpha \in F_2^{64}$;

Step 2. choose R_0 and R_0^* satisfying $R_0 \oplus R_0^* = \alpha$;

Step 3. compute (R_0, K_1) and $E(R_0^*, K_1)$;

Step 4. let $L_0 = X_1 \oplus E(R_0, K_1)$, $L_0^* = X_2 \oplus E(R_0^*, K_1)$;

Step 5. output (L_0, R_0) and (L_0^*, R_0^*).

Since $E(L_0 \oplus E(R_0, K_1), K_1) \oplus E(L_0^* \oplus E(R_0^*, K_1), K_1) = E(X_1, K_1) \oplus E(X_2, K_1) = \alpha = R_0 \oplus R_0^*$, (L_0, R_0) and (L_0^*, R_0^*) satisfy

$$R_0 \oplus E(L_0 \oplus E(R_0, K_1), K_1) = R_0^* \oplus E(L_0^* \oplus E(R_0^*, K_1), K_1).$$

5 Concluding remarks

In this paper, we have analyzed DFC, LOKI97 and MAGENTA respectively. Since cryptographic properties of the confusion permutation of DFC are weak, its security depends on its subkey addition completely. It seems that LOKI97 is not stronger than LOKI89 and LOKI91. Because of characteristics of the round function, LOKI97 is susceptible to linear cryptanalysis for some keys. MAGENTA is a Feistel cipher with only 6 rounds and its key schedule is very simple. Based on these limitations, we present a chosen plaintext attack against it using 2^{64} chosen plaintexts. As a result, we do not think DFC, LOKI97 and Magenta are qualified candidates for the AES. Although some similar results were presented in the Proceedings of second AES conference, we had mentioned our results in Ref.[10] before.

References

1. RSA Data Security Inc., http://www.rsa.com
2. Serge Vaudenay, "Decorrelated Fast Cipher", http://www.nist.gov/aes.

3. L.Brown and J.Pieprzyk, "LOKI97", http://www.nist.gov/aes.

4. M.J.Jacobson and Jr.K.Huber, "MAGENTA", http://www.nist.gov/aes.

5. E.Biham and A. Shamir, Differential cryptanalysis of DES-like cryptosystems, *Journal of Cryptology*, 4(1): 3-72, 1991.

6. Mitsuru Matsui, Linear Cryptanalysis Method for DES Cipher, *volume 765 of Lecture Notes in Computer Science*, pages 368-397. Springer-Verlag, 1993.

7. Eli Biham and Adi Shamir, Differential Cryptanalysis Snefru, Kharfe, REDOC-II, LOKI and Lucifer, *volume 576 of Lecture Notes in Computer Science*, pages 156-171. Springer-Verlag, 1991.

8. Toshio Tokita, Tohru Sorimachi and Mitsuru Matsui. Linear cryptanalysis of LOKI and S2DES, *volum 917 of Lecture Notes in Computer Science*, pages 363-366. Springer-Verlag, 1994.

9. J.Pieprzy, Bent Permutations, *volume 141 of Lecture Notes in Pure and Applied Mathematics*, Springer-Verlage, 1992.

10. Wenling Wu, Dengguo Feng and Sihan Qing, Brief Commentary on the 15 AES Candidate Algorithms Issued by NIST of USA, *Journal of Software*, 10(3): 225-230, 1999.

Issues in the Design of a Language for Role Based Access Control

Michael Hitchens[1] and Vijay Varadharajan[2]

[1]Basser Department of Computer Science, University of Sydney, NSW, Australia
michaelh@cs.usyd.edu.au
[2]School of Computing and Information Technology, University of Western Sydney,
Nepean, NSW, Australia
vijay@cit.nepean.uwd.edu.au

Abstract. In this paper, we describe a language based approach to the specification of authorisation policies that can be used to support the range of access control policies in commercial object systems. We discuss the issues involved in the design of a language for role based access control systems. The notion of roles is used as a primitive construct within the language. This paper describes the basic constructs of the language and the language is used to specify several access control policies such as role based access control, static and dynamic separation of duty, delegation as well as joint action based access policies. The language is flexible and is able to capture meta-level operations and it is often these features which are significant when it comes to the applicability of an access control system to practical real situations.

1. Introduction

In a computing system, when a request for a certain service is received by one principal (agent) from another, the receiving principal needs to address two questions. First, is the requesting principal the one it claims to be? Secondly, does the requesting principal have the appropriate privileges for the requested service? These two questions relate to the issues of authentication and access control (authorisation). Traditionally the work on access control has classified security models into two broad categories, discretionary and non-discretionary (mandatory) models. Typically, discretionary access control (DAC) models leave the specification of access control policies to individual users and control the access of users to information on the basis of identity of users. In mandatory access control (MAC) models the standard approach is to have the access defined by a system administrator and employ attributes such as classifications and clearances [14]. Recently, there has been extensive interest in Role Based Access Control (RBAC) [19] even though the idea of the use of roles for controlling access to entities and objects is as old as the traditional access control models. In RBAC models the attributes used in the access control are the roles associated with the principals and the privileges associated with the roles.

An important aspect of an access control model is the type of polices that the model can support. The model must be flexible enough to support a variety of access control requirements as required in modern application environments. Many of the proposed models are often inflexible because they assume certain pre-defined access policies and these policies have been built into the access control mechanisms. In fact, in some sense, DAC, MAC and RBAC are all mechanism oriented. They all fall into the trap of using access control mechanisms both for policy expression as well as in policy enforcement. By separating out the policy expression from the mechanisms used to implement and to enforce the policy, a number of advantages can be achieved in practice. On the one hand, a mechanism can be used to support a number of policies while on the other hand, a policy may be supported by multiple mechanisms. One mechanism can be used to specify and support a number of policies.

It has been claimed elsewhere that role based access control better suits the needs of real world organisations than MAC or DAC based approaches [17]. While we support this view, we believe that a number of issues in the design and implementation of RBAC systems have not been adequately addressed in previous work. Much of the work on RBAC models and systems has not addressed the issue of how to express policies in a real world system. We strongly believe that a language-based approach to authorisation is required to support the range of access control policies required in commercial systems. While some language-based proposals have been presented, such as in [21], [8], these tend to either lack expressiveness or be highly formal.

It has been recognised that RBAC systems need to deal with permissions which reflect the operations of the application (such as credit and debit in a financial system) rather than the more traditional approach of a fixed set of operations such as read, write and execute [18]. However the effects of this on the design of such systems is rarely addressed. This is especially important as choosing this level of granularity of access control has obvious parallels with the ideas of object-oriented design.

Several RBAC proposals tend to avoid the question of ownership. Indeed, it has been claimed that RBAC is simply a form of MAC [6]. Expecting the system manager to make all policy decisions in a distributed system is impractical and probably unnecessary in most cases. This has also been recognised, with proposals such as [17] recommending that the system manager delegate authority over various objects to other users. These delegated authorities are then exercising control equivalent to that of ownership in DAC systems. It has also been noted that, in practice, users of RBAC systems still wish to maintain individual control over some objects. Given this, the concept of ownership in relation RBAC deserves consideration.

If a language based approach to RBAC is adopted, then we also need to consider other issues, such as how a history of actions can be maintained. The importance of this, and the related idea of history based constraints, has been recognised, [25], but the issue of how to include this ideas in policy expressions is less commonly addressed. Such information is often necessary for access policies such as separation of duty [15]. Consideration also needs to be given to how users are represented and assigned to roles.

We have developed a language for RBAC, called Tower, which is based on the principles we explore in this paper. Tower can be used to express RBAC policies for an object-oriented system. The conclusions reached in the following discussions are all reflected in the design and facilities of Tower. A brief introduction to Tower is given at the end of the paper, space precluding a detailed examination of the language. For more details the reader is referred to [7].

In the next section, we present our arguments for using a language to express RBAC policies. Section 3 outlines the basic constructs required for role based access control in object systems. The subject of ownership and how it can be integrated into an RBAC model is covered in section 4. Section 5 describes an approach to handling history and the use of attributes. This is followed, in section 6, by a brief description of Tower. We conclude with a brief comparison of our work with other proposals.

2. A Language Based Approach

Roles are intended to reflect the real world job functions within an organization. The permissions that are attached to the roles reflect the actual operations that may be carried out by members of the role. The policies that need to be expressed in an RBAC system can therefore be seen to have the potential to be both extensive and intricate. The inter-relationships between structures of an RBAC system, such as role, users and permissions, likewise have the potential to become complicated.

Languages, in various forms, have long been recognized in computing as ideal vehicles for dealing with the expression and structuring of complex and dynamic relationships. Therefore it seems sensible to at least attempt to employ a language based approach for expressing RBAC. If the language is well designed, this will deliver a degree of flexibility superior to other approaches. Flexibility is a necessary requirement if the RBAC system is to be capable of supporting a wide range of access policies. Other work has recognised to possibility of using a language to express RBAC policies, for example [25].

While in theory a general purpose language could be used, a special purpose language allows for optimizations and domain specific structures which can improve conceptual design and execution efficiency. In particular, the notion of roles and other domain specific concepts should be available as primitive constructs within the language. This would simplify the expression of access control policies. The permissions associated with a role tend to change less often than the people who fill the job function that the role represents. Being able to express the relationships between permissions and roles within the structure of the language would make the administration of the access control system simpler and this is a major advantage when it comes to management of authorization policies. In fact, key aspects of any authorization model are the ease with which one can change, add and delete policy specifications and the specification of authorities that are able to perform such operations. A language that does not support the necessary construct will not fulfill this requirement.

In some sense the question of who can modify the policy setting is what determines whether something is discretionary or mandatory. In general, we feel that the traditional notions of discretionary and mandatory are not very helpful in that a policy may be discretionary to some and mandatory to others. For instance, consider a manager of a project group and a number of group members. A policy can be mandatory to the group as it is set by the manager but is discretionary to the group manager. Similarly at a higher level, a policy may be mandatory to the group manager but is discretionary to the laboratory manager above. This is typical of many organizations and is often true in large distributed systems. It is often the flexibility and management of the operations of the access control system itself which are significant when it comes to the applicability of such a system to practical real situations. The use of a language based policy approach helps us to better structure such policies. Some recent work such as [8] and [1] have considered the use of a logic based language in the specification of authorization policies. This paper proposes a language theory based approach and is primarily targeted towards object based systems.

Of course, another reason for employing a special purpose language is that a general purpose language will include many constructs not required for access control. This would impinge on the efficiency, safety and usability of such a system.

We envisage that such a language would be directly implemented, that is, it would not be implemented on top of another access control mechanism, such as access control lists. This could be achieved in systems which support mechanisms in the style of CORBA interceptors [12,13]. Given the power of such a language it would be difficult to implement it using a significantly less expressive access control mechanism.

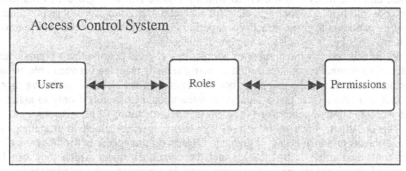

Figure 1: Classical Role Based Access Control Model

3. The Structures of an RBAC Language

Typically, a role based access control model has the following three essential structures [18]:

- users: which correspond to real world users of the computing system
- permissions: a description of the access users can have to an object
- roles: a description of the functions of users within an organization

Permissions and users are mapped to roles, as shown in figure 1. The relationships are many to many.

3.1 Permissions and Privileges

The "classical" role based access model takes a simplistic view of the entities to which access is being controlled. When its application to object oriented systems is considered, it can be noted that the *object*, the basic construct of such systems, is not directly represented within the classical model. In an object-oriented system operations on objects are represented by the methods of the objects. In the classical model, permissions will therefore specify the object and the method to which access is allowed. For example, [4], defines a permission (there called a privilege) to be a pair (am,o), where o is a particular object and am is an access mode identifying a particular operation on that object. We choose to examine access control at a fine level of granularity, so an operation in an object-oriented system would correspond to a method of the object.

If a user has access to a particular object, it is likely that the user will have access to more than one method of that object. While this will not always be true, it will generally hold. This could be handled by having a separate permission for each method. Given the number of methods to which users will have access, this is obviously cumbersome. More promising is the option of allowing a permission to control access to a number of methods. The language should be able to handle both these situations.

Another important concept in object-oriented systems is that of a *class*. Users will often have access to a number of objects in a class and may have identical access to such objects. Similar options to those for dealing with multiple methods are available. A separate permission for each object could be employed. Again this could lead to an unwieldy number of permissions. As we have chosen to police access at the method level, and it is likely that objects to which the same access is granted to the same methods will belong to the same class (or superclass), handling such objects within a single permission appears attractive. A permission should therefore be able to cover a number of objects as well as a number of methods.

While this restricts the number of permissions, it gives permissions a complicated internal structure. Permissions would be many to many mapping between objects and methods. Given that the mappings between entities in the classical model are many to many, it seems clear that a separate construct could be useful, if only to assist in the representation of policies and the clear internal structuring of permissions.

This construct, which we refer to as a *Privilege*, specifies a method or methods of those objects to which access is granted. Objects do not appear in privileges, which are concerned solely with methods and the conditions under which they can be accessed. Privileges are mapped to permissions, in a many-to-many relationship,

A permission specifies an object or objects to which a user has access. It is important that the set of objects governed by a permission be flexible. The language must allow a variety of means for expressing this set, including the objects being explicitly named[1], referenced by a named set, or the reference being to a complete. Using named sets allows grouping of objects, and such groups may change dynamically.

A permission (together with its associated privileges) therefore specifies a set of objects, a set of common methods of those objects and the conditions under which those methods can be accessed.

The structures and the mappings between the components of our RBAC model are shown in figure 2. This model takes into account the objects, which are the chief structuring concept in an object-oriented system. The dashed lines in figure 2 represent ownership relations, which are discussed in section 3.4.

3.2 Users and Sessions

Another basic structure in a RBAC language should be that of the *user*. The elementary requirements this structure are that it specify the user (such as a name and

---

[1]The permission may not actually include the name (as in directory path) of the object, but could hold a pointer to the object or other system dependent construct.

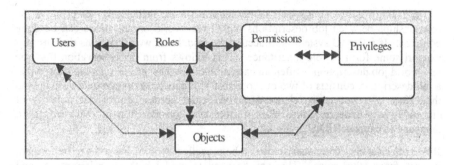

Figure 2: Object-Oriented Role Based Access Control Model

a user identity) and the roles for which the user is authorized. The language should also allow various attributes of the user to be specified (for example, place of work).

As displayed in figure 1, users will have access to multiple roles that they can assume as their authorized roles. This is slightly misleading, in that users will not always want to have access to all their objects. For example, a user may wish to run some untrusted code with only a small subset of their normal access.

For a particular process or login session[2], a user may wish (or be required to have) their access governed by a subset of their authorized roles. We shall refer to this subset as the *active* roles. In a session, allowed accesses to objects are limited to those within the current active roles for that session.

If a user is *authorized* for a given role then a user may adopt that role as an *active* role. A role is not used for access checking until it becomes active. When a role becomes active it does so for a particular *session* of the user. Some systems allow only one role to be active at any one time. Others allow any number of active roles and, indeed, the user may have different roles active for simultaneously existing sessions.

There is obviously a need for a *session* structure that records the active roles, i.e. the ones actually used to determine whether accesses attempted in that session will be allowed.

3.3 Roles

Roles delimit the functions of users within organisations by prescribing the access users have to objects. It must be possible within the *role* structure of the language to specify a set of permissions. There are two other important considerations in determining what must exist within a role. Many RBAC models include the concepts of role hierarchy and constraints. Role hierarchies serve various purposes – they may allow roles to be constructed from other roles [19] or they may define supervision (or subsidiarity) structures [11, 5]. Constraints place restrictions on how users are assigned to roles. It is well understood that it must be possible to specify constraints on users taking on roles (for example, users may not be allowed to take on one role if another role they already have specifically disallows the other role).

---

[2]As systems vary significantly in their treatment of processes and login sessions, we shall simply refer to sessions and leave the details to implementors.

Regarding roles as simply sets of permissions is not sufficiently powerful, as the relationships between job descriptions within real world organisations can be quite complex. If a RBAC system is to accurately model real world systems then it must support some form of role inheritance. This follows from the observation that one real world job description is often an extension of another (an *isa* relationship) or that a job description consists of two or more sets of functionality (aggregation) [11]. As the effect of supervision relationships on actual access control decision is not currently well understood we omit it from our consideration of the needs of a language to express RBAC.

Role Inheritance. Role inheritance can be modelled by allowing roles to be formed, in part, from other roles. This can be used to model both *isa* relationships and *aggregation* relationships. In the case of aggregation (and possibly *isa*) roles higher in the hierarchy will not correspond to actual job descriptions (to which users are assigned) but exist solely for structuring purposes. Such roles can be thought of as *virtual* roles in that no user is ever assigned to such a role. Such roles can be used to implement another concept in role structuring – template roles [4, 10]. Template roles are roles which are not fully bound to all their necessary values (such as exactly what permissions they contain). A virtual role could be a partly defined role, which can serve as a template for any number of actual roles.

One obvious consequence of building roles from permissions and other roles, which themselves consist of yet other permissions and roles (i.e. inheritance) is the possibility of conflicts between the permissions (and their privileges) of the various roles. Consider the simple example of role *branch_manager* which inherits from both role *accounts_manager* and role *loans_manager*. Both *accounts_manager* and *loans_manager* inherit from role *teller*.

There are two points that require discussion

- what is the effect of the *branch_manager* having the permissions of the *teller* twice (once through *accounts_manager* and once through *loans_manager*)?
- what if the permissions unique to *accounts_manager* and *loans_manager* are in conflict at some point?

The answer to first question should be that there is no effect. The operations allowed by an active role, its permissions and privileges should not depend upon how they were retrieved internally by the RBAC system during access checking. This is the classic answer to this question when multiple inheritance is allowed.

The second question revolves around the forms of conflicts that are possible in the language. If the language allows negative permissions (i.e. operations a user is specifically barred from carrying out) then a conflict resolution mechanism is required. If the language only allows only positive permissions, then a decision must be made as to whether the most restrictive or most generous permission will apply and this is a policy issue.

For example, the accounts manager may only be allowed to examine loan balances during business hours, but the loans manager may be able to examine them on Saturdays as well. If the branch manager is attempting to check a loan balance on a Saturday, the permission inherited from the accounts manager's role may be checked first. Although this permission does not allow the action, the search continues and eventually the appropriate permission, inherited from the loans manager's role, is checked. As the branch manager is inheriting all the functions of the loans and accounts managers, it follows that the least restrictive form should apply. In this case, the second option of most generous permission applies. Similar argument applies regardless of whether the conditions conflict in external factors (such as different limits to time or place of access) or internal factors (such as status of the object or access control system).

The other requirement for role inheritance within the language is that it should be possible for the inheritance to be partial. That is, a role inherits some (but not all) of the permissions of another role. It has been noted that in practice [24] the real world organisations that roles are intended to model do not always display total inheritance. It must be possible to specify that one role inherits all the permissions of another, less certain stated permissions. The converse (explicitly stating those permissions to be inherited) does not really require an ancestor role at all, although it may be useful for modelling purposes. Real world job descriptions may simply be related by some overlap of job functions, rather then the strict subset relationship implied by simple inheritance. It may be useful, for modelling purposes, to create roles which contains the common functions of job descriptions and have the roles which correspond to the actual job functions inherit from those roles. As the ancestor role does not correspond to any actual real world job, it can be nominated as a virtual role and no user may then have it as an authorised role.

Role Constraints. Constraints on roles limit the possible roles that may be active and or authorised for a user. Constraints between roles must be checked at two points. Firstly, when the role is authorised for the user. Secondly, whenever an attempt is made to make the role active for a particular session. While a user may be authorised for two roles, it should be possible to prevent the user being simultaneously active in both roles. This decision on whether a role can be made active for a session may be based solely on the roles already active for that particular session, or on the roles active for all current sessions of the user. This requires that the language allow constraints to be expressed at three levels – authorised roles, active roles within a session and active roles within all simultaneous sessions for a principal.

In general, constraint decisions can be based on a number of factors, such as which roles are already authorised or active for the user or session or how many users are already authorised or active in the role. It should be possible to stop a role being authorised or active based on other role assignment of the user or session (mutual exclusion). It should also be possible to require that another role is authorised or active for the user before a role assignment is made (prerequisite roles). The more expressive power the language allow for constraints the richer the possible set of constraints.

4. Object Ownership

In most role based access control systems, there appears to be an implicit (or at times explicit) assumption that objects are explicitly referred to in the permissions. While this is often what is required, it can lead to problems. Specifically, what happens when a new object is created? If objects are explicitly named within permissions, then permissions have to be updated or created every time a new object is created. This is an unwanted and unnecessary overhead. It also violates the general requirement that the access control system interfere as little as possible with the rest of the system. However, if objects are not explicitly named within permissions, how can one determine to which objects a permission is controlling access? While the possibility of specifying a class, rather than a group of objects, within a permission partially solves this problem, there are many occasions when users should not have access to all objects of a given class.

The concept of ownership, found in other access control models, can be used to provide a solution to this problem. This would allow permissions to be written so that access to objects was granted only to the owners of the objects. Such permissions could be written in advance of the objects coming into existence and would not need to refer to the objects by name. Ownership can be checked within the permissions and privilege, allowing dynamic creation of objects without each such creation requiring update of the permissions and privileges. While the information for a newly created object (who or what owns it) must be entered into the access control system, this is no more information required than in any other approach to access control.

The immediate question that arises is where to vest the ownership – in users or in roles? The obvious answer, for a role based system, may be in the roles. However, consider the following example. A system may have a basic role, *user*, which includes a permission giving access to the factory object for word processing documents. If ownership of word processing objects (documents) was vested in the role, then all members of the role *user* (i.e. probably every user of the system) could access all word processing objects. This situation would (at least in general) be undesirable. More sensible would be that ownership of a new word processing object would be vested in the user that requested its creation.

However, restricting ownership to users only is also undesirable. For example, consider the role *budget_working_group*. It is likely that access to spreadsheet documents created by members of this role should be automatically granted to all other members of the role, not just the particular member who created the document. The granting of such access should be on creation of the document and not require manual input for every new document. It would not be feasible to say that all spreadsheets are owned by the role through which access was granted to their creating object, as this would preclude private ownership of such objects. As a further example, consider that *user* and *budget_working_group* may exist on the same system and that the *budget_working_group* may produce word processing documents as well as spreadsheets.

Another possibility is to use the concept of role inheritance to guide ownership, if ownership is to be vested solely in roles. Ownership is then granted to the role that is actually the active role of the user, not any intermediate roles from which the active role has inherited. Therefore, if *budget_working_group* was the active role of the user[3], then ownership of any objects created would be vested in *budget_working_group*. However, allowing "private" ownership of objects would then require the creation of "private" roles for each individual user, each of which inherits basic permissions from role *user*. Each user could retain ownership of private objects by having this private role as their active role. The result would be that no user would ever have *user* as their active role – which, at the very least, does not follow the spirit of role based access control.

---

[3] For purposes of the example in this paragraph we assume that each user can have only one active role at a time. This is unnecessarily restrictive, but allows a more clear presentation of the example.

It is clear that it must be possible to vest ownership of an object in one or more users, one or more roles or some combination thereof. A user owning an object has a clear meaning – they may alter they access control policies which govern that object. With a user exercising ownership through membership of the role which directly owns the object the situation is less clear. It could be required that the user attempting an operation requiring ownership have the owning role as one of their currently active roles. Less restrictively the user could simply have that role as one of their authorised but but not necessarily currently active roles. The latter case restricts the possibility of sand-boxing operations and hence we advocate the former. The final argument in favour of allowing user-based ownership is that practical experience with role based systems has found that it is impractical to do away with them. Such permissions should then be encompassed within the RBAC system, as otherwise a separate access control system would be required to handle them. This would then produce obvious problems in coordinating and understanding the two access control systems.

Whether the information regarding ownership is stored with the object or elsewhere (and accessible to the access control system) or by the access control system (and if so exactly how) is implementation dependent.

5. History, Variables and Attributes

Access control decisions will often need to be made based on the previous actions of the user attempting access (or of other users with the object in question or even other objects). This is as true for RBAC systems as any other and various proposals for dealing with this have been made, e.g. in [21], [9]. Most of these are fairly restricted mechanisms, recording simply the past occurrence of actions (or transactions).

In taking a language-based approach we have the option of a more flexible mechanism. One of the important features of most programming languages is the manner in which they store information. We propose that a language for RBAC should allow the declaration and use of variables to hold information useful for such purposes as tracking the history of user actions. Access control decisions can be based upon the values of such variables; we refer to such variables as meta-variables, to separate them from the variables of the programming languages used to write the objects being protected[4].

The values of meta-variables can be checked in the condition clauses of the various structures of the language, such as roles and privileges. Meta-variables can be used to implement attributes [6], history-based constraints [25] and activiation criteria [5]. Employing variables (the values of which can change) will allow the state of the access control system to automatically change in response to user actions; hence such variables are particularly useful when it comes to modelling dynamic access control. While classical role based access control is easily able to model static separation of duties, modelling dynamic separation of duties is often less straightforward [3]. While the need for dynamic separation of duties is commonly acknowledged, the details of how it is to be achieved in practical implementations are often omitted. A common example of dynamic separation of duties is the initiation

---

[4] The term "meta" is used in this paper to refer to operations and values within the access control mechanism itself, as opposed to those of the entities to which access is being controlled.

and authorisation of a payment. While members of a given role, such as *accountant*, may be allowed to initiate and authorise payments, a given member may not be allowed to both initiate and authorise a particular payment. Assume that the payment objects do not record the identity of the initiator and do not enforce the separation of duties. It is then the responsibility of the access control system to record this information and enforce the separation. When the payment is to be authorised, the identity of the authorising entity can be checked against that of the initiator. This is achieved using meta-variables. As another example, meta-variables are useful when it is required to limit the number of times a permission can be used and to alter the state of roles and permissions and to record parameters to actions.

5.1. Attributes

Roles are intended to reflect the real world job functions within an organisation. In the classical RBAC model, roles consist of privileges where privileges are simple expressions of access to objects. This does not fully encompass the real world description of a job function. Job functions often encompass attributes such as physical ones (e.g. location) and job classification (e.g. probationary period). Attributes can also be used in handling Mandatory Access Control and Lattice based controls (such as a those in [2] and [16]).

As a simple role example, consider a wholesale company with a number of warehouses. There may be a role *stock_inventory_clerk*. Users in this role have the responsibility for recording movement of goods into and out of the warehouses. However, there is no need for (and possibly good arguments exist against) such clerks being able to alter the values at warehouses other than the specific ones at which they work. While this could be handled by having a separate role for each location and including within each such role permissions for the objects at each location, this simply results in a large number of virtually identical roles. This can be avoided by giving each user and object a location attribute and checking (in the conditions of prvileges) at time of access, whether the values of the attributes match.

It may be thought that the attributes that specify the fine details of a job description are unnecessary. For example, whether a clerk is in a probationary period or not could be handled by having two separate roles. However, the number of such attributes and their interaction, in practice, can again lead to an explosion in the number of roles. An example quoted [6] is for 3125 separate sub-categories of bank teller arising from only different attributes. It is obviously far easier to handle these using attributes than by using multiple roles.

We believe that it is impossible to precisely define in advance all the possible attributes that could be of interest to the designers of security systems. However we believe that it is possible to define a useful set of attribute types. This is inline with programming language design, where language designers do not attempt to pre-set the variables available to programmers. Therefore an RBAC language should allow attributes of various names and types to be associated with users and objects.

5.2. Ownership and Attributes

Ownership, as discussed in section 3.4, can be thought of as a special attribute. It can be tested in a condition, as any other attribute can. This allows users to delegate access to other users on the basis of the objects they own, if they wish. For example, a user could delegate access to their secretary on the basis of the objects they own. Of course, this is a powerful form of access and any user who did not limit such access in other ways, for example by setting a lifetime on the permission, deserves any problem they encounter. Lifetimes could be set by simply checking the current time within the permission and ensuring that it is not beyond a cut-off point.

6. Access Control Language: Tower

In this section, we briefly discuss some of the major features of a language, called Tower, which implements the constructs discussed above. Using a language allows more flexible expression of access control policies then other approaches as mentioned earlier. RBAC policies are expressed as Tower code. The total compiled body of Tower code forms an executable system that polices access to a system. Tower code is compiled to check for syntactic correctness and then submitted to the system. When code is submitted it may cause new meta-variables and structures to come into existence and may also alter the value of existing data.

Requests to the access control system are of the following form:

    access(s,o,m,p)

where

    s is a session identifier, which can be used to identify a user
    o identifies an object being accessed
    m identifies the method being invoked
    p is the set of parameters for the invocation

The access control system will report whether the method invocation is to be allowed or disallowed, according to the following algorithm:

for each active role for the session
  for each permission in the role
    if the permission applies to the object
      for each privilege in the permission
        if the method being invoked is one of those listed for the privilege and
            the privilege's condition evaluates to true
          execute the action of the privilege
          allow the access

Note that the ordering of roles, permissions and privileges are significant in this algorithm; the ordering determines which action section will be executed if more than one is possible.

The most important structures in Tower are the definitions of users, roles, permissions and privileges. The details of these structures are discussed in subsequent sections but some general comments can be made. Each structure is declared and is given a name. The name is used to identify the structure throughout the access control system. Therefore each of these names must be unique and these can be block structured for scoping. A new structure instance may be created and assigned to a structure variable. The closure of a structure includes any variables declared in the same scope. The structures are immediately available upon creation for evaluating access requests. They may also have their values modified in code which is subsequently executed. In this paper we do not specify the management interface of the access control system. We envisage that both users and administrators can enter policies (in the form of Tower expressions) into the system.

Whether this is in a form similar to the Adage VPB [25] or by some other means is not relevant to the design of the language itself.

Most previous work on RBAC has not considered in any depth the problem of controlling the name space within the access control system itself. The roles and permission appear to implicitly reside within a flat name space. While this is convenient in making all role and permission globally accessible, it presents a number of practical problems, common to any flat name space. These problems include uniqueness of names and creating an understandable structure.

Solutions to the problems of flat name spaces are well known. In Tower we adopt one based on the hierarchical structuring of name spaces common in operating systems combined with block structuring in languages such as Pascal. Declarations of structures and variables are made within a *module*. Modules can be named, and these names (at the outermost level) identify the module at the level of the access control system and so must be unique. Tower is a block-structured language, in that modules can be declared with a module. It is not the objective of this paper to describe the complete syntax of Tower; space limitations prevent this. However, the following examples hopefully convey some of its flavour.

The first is static separation of duties. Consider an object-oriented system that has a class *container_class* with two operations *create* and *certify*. We wish total separation between the users who can carry out these two operations. However, if the operation is allowed, it is allowed for all instances of the class. The following Tower code creates the necessary roles, permissions and privileges. Once the roles have been created users may be assigned to them.

```
create_privilege := privilege
   {create}
end_privilege
certify_privilege := privilege
   {certify}
end_privilege
create_permission := permission
   container_class
   objects {container_object}
   privileges {create_privilege}
end_permission
certify_permission := permission
   container_class
   objects {container_object}
   privileges {certify_privilege}
end_permission
creator_role := role
   authorised exclude certifier_role
   permissions {create_permission}
end_role
certifier_role :=role
   authorised exclude creator_role
   permissions {certify_permission}
end_role
```

Only one constraint expression is actually necessary. For completeness, a constraint expression is included in both. Note that Tower structures are set based, so that privileges may hold many permissions, permissions many privileges and roles many permissions.

The second example is dynamic separation of duty. Consider a class of cheque objects, which may be accessed by members of the role *accountant*. However, the same user may not both issue and authorise the same cheque.

```
begin
    issuing_user* : userid
    issue_privilege :=privilege
        issuing_user := user
        {issue}
    end_privilege
    authorise_privilege := privilege
        issuing_user <> user
        {authorise}
    end_privilege
    cheque_permission := permission
        cheque_class
        privilege
{issue_privilege,authorise_privilege}
    end_permission
    accountant := role
        permissions {cheque_permission}
    end_role
end
```

The first line within the module is a meta-variable declaration. The '*' after the declaration causes one copy of the variable *issuing_user* to be created for **each** object covered by the *cheque_permission* and its tokens. If access is attempted to an object via these structures and a corresponding instance of the variable does not exist then it is automatically generated. The value of the variable is set in *issue_privilege* and checked in *authorise_privilege*. This allows the access control system to ensure that the same user does not both issue and authorise a cheque. The role *accountant* does not need to be declared within the block, but placing it within the block aids readability.

Let us know consider s a delegation example. Delegation within Tower is handled by dynamically assigning and de-assigning roles. While roles are usually thought of as broad concepts covering complete job descriptions, they can also be used in a much more fine-grained manner. The permissions representing the delegated authority can be placed in a new role. This role can be added to the authorised role of the user (which may represent a real world user or some active system entity) to whom the authority is to be delegated. When the delegated actions are completed, the role can be removed.

Consider the following delegation situation [22]. A departmental manager has access to view and modify the overall departmental portfolio object DP. The department may have several projects, each of which has an individual portfolio object DPi. A project manager can only view or modify his or her own portfolio object. A project manager, *pm*, can only view or modify another project's portfolio if and only if the departmental manager, *dm*, has delegated the appropriate privilege to it. That is, in this case, the project manager is acting on behalf of the departmental manager. This could be handled in Tower by *dm* executing the following:

```
delegated_rights := permission
   ...
end_permission
delegation_role := role
   permissions {delegated_rights}
end_role
pm := pm + delegation_role
```

The departmental manager (after executing the above code) has created a new role holding the delegated privileges and this role has been added to the set of authorised roles for the project manager. As the departmental manager is probably not the owner of the user structure for the project manager, the last line will not take effect until the project manager (assuming he/she owns his/her own user structure) gives permission for the update. As the department manager retains ownership of the new role and permission, the project manager can not pass on the delegated rights without the department manager's agreement.

However, the department manager may wish the project manager to be able to further pass on the delegated permission without referring back to the department manager. This can be handled in a number of ways. The department manager could transfer ownership of the new structures to the project manager. The project manager could then distribute them freely. However, the project manager could also alter them before distribution. Even though such alteration would have to obey the access restrictions implied by ownership of objects, this may still be more then the department manager desires. In such a case the department manager could transfer ownership of only the *delegation_role*. This would allow the project manager to add this role to other user's roles or user structures, without being able to alter the encapsulated permission(s).

Using Tower we have been able to specify a variety of other access control policy examples. In particular polices involving role hierarchies, separation of duties, the Chinese Wall [2], joint actions[23] and policies limiting access to fixed number of accesses have been specified.

7. Brief Comparison with Other Work

The language described above is far from the first attempt at expressing role based access control. Other proposals have been put forward which allow the access control policies to be expressed in a systematic manner for role based or related systems. These proposals range from the formal one such as in [8] to more practical ones such as [21] and [24] to related mechanisms such as in [9]. While formal languages such as the ASL in [8] can have good expressive power, they suffer from a resistance amongst real users, due to their highly intricate nature. For example, these languages often depend upon their users having a reasonable level of understanding of logical principles. This is not always found amongst real world users, even those entrusted with the management of access control policies for a system. Their syntax will often contain symbols not commonly used, limiting their appeal. While it is true that such languages are not generally written for widespread use, this simply perhaps strengthens the argument that a different approach should be used for the expression of access control policies in real world systems. Another drawback of some such proposals is the attempt to be too general. While it may be useful in a theoretical language to be able to cover a number of access control approaches, in the real world, it is more important to be able to address those that are used in practice and to develop tools tailored to support them. The language described in this paper is intended to address this practical issue and does not assume an overly high level of theoretical ability.

Furthermore, in the language that has been proposed in this paper, we have considered meta-variables and role constraints, which few of the other proposals have included. For example, [21] attempts to use an assortment of predefined functions to fulfil the functionality that we address using meta-variables. A set of predefined functions is highly unlikely to present a sufficient degree of flexibility and capability. It is far better to provide the user needing to specify the access control policies with flexible mechanisms (such as those we provide) and allow them to construct structures for expressing their policies. Finally, some other earlier work such as the one in [9] is limited in their expressiveness; it does not address concepts such as time or object attributes and the syntax is also somewhat limited targeted at specific operating systems.

Consideration has been given elsewhere to the use of RBAC structures in the management of RBAC policies [20]. We believe that the structures of Tower could be used in this manner but leave detailed consideration of this to future work

8. Concluding Remarks

In this paper, we have proposed a language based approach to the specification of authorization policies. We believe that such an approach is required to support the range of access control policies in commercial systems. We have discussed the issues involved in the design of a language for role based access control systems. The proposed language focuses in particular on object-oriented systems. The notion of roles is used as a primitive construct within the language. It is often the flexibility and management of the meta-level operations, which are significant when it comes to the applicability of an access control system to practical real situations. The use of a language based policy approach helps us to better structure such meta-level policies. In this paper, we have described the basic constructs of the language and used the language to specify several access control policies. In particular, we have described policy example scenarios involving role hierarchy, separation of duties both static and dynamic, Chinese Wall policy, delegation as well as joint action based access policies.

References

1. Bai, Y., Varadharajan, V.: A Logic for State Transformations in Authorisation Policies: Proceedings of the 10th IEEE Computer Security Foundations Workshop (1997) 173-183.
2. Brewer D., Nash, M.: The Chinese Wall Security Policy, IEEE Proceedings on Security and Privacy, (1989) 206-214.
3. Ferraiolo, D., Kuhn, R.:Role based Access Controls, 15th NIST-NCSC National Computer Security Conference, (1992)

4. Giuri L., Iglio, P.: Role Templates for Content-Based Access Control: 2nd ACM RBAC Workshop, (1997) 153-159.
5. Goh. C.:Towards a more Complete Model of Role: 3rd ACM RBAC Workshop, (1998) 55-61.
6. Hilchenbach, B.: Observations on the Real-World Implementation of Role-Based Access Control: National Information Systems Security Conference (1997) 341-352
7. Hitchens, M. & Varadharajan, V.,: Specifying Role Based Access Control Policies for Object Systems, submitted for publication.
8. Jajodia S., Smarati, P., Subrahmanian, V.: A Logical Language for Expressing Authorizations: IEEE Proceedings on Security and Information Privacy (1997).
9. Karger, P.: Implementing Commercial Data Integrity with Secure Capabilities: IEEE Symposium on Security and Privacy (1988), 130-139.
10. Lupu E., Sloman, M.: Reconciling Role Based Management and Role Based Access control: 2nd ACM RBAC Workshop (1997) 135-141.
11. Moffett. J. Control Principles and Role Hierarchies: 3rd ACM RBAC Workshop, (1998) 63-69.
12. Object Management Group (OMG) : Security Services in Common Object Request Broker Architecture 1996.
13. Object Management Group (OMG), CORBAservices: Common Object Services Specification, OMG Document 97-07-04. (1997)
14. Pfleeger, C.P.,: Security in Computing, 2nd ed., Prentice Hall, 1997.
15. Sandhu, R.: Transaction Control Expressions For Separation of Duties: Fourth Aerospace Computer Security Applications Conference (1988) 282-86.
16. Sandhu, R.:Lattice-Based Access Control Models: IEEE Computer 11 (1993) 9-19.
17. Sandhu, R., Feinstein, H.: A Three Tier Architecture for Role-Based Access Control: 17th National Computer Security Conference (1994), 34-46.
18. Sandhu, R., Coyne, E., Feinstein H., Youman, C.: Role-Based Access Control: A Multi-Dimensional View, 10th Annual Computer Security Applications Conference (1994), 54-61.
19. Sandhu,R., Coyne,E.J., Feinstein, H.L.: Role based Access Control Models, IEEE Computer, 2 (1996) 38-47.
20. Sandhu. R.: Role Activation Hierarchies: 3rd ACM RBAC Workshop, (1998) 33-40.
21. Simon R., Zurko, M.: Separation of Duty in Role-Based Environments: 10th Computer Security Foundations Workshop, (1997) 183-194.
22. Varadharajan, V., Allen, P., Black, S.: Analysis of Proxy Problem in Distributed Systems, IEEE Proceedings on Security and Privacy (1991).
23. Varadharajan V., Allen, P.: Joint Action based Authorisation Schemes, ACM Operating Systems Review 7 (1996).
24. Varadharajan V., Crall, C., Pato, J.: Authorisation for Enterprise wide Distributed Systems: Design and Application, IEEE Computer Security Applications Conference, (1998).
25. Zurko, M., Simon R., Sanfilippo, T.: A User-Centred, Modular Authorisation Service Built on an RBAC Foundation: IEEE Symposium on Security and Privacy, (1999).

Extending Erlang for Safe Mobile Code Execution

Lawrie Brown[1], Dan Sahlin[2]

[1] School of Computer Science, Australian Defence Force Academy, Canberra 2600, Australia. Lawrie.Brown@adfa.edu.au
[2] Computer Science Laboratory, Ericsson Utvecklings AB, S-125 25 Älvsjö, Sweden. dan@cslab.ericsson.se

Abstract. This paper discusses extensions to the functional language Erlang which provide a secure execution environment for remotely sourced code. This is in contrast to much existing work which has focused on securing procedural languages. Using a language such as Erlang provides a high degree of inherent run-time safety, which means effort can be focused on providing a suitable degree of system safety. We found that the main changes needed were the use of unforgeable (capability) references with access rights to control the use of system resources; the provision of a hierarchy of execution nodes to provide custom views of the resources available and to impose utilisation limits; and support for remote module loading. We then discuss prototype implementations of these changes, used to evaluate their utility and impact on visibility for the users of the language, and mention work in progress using this foundation to specify safety policies by filtering messages to server processes.

1 Introduction

We present some extensions to the functional language Erlang [4] which provide a secure execution environment for mobile code. Supporting safe mobile code is an area of considerable recent research, particular with the increasing use of Java and similar languages (see surveys such as [14, 29, 1]). However much of the work has focused upon securing traditional procedural languages or their derivatives. This paper considers the changes needed to secure a production use functional language, Erlang. We believe that using a functional language can provide significant benefits in productivity and maintenance of large applications, and that it would be beneficial to provide these benefits for applications requiring the use of remotely sourced code. Further, we believe that functional languages have a very high degree of intrinsic run-time safety, which means the changes required to provide a safe execution environment are much smaller than those needed to secure a procedural language. This paper will discuss the results of our initial prototypes of a safer Erlang, used to investigate these assertions.

We will use the term *mobile code* to refer to any code sourced remotely from the system it is executed upon. Because the code is sourced remotely, it is assumed to have a lower level of trust than locally sourced code, and hence needs to be executed within some form of *constrained* or *sandbox* environment to protect the local system from accidental or deliberate inappropriate behaviour. A key assumption is that the local system is trusted, and provides adequate access control mechanisms. We also assume that appropriate protection of inter-node communications, using either encryption techniques (eg IPSEC, SSL) or physical isolation, is provided.

In his taxonomy of issues related to distributed hypermedia applications, Connolly [11] in particular distinguishes between *Run-time Safety* — which is concerned with whether the runtime system will guarantee that the program will behave according to its source code (or abort); and *System Safety* — which is concerned with whether it is possible to control access to resources by a piece of executing code. A similar emphasis on constraining resource access or certifying code is identified by Hashii et al. [14], Rubin and Geer [24], or Adl-Tabatabai et al. [1] amongst others.

Traditionally the necessary protection and access control mechanisms needed for system safety have been supplied using heavyweight processes with hardware assisted memory management. These have a long and successful history as a proven means of providing safety in traditional computing environments. However, we believe that in applications with rapid context changes, such as would be found when loading and running code from a wide range of sources of varying trust levels, such mechanisms impose too great an overhead. Further they also tend to restrict the portability of the code sourced [28, 29]. Here we are interested in the use of lightweight protection mechanisms, which typically rely on the use of language features and an abstract machine to provide the necessary safe execution environments for portable code. This has been the approach of most recent research into safe mobile code, and may be regarded as a combination of the Sandbox and Proof-Carrying code approaches, see for example [24].

The traditional focus for safe mobile code execution has been on securing procedural languages and their run-time environment. The best known example is the development of Java [6, 29] from the production C++ language. Other examples include: SafeTCL [23], Omniware [19, 1], and Telescript (see overviews by Brown [8], Thorn [28]). With these languages, much effort has to be expended to provide a suitable degree of run-time safety in order that the system safety can then be provided. In part this has been due to the ease with which types can be forged, allowing unconstrained access to the system. A number of attacks on these systems have exploited failures in run-time safety (cf Dean et al. [13], McGraw and Felten [20], Oaks [22], or Yellin [31]).

We are interested in providing safe mobile code support in a functional language. This is motivated by evidence which suggests that the use of a functional language can lead to significant benefits in the development of large applications, by providing a better conceptual match to the problem specification. This has been argued conceptually by Hughes [15] for example. Further, significant benefits have been recently reported with large telecommunications applications written in Erlang, see Armstrong [3].

In addition, we believe a dynamically typed, functional language can provide a very high degree of intrinsic run-time safety, since changing the type interpretation should be impossible (except perhaps via explicit system calls). This is noted by Connolly [12], who observes that languages like Objective CAML and Scheme48 provide a high degree of run-time safety, though they need further work to provide an appropriate level of system safety. This should greatly reduce the work required to support safe mobile code execution with such languages. We do need to assume that the basic run-time system correctly enforces type accesses, though the language semantics make checking and verifying this considerably easier.

The work on Objective CAML [18] is perhaps closest in some respects to the Safe Erlang system we discuss. However whilst Objective CAML has the necessary features for run-time safety, its system safety relies on the use of signed modules created by trusted compilation sites. Our approach however, provides appropriate system safety by constraining the execution environment into nodes and controlling resource access, so that untrusted imported code is unable to access resources beyond those permitted by the policy imposed upon the node within which it executes.

1.1 Erlang

For the work discussed in this paper, the authors, initially independently, decided to concentrate on the language Erlang [4, 3, 5, 25, 30]. Erlang is a dynamically typed, single assignment language which uses pattern matching for variable binding and function selection, which has inherent support for lightweight concurrent and distributed processes, and has error detection and recovery mechanisms. It was developed at the Ericsson Computer Science Laboratory to satisfy a requirement for a language suitable for building large soft real-time control systems, particularly telecommunications systems. The Erlang system has now been released as an open source product[26], and is freely available for download. Most of the Erlang system is written in Erlang (including compilers, debuggers, standard libraries), with just the core run-time system and a number of low-level system calls (known as BIFs for BuiltIn Functions) written in C. Distributed use is almost transparent, with processes being spawned on other nodes, instead of locally. An Erlang node is one instance of the run-time environment, often implemented as a single process (with many threads) on Unix systems. There are currently several different implementations of the Erlang run-time system, with different performance tradeoffs. The focus of the research described here though, is on the language, and the features it provides. The prototypes described use the original JAM implementation of Erlang. In future, an alternative may be more appropriate.

Erlang is currently being used in the development of a number of very large telecommunications applications, and this usage is increasing [3]. In future it is anticipated that applications development will be increasingly outsourced, but that they will be executed on critical systems. Also that there will be a need to support applications which use mobile agents, which can roam over a number of systems. Both of these require the use of mobile code with the provision of an appropriate level of system safety. The extensions we have proposed would, we believe, provide this.

2   Current Limitations of Erlang

Using Erlang as the core language for a safe execution environment provides a number of advantages. For instance, *pure* functional language use, where functions manipulate dynamically typed data, and return a value, is intrinsically safe (apart from problems of excessive utilisation of cpu, memory etc). It is only when these functions are permitted to have side-effects that they may threaten system safety. Side effects are possible in Erlang when a function:

accesses other processes by sending and receiving messages or signals, viewing process information, or changing its flags.

accesses external resources outside the Erlang system (files, network, hardware devices etc), using the open_port BIF.

accesses permanent data in databases managed by the database BIFs.

Thus, a safer Erlang requires controls on when such side-effects are permitted.

In Erlang, a *process* is a key concept. Most realistic applications involve the use of many processes. A process is referred to by its *process identifier (pid)*, which can be used on any node in a distributed Erlang system. Given a pid, other processes can send messages to it, send signals to it (including killing it), or examine its process dictionary, amongst other operations. Erlang regards external resources (devices, files, other executing programs, network connections etc) also as processes (albeit with certain restrictions, in much the same way that Unix regards devices as a special sort of file). These processes are called *ports* and are referred to by their port number, which is used like a pid to access and manipulate the resource.

Consider the following code excerpt from an account management server, which when started, registers itself under the name **bank**, and then awaits transactions to update the account balance:

```
-module(bankserver).
-export([start/1]).
start(Sum) ->                   % start account server
    register(bank,self()),      % register as 'bank'
    account(Sum).               % process transactions
account(Sum) ->                 % transaction processing loop
    receive                     % await transaction
        {Pid, Ref, Amount} ->   % account update msg received
            NewSum = Sum+Amount,  % update sum
            Pid ! {Ref,NewSum}, % send response back
            account(NewSum);    % loop (recurse)
        stop -> nil             % end server msg received
    end.
```

This could be started with a balance of 1000, and updated, as follows:

```
...
% spawn a bank account process with initial Sum 1000
Bank_pid = spawn(BankNode,bankserver,start,[1000]),
...
Ref = make_ref(),              % make a unique ref value
Bank_pid ! {self(),Ref,17},    % send msg to server
receive                        % await reply from server
    {Ref,New_balance} ->       % reply says updated ok
    ...
end,
...
```

In standard Erlang a pid or port identifier used to access processes or external resources is both forgeable and too powerful. Apart from legitimately obtaining a pid by being its creator (eg Bank_pid = spawn(BankNode, bankserver, account, [1000]) in the example above which creates a new process and returns its pid), receiving it in a message (receive Bank_pid -> ok end,), or looking it up via a registered name (Bank_pid = whereis(bank)); it is also possible to obtain a list of all valid pids on the system (AllPids = processes()), or to simply construct any arbitrary pid from a list of integers (FakePid = list_to_pid([0,23,0])). These latter features are included to support debugging and other services, but open a significant safety hole. Further, once a valid pid has been obtained, it may be used not only to send messages to the referenced process (FakePid!{self(),Ref,-1000}), but to send it signals, including killing it (exit(FakePid,kill)), or inspect its process dictionary (process_info(FakePid)). There is no mechanism in the current specification of the Erlang language, to limit the usage of a pid (to just sending messages, for example).

Another limitation of the current Erlang system from a safety perspective is the fact that a given Erlang system (that is, one instance of the run-time environment) forms a single node. All its processes have the same privileges and the same access to all resources (file system, modules, window manager, devices) managed by the system. There is no mechanism to partition access within a system, so that it may be mediated via a trusted server process. The only current method for providing this is to run a separate system in a separate heavyweight process, at a considerable cost in system (cpu, memory etc) resources. There is also no means to limit the resources utilised by processes, apart from imposing restrictions on the entire system.

Lastly, there is a need to provide a remote module loading mechanism in order to properly support mobile agents. Whilst Erlang currently supports distributed execution and remote spawning of processes, the module containing the code to be executed must exist on the remote system in its default search path. Further, any modules referenced in the loaded module will also be resolved on the remote system. The code loading must, however, be implemented in such a way that the remote code server cannot be tricked into sending code that is secret. Note that pid and port identifiers are globally unique, so they may be passed in messages between nodes whilst maintaining their correct referent.

3 Extensions for a Safer Erlang

In order to address the deficiencies identified above, we believe that extensions are required in the following areas:

capabilities may be used to impose a finer granularity of control on the use of process (and other resource) identifiers, making these unforgeable with a specified set of rights on the use of any specific instance.

nodes that form a hierarchy within an Erlang system (one instance of the runtime environment), to provide a *custom context* of services available, to restrict the use of code with access to external resources, and to impose utilisation limits (on cpu, memory etc).

remote module loading may be used to allow execution of code modules sourced on another system, retaining knowledge of the source system so that subsequent module references can also be sourced from that system, rather than the local module library.

3.1 Capabilities

We define a *capability* as a globally unique (in time and space), unforgeable name for (a specific instance of) some resource, and a list of rights for operations permitted on the resource by holders of the capability. We use such capabilities to replace the existing forgeable, unconstrained identifiers for nodes, ports, processes etc. Note that all access to external resources (files, network connections, other programs etc) is provided via a port. There are no capabilities which refer to a file directly, for example, only to some access currently provided on some node.

The same resource may be referred to from different capabilities giving the owners of those capabilities different rights. We are using capabilities to ensure that these identifiers cannot be forged, and to limit their usage to the operations desired. A capability is always created (and verified upon use) on the node which manages the resource which it references, and these resources never migrate. This node thus specifies the *domain* for the capability, and is able to select the most appropriate mechanism to provide the desired level of security. Further, the resources referenced are never reused (a new process, even with the same arguments, is still a new instance, for example), so revocation is not the major issue it traditionally is in capability systems. In our usage capabilities are invalidated when their associated resource is destroyed (eg. the process dies, or a port accessing a file is closed). Other processes may possess invalid capabilities, but any attempt to use them will raise an *invalid capability* exception. Most particularly, if a node is destroyed, then **all** the capabilities it created are now invalidated. Any replacement node will create and use new capabilities, even if performing the same task, or accessing the same external resource.

The use of capabilities to control resource access echoes it use to provide safe resource use in systems such as Amoeba [27]. However Amoeba was a general operating system, which had to use heavyweight protection mechanisms to isolate processes running arbitrary machine code from each other, and to provide custom contexts. Here we rely on the language features and an abstract machine to provide lightweight protection mechanisms at much lower cost in system resource use.

Capabilities may be implemented in several ways. Since we are concerned with their use in a mobile code system, hardware implementations, for example, are not relevant. We focused on the following types of capabilities as being most appropriate:

Encrypted (hash) Capabilities use a cryptographic hash function (cf [17]) to create a encrypted check value for the capability information, which is then kept in the clear. Only the node of creation for the capability can create and check this hash value. The overhead of validating the check value can be minimised if any *local* encrypted capabilities are checked once on creation (or import) and then simply flagged as such (say by amending the hidden data type tag to indicate that it has been verified). Subsequent use of the capability then incurs no overhead. Further, for *remote* capabilities, any delays due to cryptographic overheads are likely to be swamped by network latencies. Each node would keep the key used to create and validate its capabilities secret, and this key could be randomly selected when the node is initialised. Any previously created capabilities must refer to no longer extent (instances of) resources (from a previous incarnation of the node), so there is no requirement to continue to be able to validate them. This approach could be attacked by attempts to guess the key used, and verifying the guess against intercepted capability data. The likelihood of success will depend on the type of hash function used, so some care is needed in its selection to avoid known flaws in some *obvious* modes of use [7].

Password (sparse) Capabilities [2, 27] use a large random value (selected sparsely from a large address space) to protect the capability, with the node of creation maintaining a table of all its valid capabilities. This table is checked whenever it is presented with a capability, and capabilities may be revoked by removal from this table. One disadvantage of this approach is the size this table may grow to, particularly for long running server processes, or when user defined capabilities are used, where it is impossible to know when they have no further use. Another is that large tables may take some time to search, though careful selection of the table mechanism can reduce this to a minimum. It is possible to try and forge such a capability, but success is statistically highly improbable, and attempts should be detectable by abnormally high numbers of requests presenting invalid capabilities.

There is thus a tradeoff between these alternatives — trading some level of security with encrypted capabilities for space with password capabilities. The best alternative is likely to depend on the target environment, since which of these tradeoffs is appropriate depends on the particular application.

Experience with the prototypes has shown that it is important for efficient execution that all information needed to evaluate guard tests or pattern matches be present locally in the capability. This information must include the type (eg node, port, process) and the value (to test if two capabilities refer to the same object). Because different applications may wish to choose between the security tradeoffs, we decided to support both hash and password capabilities, chosen on a node by node basis, in an interoperable mechanism, where only the node of creation need know the actual implementation. Thus, we have chosen to use capabilities with the following components:

<Type,NodeId,Value,Rights,Private>, where:

Type the type of resource referenced, eg. module, node, pid, port, or user.

NodeId the node of creation, which can verify the validity of the capability or perform operations on the specified resource.

Value the resource instance referenced by the capability (module, n- ode, process identifier, port identifier, or any Erlang term, respec- tively)

Rights a list of operations permitted on the resource. The actual rights depend on the type of the capability. For a process capability these could include: [exit,link,register,restrict,send].

Private an opaque term used by the node of creation to verify the validity of the capability. It could either be a cryptographic check value, or a random password value: only the originating node need know.

A capability may be restricted (assuming it permits it). This results in the creation of a new capability, referencing the same resource, but with a more restricted set of rights. Using this, a server process can, for example, register its name against a restricted capability for itself, permitting other processes to only send to it. eg. register(bank, restrict(self(),[send,register]))

Capabilities would be returned by or used instead of the existing node names, pids, or ports, by BIFs which create or use these resources.

3.2 Nodes

We have provided support for a hierarchy of **nodes** within an Erlang run-time system (an existing Erlang **node**). These provide a *custom context* of services available, restrict the use of code with side-effects, and impose utilisation limits (eg on cpu, memory etc). Functionally, these would be similar to existing Erlang **nodes** with some additional features.

A **custom context** for processes is provided by having distinct:

registered names table of *local* names and associated capability identifiers, used to advertise services. These names are not visible on other nodes.

module alias table which maps the name used in the executing code to the name used locally for the loaded module. This module name *aliasing* mechanism is used to support both redirection of names to *safer* variants of modules, as well as to provide unique names for modules loaded from remote systems (for agents or applets). This table is consulted to map the module name on all external function calls, applys, and spawns.

capability data which is all the information (eg. keys or table) necessary to create and validate unforgeable capabilities for the node.

Restrictions on **side-effects** are enforced by specifying whether or not each of the following are permissible for all processes which execute on the node:

open_port for direct access to external resources managed by the local system.

external process access for access to processes running on other Erlang systems, which could provide unmediated access to other resources, or reveal information about the local system to other nodes.

database BIFs usage for access to permanent data controlled by the local database manager.

When disabled, access to such resources would have to be mediated by server processes running on the local system, but in a more privileged node, trusted to enforce an appropriate access policy for safety. Typically these servers are advertised in the registered names table of the *restricted* node.

Utilisation limits can be imposed by a node for all processes executing within the node, or any descendent child nodes of it. Limits could be imposed on some of cpu usage, memory usage, max no reductions; or perhaps on combinations of these.

The general approach to creating a controlled execution environment is as follows. First, a number of servers are started in a node with suitable privileges to provide controlled access to resources. Then a node would be created with *side-effects* disabled. Its *registered names* table would be pre-loaded to reference these servers, its *loaded modules* table pre-loaded with appropriate local library and safe alias names; and appropriate utilisation limits set. Processes would then be spawned in this node to execute the desired *mobile code* modules, in a now appropriately constrained environment. eg. `BankNode=newnode(node(), ourbank, [{proc_rights,[]}])`

The nodes are structured in a hierarchy, the root of which corresponds to some instance of the Erlang run-time system, and which has full access to the system resources. For each application which requires a distinct run-time environment, a sub-node can be created with suitable utilisation limits. It can start whatever servers are required for that application, and then create further, restricted sub-nodes, as required by it. In this way, various applications can be isolated from each other, with a share of the system resources, and their own servers with appropriate restrictions.

3.3 Remote Module Loading

Our prototypes do not directly support mobile processes as once started, a process may not migrate. However, a new process may be spawned off at a remote node. We have included extensions which ensure that subsequent module references from within a remotely loaded module are also sourced from the originating node, rather than being resolved from the local module library. This is to ensure that a package of multiple modules will execute correctly, remotely, and not be confused with local modules that may have the same name.

Supporting this required an extension of the apply BIF handler so that it checks whether the originating module is local or remote, and proceeds accordingly to interpret the module requested in context, querying the code server on the remote node for the module, if necessary. Some care is needed in managing the acquisition of an appropriate capability for requested module. This is issued by the remote code server upon receipt of a request which includes a valid capability for the parent module, and is then used to request the loading of the requested module.

4 Prototypes

A number of prototypes of a safer Erlang environment, with differing goals and features supported, have been trialed.

SafeErlang was developed by Gustaf Naeser and Dan Sahlin in during 1996 [21]. The system supports a hierarchy of subnodes to control resource usage and to support remotely loaded modules; *encrypted capabilities* for pids and nodes to control their usage; and remote module loading. Whilst this prototype was successfully used to trial a mobile agent application, limitations were found with the complexity of its implementation, the incomplete usage of capabilities for all resource items (in particular for ports), and the use of fully encrypted capabilities and the consequent need to decrypt them before any information could be used.

In Uppsala, as a students project, a design for safe mobile agents, was implemented in 1996 [16]. Distributed Erlang was not used in that system, which instead was based on KQML communication. Safety was supported by protected *secure domains* spanning a number of nodes, where it was assumed that all nodes within a single domain were friendly.

More recently the *SSErl* prototype was developed by Brown [9] whilst on his sabbatical in 1997 at SERC [1] and NTNU[2] to address the perceived deficiencies of the previous prototypes. It supports a hierarchy of nodes on each Erlang system which provide a *custom context* for processes in them, the use of both *hash* and *password capabilities* for pids, ports, and nodes to constrain the use of these values; and remote module loading. This prototype has evolved through a number of versions in the process of refining and clarifying the proposed changes.

---

[1] Software Engineering Research Centre, RMIT, Melbourne, Australia
[2] Norwegian University of Science and Technology, Trondheim, Norway

Both of the latter prototypes implement the language extensions using glue functions for a number of critical BIFs. These are substituted by a modified Erlang compiler (which itself is written in Erlang). The glue routines interact with *node* server processes, one for each distinct node on the Erlang system. Most of the SSErl glue functions have the form:

- possibly resolve a registered name to capability
- check with the node manager to see if the operation is permitted by the capabilities rights, and if not, to throw an exception
- otherwise some key information is returned (eg a real pid or capability)
- perform the desired operation (in the user's process)

For example, the k_exit glue routine looks as follows:

```
k_exit(CPid,Reason) ->
    Pid = node_request(check,CPid,?exit), exit(Pid,Reason).
```

Both support a hierarchy of **nodes** within an Erlang run-time system (an existing Erlang **node**). Each node is represented by a node manager process, which manages the state for that node, and interacts with the glue routines to manage resource access.

SSErl capabilities are a tuple with the components identified previously: {Type,NodeId,Value,Rights,Private} which specify the type of resource the capability references, the node managing that resource, the resource instance, the list of access rights permitted on the resource, and the opaque validation value (crypto hash or password) for the capability.

To support these new features, SSErl provides some new BIFs:

check(Capa,Op) checks if the supplied capability is valid and permits the requested operation, throwing an exception if not. This is not a guard test as it must consult the originating node to validate the capabilities check value.

halt(Node) halt a node along with all its nodes and processes.

make_capa(Value) create a user capability with value given.

make_mid(Module) create a module (mid) capability for the named module.

newnode(Parent,Name,Opts) creates a new node as a child of the Parent, with the specified context options.

restrict(Capa,Rights) creates a new version of the supplied capability, referring to the same resource, but with a more restricted set of rights.

same(Capa1,Capa2) guard testing whether the supplied capabilities refer to the same resource, without verifying the check value (for efficiency reasons).

Except when explicitly configuring the execution environment, the new features are mostly invisible to existing user programs. The SSErl prototype successfully compiles many of the standard library modules (only those interacting with ports require some minor, systematic, changes necessitated by the protocol currently used). It is now being used to trial some demonstration applications.

These prototypes have demonstrated that the Erlang language can be successfully extended to support safe execution environments with minimal visible impact on most code. In the future we anticipate that these extensions will be incorporated into the Erlang run-time environment. This should remove some unavoidable incompatibilities (such as with ports) found in the prototypes, as well as ensuring that these safety extensions cannot be bypassed.

4.1 Safe Erlang Applications

Our work has mainly concentrated on design and implementation issues. However, a couple of simple applications have been successfully implemented within the prototypes, validating the basic approach we have proposed.

In 1998, Otto Björkström, implemented a distrubuted game where a number of players share a common board, each taking their turn in order.

The board itself is implemented with a server, and to enter the game a player only needs to send a message to the server containing a reference to a local protected node where the server will spawn off a process representing the player. Code loading is thus made transparent. The local process is quite restricted, but may draw any graphics within a certain window on the player's screen.

Being able to draw any graphics on a screen makes the user vulnerable to a "Trojan Horse" attack as the process might draw new windows asking for sensitive information such a passwords. We have been contemplating drawing a distinctive border around windows controlled by remote code to warn the user about this potential hazard.

The second application implemented concerned the remote control of a telephone exchange. Here no code was spawned, and the essential functionality was to prevent anybody else from taking over control of the exchange. In fact, a system without encryption, just having an authentication mechanism would be sufficient for this application. In countries where use encryption is restricted, this might be an interesting mode of operation.

5 Further Work - Custom Safety Policies

The extensions described above provide mechanisms necessary to impose an appropriate safety policy, without enforcing any specific policy. Work is now continuing on providing an appropriate level of *system safety* by executing the mobile code in a constrained environment created using these extensions, trusted to prevent direct access to any external resources by the executing mobile code, and then applying filters to messages between the mobile code and server processes which mediate the access to resources. This allows the use of standard servers, but with distinct policies being imposed for each instance. By restricting the security problem domain to just verifying that the messages conform to the desired policy, the problem should be considerably more tractable than approaches based on proofing the imported code directly, for example. It is also much more efficient than approaches which attempt to validate all function calls. The use of such a safety check function on client-server messages is described in Brown [10].

6 Conclusions

This paper describes proposed extensions to Erlang to enhance its ability to support safe execution of remotely sourced code. The changes include the provision of a hierarchy of nodes to provide a *custom context*, restrictions on *side-effects*, and resource limits; and the use of unforgeable references (capabilities) with associated rights of use for modules, nodes, processes, ports, and user defined references. Further extensions are required to support the remote loading of code with module references interpreted in context. Two experimental prototypes, used to evaluate the utility of these extensions, are then described. Finally, mention is made of ongoing work using this safe execution environment as a foundation for imposing custom safety policies by filtering messages to server processes.

References

[1] Ali-Reza Adl-Tabatabai, Geoff Langdale, Steven Lucco, and Robert Wahbe. Efficient and Language Independent Mobile Programs. *SIGPLAN*, 31(5):127–136, May 1996.

[2] M. Anderson, R.D. Pose, and C.S. Wallace. A Password Capability System. *The Computer Journal*, 29(1):1–8, 1986.

[3] J. Armstrong. Erlang - A Survey of the Language and its Industrial Applications. In *INAP'96 - The 9th Exhibitions and Symposium on Industrial Applications of Prolog*, Hino, Tokyo, Japan, Oct 1996. http://www.ericsson.se/cslab/erlang/publications/inap96.ps.

[4] J. Armstrong, R. Virding, C. Wikstrom, and M. Williams. Concurrent Programming in Erlang. Prentice Hall, 2nd edition, 1996. http://www.erlang.org/download/erlang_book_toc.html.

[5] Joe Armstrong. The Development of Erlang. In *Proceedings of the ACM SIGPLAN International Conference on Functional Programming*, pages 196–203. ACM, 1997.

[6] Ken Arnold and James Gosling. *The Java programming Language*. Addison-Wesley, 2nd edition, 1998. 0201310066.

[7] M. Bellare, R. Canetti, and H. Krawczyk. Keyed Hash Functions and Message Authentication. In *Advances in Cryptology - Proceedings of Crypto'96*, volume 1109 of *Lecture Notes in Computer Science*, pages 1–15. Springer-Verlag, 1996. http://www.research.ibm.com/security/keyed-md5.html.

[8] L. Brown. Mobile Code Security. In *AUUG 96 and Asia Pacific World Wide Web 2nd Joint Conference*, pages 46–55, Sydney, Australia, Sept 1996. AUUG. http://www.adfa.edu.au/~lpb/papers/mcode96.html.

[9] L. Brown. SSErl - Prototype of a Safer Erlang. Technical Report CS04/97, School of Computer Science, Australian Defence Force Academy, Canberra, Australia, Nov 1997. http://www.adfa.edu.au/~lpb/papers/tr9704.html.

[10] L. Brown. Custom Safety Policies in SSErl. Technical note, School of Computer Science, Australian Defence Force Academy, Canberra, Australia, Jun 1999. http://www.adfa.edu.au/~lpb/research/sserl/sspol99.html.

[11] Dan Connolly. Issues in the Development of Distributed Hypermedia Applications, Dec 1996. http://www.w3.org/OOP/HyperMediaDev.

[12] Dan Connolly. Mobile Code, Dec 1996. http://www.w3.org/MobileCode/.

[13] D. Dean, E.W. Felten, and D.S. Wallach. Java Security: From Hotjava to Netscape and Beyond. In *Proceedings IEEE Symposium on Security and Privacy*. IEEE, May 1996. http://www.cs.princeton.edu/sip/pub/secure96.html.

[14] Brant Hashii, Manoj Lal, Raju Pandey, and Steven Samorodin. Securing Systems Against External Programs. *IEEE Internet Computing*, 2(6):35–45, Nov-Dec 1998.

[15] John Hughes. Why Functional Programming Matters. *Computer Journal*, 32(2):98–107, 1989. http://www.cs.chalmers.se/~rjmh/Papers/whyfp.ps.

[16] I. Jonsson, G. Naeser, D. Sahlin, and et al. Adapting Erlang for Secure Mobile Agents. In *Practical Applications of Intelligent Agents and Multi-Agents: PAAM'97*, London, UK, Apr 1997. http://www.ericsson.se/cslab/~dan/reports/paam97/final/paam97.ps.

[17] H. Krawczyk, M. Bellare, and R. Canetti. HMAC: Keyed-Hashing for Message Authentication. Rfc 2104, IETF, Feb 1997.

[18] Xavier Leroy. Objective CAML. Technical report, INRIA, 1997. http://pauillac.inria.fr/ocaml/.

[19] S. Lucco, O. Sharp, and R. Wahbe. Omniware: A Universal Substrate for Mobile Code. In *Fourth International World Wide Web Conference*. MIT, Dec 1995. http://www.w3.org/pub/Conferences/WWW4/Papers/165/.

[20] Gary McGraw and Edward W. Felton. *Java Security: Hostile Applets, Holes, and Antidotes*. Wiley, 1997. 047117842X.

[21] G. Naeser. Your First Introduction to Safeerlang. Technical report, Dept. Computer Science, Uppsala University, Sweden, Jan 1997. ftp://ftp.csd.uu.se/pub/papers/masters-theses/0109-naeser.ps.gz.

[22] Scott Oaks. *Java Security*. O'Reilly, 1998. 1565924037.

[23] J.K. Ousterhout, J.Y. Levy, and B.B. Welch. The Safe-Tcl Security Model. Technical report, Sun Microsystems Laboratories, Mountain View, CA 94043-1100, USA, Nov 1996. http://www.sunlabs.com/research/tcl/safeTcl.ps.

[24] Aviel D. Rubin, Daniel E. Geer, and Jr. Mobile Code Security. *IEEE Internet Computing*, 2(6):30–34, Nov-Dec 1998.

[25] Dan Sahlin. The Concurrent Functional Programming Language Erlang - An Overview. In *Workshop on Multi-Paradigm Logic Programming, Joint Intl. Conf. and Symposium on Logic Programming*, Bonn, 1996. http://www.ericsson.se/cslab/~dan/reports/mplp/web/mplp.html.

[26] Erlang Systems. Open Source Erlang Distribution, 1999. http://www.erlang.org/.

[27] Andrew S. Tanenbaum, Robbert van Renesse, Hans van Staveren, Gregory J. Sharp, Sape J. Mullender, Jack Jansen, and Guido van Rossum. Experences with the Amoeba Distributed Operating System. *Communications of the ACM*, 33(12):46–63, Dec 1990.

[28] Tommy Thorn. Programming Languages for Mobile Code. *ACM Computing Surveys*, 29(3):213–239, Sept 1997.

[29] Dan S. Wallach, Dirk Balfanz, Drew Dean, and Edward W. Felten. Extensible Security Architectures for Java. In *Proceedings of the Symposium on Operating Systems Principles*, pages 116–128. ACM, 1997.

[30] C. Wikstrom.   Distributed Programming in Erlang.   In *PASCO'94 - First International Symposium on Parallel Symbolic Computation*, Sep 1994. http://www.ericsson.se/cslab/erlang/publications/dist-erlang.ps.

[31] F. Yellin.   Low Level Security in Java.   In *Fourth International World Wide Web Conference*. MIT, Dec 1995. http://www.w3.org/pub/Conferences/WWW4/Papers/197/40.html.

Detachable Electronic Coins

Chris Pavlovski, Colin Boyd, and Ernest Foo

Information Security Research Centre
School of Data Communications
Queensland University of Technology
Brisbane, Australia

chripavl@aul.ibm.com, {boyd, foo}@fit.qut.edu.au

Abstract. *We propose a non-divisible electronic cash system that combines the restrictive blind signature technique and a batch signature scheme based upon the binary tree structure. Contemporary electronic cash solutions rely on the divisibility paradigm to achieve significant computational savings, and such techniques are considered to be the most efficient. We demonstrate an alternative approach that employs batch cryptography to withdraw and spend multiple coins. This approach provides a set of alternative primitives for an electronic cash system that possesses comparable efficiency to the most efficient divisible based schemes.*

1. Introduction

Electronic cash is the virtual manifestation of physical cash, endeavoring to possess similar properties to physical cash including portability, transferability, and anonymity. Additionally, electronic cash is subject to the same attempts at fraudulent use as physical cash.

Physical cash systems essentially involve three entities: the merchant, customer and bank. In much the same way, electronic cash systems involve these entities; however, the presence of physical notes and coins is replaced by a bitstring that may be easily copied and reused. To masquerade bit strings as cash, a number of security properties must be satisfied. These include identity detection of a double spending party, the ability to conduct payments without being traced, and difficulty in forgery. Whilst there are a number of other desirable features, such as prior restraint of double spending, these properties appear to be the primary constituents of many electronic cash systems.

Electronic cash propagates through society under three prominent transaction analogies; these are the withdrawal, payment, and deposit transactions. The customer, after identification, establishes an account with a bank, and is then able to obtain electronic notes and coins from the bank by way of the withdrawal transaction. To purchase items the customer presents the electronic cash to a merchant during the payment transaction. And finally, a merchant is able to redeem hard currency for the electronic cash by conducting the deposit transaction with the bank. A key feature of any proposed electronic cash system is the ability to conduct payment in an off-line manner, i.e. without the involvement of a third party, particularly the bank. Furthermore, both the withdrawal and payment transactions should be efficient, this is quite crucial as merchants and banks represent a central processing point in a distributed cash environment.

Despite the intense research of this area [3, 8, 13, 15, 16, 14, 5], electronic cash is yet to make significant inroads into the commercial sector. A major reason for this predicament is the inability of commercial entities to conduct electronic cash transactions in an efficient manner. In this paper we propose a new electronic cash system that provides efficient withdrawal and payment, and exhibits computational advantages for deposit. The scheme differs considerably from other electronic cash schemes in that it does not rely on a divisibility for improving efficiency when dealing with multiple coins. Rather, we demonstrate an alternative approach where multiple coins are withdrawn and spent using batch protocols. We demonstrate that the efficiency exceeds previous schemes [3, 14] and is comparable to the most efficient cash protocols that employ a divisible structure [5]. The fundamental protocols are based upon the restrictive blind signature schemes introduced by Brands [3] and the batch protocols proposed in [17, 2, 21].

1.1 Contribution

An electronic cash scheme [1] that was recently proposed by us employs the concepts of batch cryptography to improve the computational efficiency of withdrawing non-divisible coins. This scheme has similarities to previous work of Tsiounis [19]. A significant drawback of the scheme is that the signature size becomes unacceptably large for an increasing batch size n, increasing as $O(n)$. For example, when withdrawing 1,000 coins in batch the associated signature, which is part of any one coin, is 20 Kbytes. We avoid this linear size increase by applying the batch signature scheme of [17] using a binary tree structure. Under our proposed new scheme, the corresponding signature residue now becomes 200 bytes!

In this paper, we view our main contribution as proposing the first practical electronic cash system that provides an efficient alternative to divisible based cash schemes. The efficiency gains are by way of improvements in computation, communications and storage using novel batch signature and verification techniques based upon a binary tree structure. Specifically, we solve the problem of an increasing signature size for batch coins, intrinsic to [1], by reducing it to $O(\log n)$. Furthermore, we introduce several additional efficiency improvements in computation and storage.

We demonstrate that our scheme is more efficient than other electronic coin schemes [3, 14] and observe the scheme to be comparable to the most efficient divisible based systems exhibiting similar e-cash properties [5].

1.2 Organisation of Paper

The next section provides background techniques used in the electronic cash scheme we present, including details of the new concepts applied in our scheme. This is followed in section 3 by a detailed description of the new cash scheme. We then assess the security and analyse the improvements in efficiency.

2. Background

We describe the various constructions that provide the foundations of the new electronic cash system. Commencing with Brands' restrictive blind signatures and representation problems, we propose a Schnorr batch tree signature structure for improved efficiency.

2.1 Restrictive Blind Signatures

Brands [3] introduced a restrictive blind signature that extends the protocol of Chaum and Pedersen [7]. The term *restrictive* is evident in the protocol as the customer is restricted to a particular form when blinding coins, more precisely he must encode his identity. In the case of [3], the customer *must* know the representation $(u_1, 1)$ of A with respect to (g_1, g_2) (see below for a discussion of the representation problem). This restriction (on the blinding of the internal structure of coins) will enforce the ability to reveal the identity of a customer if there is an attempt to double spend. Of course if the restricted form is not adhered to, the customer does not end up with a valid signature on the blinded coins.

As in all the subsequent equations, two prime numbers p and q must be used, where q is a prime factor of p - 1, and an element g of order q. All computations take place modulo p unless otherwise stated. The public and private keys are x and h, where $0 < x < q$, and $h = g^x \bmod p$. The restrictive blind signature scheme is as follows.

Stage 1. The signer generates a random values $w \in Z_q$ and sends $z = m^x$, $a = g^w$, and $b = m^w$ to the verifier.

Stage 2. The receiver generates blinding invariants s, u, and $v \in Z_q$ and computes the following (returning c to the signer):

$$m' = m^s$$
$$a' = a^u g^v$$
$$b' = b^{su} m'^v$$
$$z' = z^s$$

$c' = H(m', z', a', b')$, and returns the blinded challenge $c = c'/u \bmod q$ to the signer.

Stage 3. The signer computes the response $r = cx + w \bmod q$ to the verifier.

Stage 4.   The receiver accepts the signature as valid if the following hold:

$$g^r = a \ h^c$$
$$m'^r = b \ z^c$$

The receiver computes $r' = ru + v \bmod q$, then ends up with the message m' blindly signed by the signer with corresponding signature (z', a', b', r'), which may be verified by others.

2.2   Representation problem

The representation problem, introduced by Chaum *et al.* [4], underpins the protocols of Brands' cash system. Simply put, the representation problem is: given a value h and the pair (g_1, g_2), find a pair (x_1, x_2) such that $h = g_1^{x_1} g_2^{x_2}$. In fact any number of g_i values can be used, and it can be seen that the representation problem is a generalisation of the discrete log problem which is the case of a single g value.

A coin in Brands' scheme [3] is a pair (A,B) and during the payment protocol the customer must prove knowledge of B's representation. This is achieved by proving knowledge of (x_1, x_2), showing that he did in fact take part in withdrawing the coins. Additionally, under the restrictive blind signature technique the customer must know a representation of A with respect to (g_1, g_2). The following protocol demonstrates how one may prove knowledge of a representation (x_1, x_2) of h with respect to (g_1, g_2).

Stage 1.   The prover generates random values $w_1, w_2 \in Z_q$ and sends $z = g_1^{w_1} g_2^{w_2}$ to the verifier.

Stage 2.   The verifier responds with a random challenge $c \in Z_q$.

Stage 3.   The prover computes responses $r_1 = w_1 + cx_1 \bmod q$, and $r_2 = w_2 + cx_2 \bmod q$, and forwards them to verifier.

Stage 4.   The verifier accepts that the prover knows a representation, if the following holds:

$$z \ h^c = g_1^{r_1} g_2^{r_2}$$

2.3   Batch Cryptography

The concept of batch cryptography to improve efficiency was first introduced by Fiat [9], through a technique that combined several messages for batch signing under RSA. The concept was further applied to Diffie-Hellman key agreement [20] and to DSA batch verification [12]. Further work concerning batch verification has been discussed for both the Schnorr and RSA digital signature schemes [21]. More recently it has been demonstrated [2] that some batch verifiers are not true verifications and that a weaker property called *screening* applies.

We next describe an alternative technique for signing a batch of messages, the objective of which is to reduce the size of the generated signature. In this paper, use is made of a binary tree structure to combine several messages into one batch for signing.

2.4 Extensions to Previous Schemes

Our major extension involves a Schnorr variant of a batch signature scheme that employs a binary tree structure. A detailed description of this technique may be found in a previous paper [17]. We also discuss several improvements in the efficiency of the batch representation check. Before describing the batch tree protocol with the Schnorr signature scheme [18], we must briefly outline the batch signature scheme for an arbitrary number of messages.

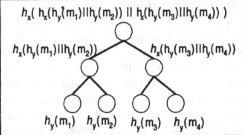

Figure 1 - Binary Tree of Messages

To generate a signature for a *batch* of messages, the hash of each message is placed at a leaf node of a binary tree. Each parent node is formed by concatenating each of its child nodes and hashing the result. This continues until the root node T is obtained where the final value is signed using the Schnorr signature scheme. In order to prevent messages representing the internal nodes from having valid signature two different hash functions are employed, h_y to hash the leaves and the function h_x to hash the internal nodes. In forming the batch signature of any individual message, we consider the unique path from the node representing the message to the root of the tree; this is found by taking the parent node of each node traversed. The batch residue for that message consists of the sibling nodes of all nodes in this path, together with their direction, which is a single bit denoted L or R. An example with four messages is shown in Figure 1. In this example the residue of m_1 consists of the two nodes $(h_y(m_2), R)$ and $(h_x(h_y(m_3)\| h_y(m_3)), R)$ while the residue for m_3 consists of the pair $(h_y(m_4), R)$ and $(h_x(h_y(m_1)\| h_y(m_2)), R)$.

More formally, for message m_i the residue $T_{res\,i}$, the sequence of nodes with their direction (dir), may be expressed as:

$$
\begin{aligned}
&c := h_v(m_i) \\
&\text{for } j := 1 \text{ to } L \\
&\qquad r_i := \text{sibling}(c) \\
&\qquad \text{if direction}(c) = \text{Left then } c := h_x(c \parallel r_i) \\
&\qquad \text{else } c := h_x(r_i \parallel c) \\
&\text{endfor} \\
&T_{res\,i} = r_1, \text{dir}(r_1), r_2, \text{dir}(r_2), \ldots, r_L, \text{dir}(r_L)
\end{aligned}
\tag{1}
$$

During verification the root node T must be recomputed to ensure that purported signature is in fact a signature on message m_i, this is given by:

$$T' := h_v\,(m_i) \tag{2}$$

```
for j := 1 to L
    r_i := sibling(c)
    if direction(r_i) = Left then c := h_x(c ∥ r_i)
    else T' := h_x(r_i ∥ c)
endfor
Output T'
```

2.4.1 Batched Schnorr Signature Scheme

The Schnorr signature scheme [18] has public and private keys x and h, where $0 < x < q$, and $h = g^x$ mod p. To sign message *m* with the private key x, outputting signature (c, r), the following steps are performed:

1. $a = g^w$ mod p, where $1 < w < q - 1$.
2. $c = h(m \parallel a)$
3. $r = w + xc$ mod q

To verify the signature (c, r), compute $a' = g^r \cdot h^c$ mod p, and check that $c = h(m \parallel a')$.

Adapting the protocol, we are able to generate *n* signatures $(C, r, T_{res\,i})$ on a sequence of *n* messages m_i, for i =1 to *n*, in the following manner:

1. $a = g^w$ mod p, where $1 < w < q - 1$.
2. Assign messages as leaves of the tree and compute root node T.
3. $C = h(T \parallel a)$
4. $r = w + xC$ mod q
5. Using algorithm (1) compute $T_{res\,i}$, for i = 1 to *n*.

Any entity is able to verify the individual signature $(C, r, T_{res\,i})$, on an associated message m_i. To verify compute $a' = g^r \cdot h^c$ mod p, compute T' using algorithm (2), and check that $C = h(T' \parallel a')$. The receiving entity is able to independently verify the signature of the sender on message m_i. The security of this scheme is proven in [17].

Examining Brands' protocol [3], we rely on the separation of m into A and B, with the identity encoded within A to detect double spending. We replace B however, with a sequence of B_i values, one for each coin, signed with the batch signature tree structure. We do this in way that offers *detachability*, rather than divisibility for the partitioning of coins.

2.4.2 Representation Screening and Improved Batch Verifiers

We introduce extensions to the payment and deposit transactions of [1], improving the efficiency for the merchant and bank respectively. In particular, we discuss whether alternative, more efficient, batch representation checks may be applied (namely those of Bellare *et al.* [2]). We show (see Appendix A) that an improved batch verifier, the Bucket test, may be useful to the merchant during payment, and that *representation screening* may be practiced by the bank during deposit, improving the efficiency of the deposit transaction.

3. Detachable Coin Scheme

The customer selects a secret number $u_1 \in_R Z_q$, computes the account number $I_c = g_1^{u_1}$, and forwards this to the bank. The bank stores the customer's details, unique account number I_c, and publishes the bank's public key $h = g^x$, for private key x. The bank also publishes g_2^x, permitting later computation of $z = (I_c g_2)^x$ by the customer. Different generators for g_2 may be used to identify both the denomination and the number of coins to be withdrawn in the cash system. In this environment, we use g_2 to indicate the number of coins that are withdrawn at any one time. This is later used by both the merchant and bank to confirm that a correct number of coins have been withdrawn and that there are not counterfeits.

In our cash system each coin is identified by (A, B_i, i), $i \in \{1...n\}$, and its signature $\sigma(A, B_i) = (T_{res\,ij}, z', a', b', r')$. A batch of coins is identified by $(A, B_j, ... B_k, i_j ... i_k)$ and signature $\sigma(A, B_j, ... B_k)$. The coins are valid when the following equations are satisfied (with suitable hash function $\mathcal{H}$):

$$g^r = a \cdot h^{\mathcal{H}(A, T, z', a', b')}$$
$$A^r = b \cdot z^{\mathcal{H}(A, T, z', a', b')}$$

The merchant can be sure that the customer presenting the coins took part in the withdrawal protocol when the customer is able to prove knowledge of the coin's secret representation pair $(x1, x2)$ with respect to (g_1, g_2).

3.2 Withdrawal Protocol

The withdrawal protocol enables the customer to receive a signature on a batch of coins, in a way that enforces the encoding of the customer's identity for each coin in the batch. For simplicity, we follow the notation of Brands [3] and derivations thereof [1]. In the interests of efficiency, we also assume that the customer precomputes the electronic coin tokens B_i prior to interaction with the bank; we describe this step first.

The customer manufactures a sequence of n electronic coin tokens, the number n determined by the bank's advertised generator g_2, encoding his identity and maintaining a representation with respect to (g_1, g_2). Choosing blinding invariant s, and initial secret representation values $x1_1, x2_2 \in_R Z_q$, remaining values are computed as $x1_i = h(x1_{i-1})$ and $x2_i = h(x2_{i-1})$ for $i = 1$ to n. The customer builds a representation of each coin by computing $A = (I_c g_2)^s$, $B_i = (g_1^{x1i} g_2^{x2i})$. Each coin (A, B_i) encodes the identity I_c, and is known to the customer by the unique representations $(u_1 s, s)$ and $(x1_i, x2_i)$ with respect to (g_1, g_2). The coins are then marshalled into a tree to build the unique structure representing the batch of coins. Commencing with a hash of each coin token B_i assigned to the leaf nodes, the tree is built until the root node T, and unique presentation of the coin batch is constructed. At this point the customer has encoded the identity into several batch coins for an interactive signature from the bank, of which the representation is known. The internal node values are not discarded as they may be used later during payment when forming the signature for the unique selection of coins.

Before commencing the on-line phase of the withdrawal protocol, it is assumed that the customer has proven his identity to the bank over an authenticated channel. The bank chooses a secret value $w \in Z_q$, and sends to the customer (blinded) values $a = g^w$, $b = m^w$. The customer, using the previously manufactured sequence of coins, now performs a number of steps that involve generating the batch of blinded coins for signatory endorsement. In order to hide the true identity of the customer from the bank during signing, the customer chooses additional blinding invariants u and v $\in Z_q$, computes new values $a' = a^u g^v$, $b' = b^{su} A'$, and $z' = z^s$. A challenge $c' = \mathcal{H}(A, T, z', a', b')$ is formed and the blinded translation $c = c'/u$ is then forwarded to the bank.

The bank responds to the challenge with $r = cx + w \bmod q$ and deducts the customer's account by the amount withdrawn. The customer confirms the presence of a correctly formed signature by checking $g^r = a\, h^c$ and $(I_c g_2)^r = b\, z^c$. The signature on the batch of coins now becomes the tuple $(T_{res i}, z', a', b', r')$, where r' is derived as $r' = ru + v \bmod q$. The following steps summarise the protocol for withdrawal, (where stage 1 may be performed prior to the on-line phase consisting of stages 2 onwards):

Stage 1.   The customer randomly chooses $s, x1_i, x2_i \in_R Z_q$, encoding the identity $A = (I_c g_2)^s$, and constructing a unique representation of each coin as $B_i = (g_1^{x1i} g_2^{x2i})$, where $x1_i = h(x1_{i-1})$ and $x2_i = h(x2_{i-1})$. The coins, forming the leaf nodes, are now marshalled into a binary tree applying function h_y to leaf nodes and h_x to internal nodes until the root node T is constructed.

Stage 2.   The bank chooses $w \in Z_q$, computes and sends $a = g^w$, $b = m^w$ to the customer.

Stage 3.   Choosing blinding invariants u and $v \in Z_q$, new values of a, b and z are computed as $a' = a^u g^v$, $b' = b^{su} A'$, $z' = z^s$. The value $c' = \mathcal{H}(A, T, z', a', b')$ is computed with blinded challenge $c = c'/u \bmod q$ returned to the bank.

Stage 4.   The bank returns $r = cx + w \bmod q$ to the customer.

Stage 5.   The verifier accepts the signature as valid if the following hold:

$$g^r = a\, h^c$$
$$(I_c g_2)^r = b\, z^c$$

If the verification holds the customer computes $r' = ru + v \bmod q$. Each coin forming the batch may be individually identified by $\{A, B_i, i, T_{res i}, z', a', b', r'\}$, where $\{T_{res i}, z', a', b', r'\}$ is the signature on the unique pair (A, B_i). Recall that $T_{res i}$ is composed of each highest level node not in the path of the leaf node identified by B_i. Each coin may be detached from the tree and individually spent with different merchants. Multiple coins may also be spent with the same merchant during the same protocol exchange. For instance, three coins may be identified with the following bitstring sequence $\{A, B_1, B_2, B_3, i_1, i_2, i_3, T_{res 123}, z', a', b', r'\}$.

The customer has encoded the identity within each pair (A, B_i), such that double spending will reveal the identity I_c of the customer. As in the scheme of [1], the sequence of batch coins may be linked between different transactions; however, coins from different batches cannot be correlated. This linkability does not reveal the customer's identity, rather it allows the bank to identify which transactions are from the same customer (or rather, which batch of coins).

3.3 Payment Protocol

When a customer wishes to spend valid coin(s) at a merchant shop, the batch sequence $(A, B_j, ... B_k, i_j ... i_k)$ and corresponding signature $\sigma(A, B_j ... B_k)$ are forwarded to the merchant. Note that the number of coins to be spent is defined by the number of B constructions forwarded. The coins may be spent individually or in batch manner, and may be forwarded in no particular order. For each set of coins to be spent in one transaction, the unique batch residue $T_{res\,jk}$ (for leaf nodes j to k) must be constructed by selecting the appropriate intermediate nodes from the binary tree. Note that it is not necessary to recompute these values as they may be stored after the execution of algorithm (1).

On receipt of the coins, the merchant computes the challenge $d = \mathcal{H}(A, B_i ... B_j, i_j ... i_k, I_m, t)$, where t represents a date/time stamp and I_m is the identity of the merchant.

Using the customers' secret value u_1, and blinding invariant s, the customer demonstrates knowledge of the coins representation $(x1, x2)$ with respect to (g_1, g_2) by computing responses $r1_i$ and $r2_i$ for each coin i to be spent in the transaction:

$$r1_i = du_is + x1_i, ..., r1_j = du_is + x1_j$$
$$r2_i = ds + x2_i, ..., r2_j = ds + x2_j$$

On receipt of the response values, the merchant proceeds to verify the bank's signature on the coins and checks that the customer knows a representation of the coins with respect to (g_1, g_2).

The bank's signatures on the blinded coins are checked by computing T' using algorithm (2), and then confirming that $g^r = a'\cdot h^{\mathcal{H}(A.\,T.\,z'.\,a'.\,b')}$ and $A^r = b'\cdot z'^{\mathcal{H}(A.\,T.\,z'.\,a'.\,b')}$. The merchant confirms that the customer knows representation of the coins by first selecting the optimal batch verifier (the bucket test is selected as the batch size grows beyond 200). Assuming for the moment that the batch size is small, random security values $w_j ... w_k$ are chosen and the following is checked

$$(3)\quad \prod_{i=j}^{k} g_1^{r1_k w_k} g_2^{r2_i w_i} = \prod_{i=j}^{k} A^{dw_i} B_i^{w_k}.$$ The values w_j through w_k are required to

prevent the customer from choosing, with high probability, r values that would satisfy equation (3).

If both the verification and representation check succeed, the merchant accepts the coins as valid and may then forward the goods to the customer. Coins that are detected to be from the same batch may be stored together, optimising the storage consumed by coins. In a similar manner, the customer can remove the coin token values B_i from storage. Summarising the payment protocol, we view the following steps to spend a coin sequence $(B_j \dots B_k)$ from the same batch:

Stage 1. The customer creates the unique signature residue $T_{res\,jk}$, under algorithm (1), for coins to be spent and sends $\{A, B_j, \dots B_k, i_j \dots i_k, \sigma(A, B_j, \dots, B_k)\}$ to the merchant.

Stage 2. The merchant computes and returns challenge $d = \mathcal{H}(A, B_j \dots B_k, i_j \dots i_k, I_m, t)$.

Stage 3. The customer computes responses and forwards to the merchant:

$$r_{1j} = du_1s + x1_j, \dots, r_{1k} = du_1s + x1_k$$

$$r_{2j} = ds + x2_j \dots r_{2k} = ds + x2_k$$

Stage 4. The merchant selects the optimal batch verifier. Assuming that the batch size is small, compute T' using algorithm (2), choose $w_j, \dots, w_k \in Z_q$, and accept the coins as valid if the both the signature verification and representation check hold:

$$g^r = a' \cdot h^{\mathcal{H}(A, T, z', a', b')}$$

$$A^r = b' \cdot z'^{\mathcal{H}(A, T, z', a', b')}$$

$$\prod_{i=j}^{k} g_1^{r1_k\,w_k} g_2^{r2_i\,w_i} = \prod_{i=j}^{k} A^{dw_i} B_i^{w_k}$$

The merchant stores the spent coins and transaction details consisting of $\{A, B_j, \dots, B_k, i_j \dots i_k, \sigma(A, B_j, \dots, B_k), t, r_{1j}, \dots r_{1k}, r_{2j}, \dots r_{2k}\}$. Under this scheme, coins spent from the same batch may be linked together. This however, may be controlled by the customer by increasing the number of distinct batch withdrawals, as different batches cannot be linked.

3.4 Deposit Protocol

The merchant is able to redeem hard currency from the electronic cash during the deposit transaction with the bank. A transcript of the payment, including the merchant's identity, is forwarded to the bank. Similar to the merchant, the bank performs a signature verification, but performs a more efficient representation check by screening the coins' representation. This is completed by first recomputing the merchant challenge $d = \mathcal{H}(A, B_j \dots B_k, i_j \dots i_k, I_m, t)$, and then confirming that the signature and representation equations hold.

Stage 1. The merchant sends coins $(A, B_j, \dots B_k, i_j \dots i_k) \sigma(A, B_j \dots B_k)$, representation responses $(r_{1j}, \dots, r_{1k}, r_{2j}, \dots, r_{2k})$ and merchant particulars (t, I_m) to the bank.

Stage 2. The bank recomputes the merchants challenge $d' = \mathcal{H}(A, B_j \dots B_k, i_j \dots i_k, I_m, t)$.

Stage 3. The bank reconstructs the root node T using algorithm (2), verifies its own signature on the coin, and screens the representation of the coins:

$$g^{r'} = a' \cdot h^{\mathcal{H}(A, T, z', a', b')}$$

$$A^{r'} = b' \cdot z'^{\mathcal{H}(A, T, z', a', b')}$$

$$\prod_{i=j}^{k} g_1^{r1_i} g_2^{r2_i} = \prod_{i=j}^{k} A^d B_i$$

If either the representation screen or coin verification fails then the bank rejects the coins. Before crediting the merchant's account, the bank checks that the coins have not been deposited previously. This involves searching a database for the same coins, each one uniquely identified by (A, B_j) for $j = 1$ to n (note that each batch may be identified by A). If the coins have not been previously spent then the bank credits the merchant's account and stores the values $(A, B_1, \ldots , B_n, i_j \ldots i_k), (r_{1j}, \ldots, r_{1k}, r_{2j}, \ldots, r_{2k})$, t and I_m.

It should be noted that the general attack on RSA screening [6] does not apply to representation screening as duplicate messages constitute double spent coins and may be easily detected. It follows then that the pruning step of [2] is not necessary.

3.4.1   Revealing Identity of Double Spender

In the event that the bank detects doubling spending, either the merchant is attempting to deposit the same coin twice or the customer has double spent the same coin. As in Brands' original scheme [3], it is now straightforward to identify the perpetrator. If the value t is the same as the value in the database then the merchant is guilty, otherwise the customer has double spent the coin. The bank has at its disposal knowledge of the two relationships $r_{1i} = dus + x_{1i}$, $r'_{1i} = d'us + x_{1i}$, and $r_{2i} = ds + x_{2i}$, $r'_{2i} = d's + x_{2i}$. Using $I_c = g_1^{(r_{1i} - r'_{1i})/(r_{2i} - r'_{2i})}$, the bank is able to derive u, which proves that the customer has double spent, and is able to identify the customer through the account number I_c.

4.   Security of Screening

In this section we consider the security of representation screening. We first note that all other enhancements do not alter the security assumptions of previous schemes [3, 1] and therefore the same security arguments hold. As already noted, the security of the tree batched signature scheme follows from the security of Merkle's authentication tree [17, 11].

Recall that screening may be applied to batch verify a number of signatures, and in our cash system, this involves proving knowledge of the representation pairs $(x1_i, x2)$. In the following we argue that if the merchant performs a fully qualified batch verification, using a small exponents test or buckets test, then the bank can perform a representation screening operation to confirm that the customer has knowledge of the coin's representation and retains the ability to reveal customer identity when double spending occurs. We also show that if the merchant performs representation screening during payment, then it is possible for the customer to successfully double spend a coin without detection.

4.1 Bank Screening

As long as the merchant rejects false coins during payment, the bank is merely duplicating the merchant's check to ensure responses are of the correct form ($du_i s + x1_i$, $ds + x2_i$). Suppose that during deposit a double spent coin is detected. Then the bank can perform a full verification on the coin in question with both sets of deposited parameters. Recall that the merchant's identity is stored with the deposited coin and used in the re-calculated challenge value d'. There are two possibilities:

1. The response values deposited by the merchant are correct, in which case the full verification will hold for both sets of parameters. This means that the customer has double spent and its identity will be revealed along with the secret value u as proof of double spending.
2. The response values deposited by the merchant are incorrect. In this case the merchant has cheated and is trying to deposit the coin twice. Notice that this case includes the possibility that the merchant is in collusion with the customer. Since the merchant will always be caught if he colludes, there is no incentive for him to do so.

4.2 Merchant Screening

If the merchant only performs screening during payment, rather than a true verification, then it can easily be seen that the customer is able to double spend without identification. Under the assumption that the merchant performs a representation screen, of the form $\prod_{i=j}^{k} g_1^{r1_i} g_2^{r2_i} = \prod_{i=j}^{k} A^d B_i$, during payment, the customer is able to provide an alternative set of r-values that enables him to double spend a coin. For example, the customer may compute new r-values $r1_a' = (x.r1_a)$ and $r1_b' = (y.r1_b)$ and double spend the coin B_a whose representation is $r1_a = du_i s + x1_a$ and $r_{2a} = ds + x2_a$, as follows[1]. Select $x \in_R Z_q$ and compute y such that: $(x \times r1_a) + (r1_b - y) = r1_a + r1_b$, (e.g. If $r1_a = 3$, $r1_b = 7$, and $x=2$, then $y=3$). When mixed into a batch, these values would still enable the batch representation check to succeed.

When the merchant finally deposits the coin, the bank will detect the double spent coin B_a and will see that it is the customer who has defrauded the system (due to different merchant parameter t). Now, if the customer has provided a new set of

---

[1] It is sufficient to provide a false set of r1-values, with correspondingly correct r2-values, to double spend successfully.

r-values, that satisfy the representation check, it is straightforward to see that the bank will not be able to reveal the associated identity.

5. Efficiency

We now review the efficiency of the cash scheme. We exclude the hash operations and also precomputation, assuming that the precomputation is conducted during processor idle time. We conduct an informal comparison to divisible coin based schemes, Okamoto [14] and Easy Come Easy Go cash (*EasyGo*) [5]. Applying the analysis of [5], we demonstrate greater efficiency over Okamoto's scheme and a somewhat comparative efficiency to *EasyGo*.

Comparisons to the divisible schemes are conducted on the basis that the customer withdraws 1000 coins from the bank. We borrow the analysis of [5], and set the security equivalent to a 512 bit modulus, even though today a much larger value would be desirable. Consequently, detachable coin parameters are $p = 512$ bits and $q = 160$ bits, with a 160 bit digest function. Divisible schemes employ a 512 RSA modulus N. Given that the complexity of modular exponentiation is $O(\log |q| \log^2 |p|)$ [10], we note that some of the divisible scheme's operations (namely the square roots) are approximately 3.2 (=512/160) times more expensive than our modular exponentiations. This is calculated as each square root is equivalent in effort to an exponentiation, where the exponent value is equal to N (i.e. 512 bits). It is worth noting that this difference would be increased with a larger composite modulus such as would be required today for an implementation of divisible schemes.

5.1 Detachable Coins

The increase in the size of each coin is determined by the depth of the binary tree constructed during the tree build-up stage, translating to $(k + 2r + (2^k + r)) \times |h|$ additional bits for a tree with $2^k - 1$ leaf nodes of depth k and 2r of depth $k + 1$. For example, a balanced tree of depth 11 (accommodating 1024 coins) adds 10×160 bits to the signature (200 bytes). In contrast to the scheme in [1], 1023×160 bits is required for the signature residue (~20Kbytes). This represents a significant improvement over the linear increase, reducing signature size and consequently the required communication costs.

During withdrawal both the bank and customer are required to execute the protocol once for an arbitrary number of coins, the effort for the bank consists of 2 modular exponentiations, whilst the customer performs 6 modular exponentiations. During communications 188 bytes of data is transferred. In addition to a fixed 296 bytes, the customer provides storage for each withdrawn coin consisting of 64 bytes per coin.

The payment protocol involves two modular multiplications/additions per coin spent by the customer, whilst the merchant performs 4 modular exponentiations and

2 multi-exponentiations for an arbitrary number of coins received. Communications costs involve the transmission of each coin to be spent; this includes A, coin tokens B_i, response values r_i, and signature $(T_{resij}, z', a', b'\ r')$, giving $252 + 104n$ bytes plus $20 \log(n)$ signature residue bytes.

Similar to the merchant, the bank performs 4 modular exponentiations and 2 multi-exponentiations for an arbitrary number of coins received during deposit. The small exponents test introduces an additional computation to ensure that the representation check is valid. Borrowing the analysis of [2][2], a batch size of up to 200 remains efficient using the small exponents tests. However, as the batch size increases beyond this, the Bucket test becomes the more efficient. Furthermore, where screening is sufficient for the bank no additional exponentiations are required.

By extension of [1], it follows that our scheme is more efficient than Brands' [3] for the withdrawal and payment of multiple coins. We now compare against divisible based schemes, generally accepted as the most efficient paradigm for electronic cash.

5.2 Okamoto Divisible Cash

Okamoto [14] proposed the first true practical electronic cash scheme, recently improved upon by [5]. In our scheme the withdrawal of each detachable coin is realised with 2 multi-exponentiations and 4 modular exponentiations by the customer; the bank performs 2 modular exponentiations. In communication, the bank transmits about 512 bytes of data whilst the customer transmits only 128 bytes. Okamoto's scheme involves only one modular exponentiation and a 128 byte message for both the customer and bank. As suggested in [5], to accommodate unlinkability between different divisible coins an additional 4,000 exponentiations is required during each withdrawal.

The payment stage requires between 2 and 22 (average 11) exponentiations depending on the coin denomination, for both customer and merchant. As mentioned above, these exponentiations are 3.2 times more expensive, and computationally equivalent to 35 of our exponentiations. In contrast, 5 modular exponentiations and 2 multi-exponentiations are conducted by the merchant in our scheme. On average 2304 bytes are transmitted by the customer, whilst our payment involves communication of a 252 bytes + 104 bytes per coin, and $\log(n)$ signature residue bytes.

5.3 Easy Come Easy Go Divisible Cash

During withdrawal the merchant performs 2 exponentiations and the user conducts 10 exponentiations. At payment time, the merchant performs a similar initial computation to our scheme plus the addition denomination revelation phase, on

---

[2] Whilst the analysis is applied to batch signature verification, we note that this is also applicable to the batch representation check.

average contributing to an additional 11 modular square roots, equivalent to 35 of our modular exponentiations.

Our scheme may be viewed as computationally more efficient during the on-line phase of payment, whilst the size of a detachable coin is greater; that is, an additional O(log n) signature residue bytes per coin is incurred for single coin payments. However, our scheme possesses O(n) precomputation activities.

6. Conclusions

In this paper we have presented a practical electronic cash system that is not based upon the divisible coin paradigm. We applied the concepts of batch cryptography in a novel way to obtain detachability of coins for the partitioning of cash. The scheme overcomes a number of deficiencies presented in our previous batch coin scheme, including a significant reduction in coin signature size. We also analysed and discussed several additional general efficiency improvements. We conducted an informal comparison to the divisible-coin approach, noting that the efficiency of our scheme exceeds previous divisible cash schemes [14] and is comparable to other well known schemes based upon divisibility [5].

Acknowledgements

We would like to thank Khanh Quoc Nguyen for drawing our attention to the related work of Tsiounis [19].

References

1. C. Boyd, E.Foo and C. Pavlovski, Efficient Electronic Cash Using Batch Signatures, 4th Australasian Conference on Information Security and Privacy – ACISP 99, LNCS, Vol. 1587, Springer-Verlag, pp244-257, 1999.
2. M. Bellare, J. A. Garay, T. Rabin, Fast Batch Verification for Modular Exponentiation and Digital Signatures, Proceedings of Eurocrypt '98, LNCS, Vol. 1403, Springer-Verlag, 1998.
3. S. Brands, Untraceable Off-line Cash in Wallet with Observers, Advances in Cryptology - Crypto '93, LNCS, Springer-Verlag, Vol. 773, pp302-318, 1993.
4. D. Chaum, E. Evertse, J. Graaf, "An improved protocol for demonstrating possession of discrete logarithms and some generalisations", Eurocrypt '87, LNCS, Springer-Verlag, Vol 304, pp127-141, 87.
5. A. Chan, Y. Frankel, Y. Tsiounis, Easy Come - Easy Go Divisible Cash, Advances in Cryptology - Eurocrypt '98, LNCS, Springer-Verlag, Vol 1403, pp561-574, 1998.
6. JS. Coron, D. Naccache, On The Security Of RSA Screening, PKC '99, 1999.
7. D. Chaum, T. Pedersen, Wallet databases with observers, Crypto '92, pp89-105, 1992.

8. N. Ferguson, Single Term Off-Line Coins, Advances in Cryptology - Crypto '93, Springer-Verlag, pp318-328, 1994.
9. A. Fiat, Batch RSA, Crypto '89, LNCS, Vol. 435, Springer-Verlag, pp175-185, 1990.
10. N. Koblitz, A course in number theory and cryptography, Graduate Texts in Mathematics, Springer-Verlag, Vol. 114, 1987.
11. R. C. Merkle, A Certified Digital Signature, Advances in Cryptology - Crypto '89, Springer-Verlag, pp218-238, 1989.
12. D. Naccache, D. M'Raihi, D. Rapheali, S. Vaudenay, Can DSA be improved: complexity trade-offs with the digital signature standard, Proceedings of Eurocrypt '94, pp85-94, 1994.
13. K. Q. Nguyen, Y. Mu, V. Varadharajan, One-Response Off-line Digital Coins, Proceedings of SAC '97, 1997.
14. T. Okamoto, An Efficient Divisible Electronic Cash Scheme, Advances in Cryptology - Crypto '95, LNCS, Vol. 963, Springer-Verlag, pp438-451, 1995.
15. T. Okamoto, K. Ohta, Divertible Zero-Knowledge Interactive Proofs and Commutative Random Self-Reducibility, Advances in Cryptology - Eurocrypt '89, LNCS, Springer-Verlag, Vol. 34, pp481-946, 1989.
16. T. Okamoto, K. Ohta, Universal Electronic Cash, Advances in Cryptology - Eurocrypt '91, LNCS, springer-Verlag, pp324-337, 1992.
17. C. Pavlovski, C. Boyd, Efficient Batch Signature Generation Using Tree Structures, CrypTEC '99, Hong Kong, 1999.
18. CP. Schnorr, Efficient Signature Generation for Smart Cards, Advances in Cryptology - Crypto '89, Springer-Verlag, pp239-252, 1990.
19. Y. Tsiounis, Efficient Electronic Cash: New Notions and Techniques, Ph.D. Thesis, College of Computer Science, Northern University, Boston MA. June 1997.
20. Y. Yacobi, M. Beller, Batch Diffie-Hellman Key Agreement Systems and their Application to Portable Communications, Proceedings of Eurocrypt '92, Vol. 658, pp208-217, 1992.
21. S. Yen, C. Laih, Improved Digital Signature Suitable for Batch Verification, IEEE Transactions on Computers, Vol. 44, No. 7, July 1995.

Appendix A: Alternative Batch Verifiers

Bellare *et. al.* [2] introduce the notion of screening for batch verification. More specifically, they prove that a simplified batch verification does not represent a true signature verification but rather a weaker property called *screening* applies. They show that screening ensures that only the true signer could have participated in generating the purported signatures, even though the individual signatures presented may not be correct. In respect of the proposed cash system, this means the in order to derive the product Πr_i, the prover must know the representation values $x1_i$ and $x2$.

Several alternative batch verifiers are also introduced in [2] and may be fitted to the representation check. In our system, the security values w, introduced by Yen and Laih [21], are used to strengthen the batch representation check; in the context of [2] this is termed a batch *verifier*. This prevents the customer, with high probability, from choosing incorrect r-values that satisfy the representation check. Some alternative techniques include the *Small Exponents* (roughly the approach introduced by Yen and Laih), *Random Subset*, and the *Bucket* tests [2]. To optimise efficiency we point out that an alternative batch verifier, the Bucket test, should be used as the number of coins spent in batch increases. Adapting the Bucket test, we present a revised batch representation check, performed with a smaller exponent $|m|$ < $|w|$.

Select security parameter $m \geq 2$, optimally chosen as in [2], set $M = 2^m$, and let W = $w/(m-1)$. For batch size n, repeat the following W times[3]:

1. Choose $f_i \in \{1 \dots M\}$, for $i = 1 \dots n$.
2. For $j = 1$ to M, $F_j = \{i: f_j = j\}$, (i.e., assign the set of elements to F_j).
3. For $j = 1$ to M, $R1_j = \Sigma_{i \in Fj} r1_i \bmod q$, $R2_j = \Sigma_{i \in Fj} r2_i \bmod q$, $B' = \Pi_{i \in Fj} B_i$.
4. Select new values $w_1 \dots w_M, \in \{0,1\}^m$ and confirm that the representation check holds on the following batch $(R1_1, R2_1, d, A, B'_1) \dots (r1_M, r2_M, d, A, B'_M)$:

$$\prod_{k=1}^{M} g_1^{r1_k w_k} g_2^{r2_k w_k} = \prod_{k=1}^{M} A^{dw_k} B_k^{w_k}$$

It should be noted that the above verifier is less efficient than the small exponents test for small batches and is shown to be more efficient as the batch size grows beyond 200 in size[4]. We may assume that an optimal batch verifier from amongst these possibilities is selected by the merchant during payment depending on the number of signatures to be checked.

---

[3] Each individual test gives probability of cheating $2^{-(m-1)}$, hence we need to perform this $w/(m-1)$ times.

[4] Although our batch operations differ from [2], we assume 200 given that our exponentiation costs are consistent in form across batch verifiers.

Linear Secret Sharing with Divisible Shares

Josef Pieprzyk

Centre for Computer Security Research
School of Information Technology and Computer Science
University of Wollongong
Northfields Avenue
Wollongong 2522
Australia
josef@cs.uow.edu.au

Abstract. Shamir secret sharing is being considered in the broader context of linear secret sharing. It is shown that any Shamir scheme built over $GF(q^v)$ can be converted into its linear equivalent defined over $GF(q)$. A notion of uniform perfectness is introduced and it is proved that Shamir schemes built over $GF(q^v)$ are not uniformly perfect. Probabilistic linear secret sharing is next studied and bounds on probability that the resulting secret sharing is uniformly perfect are given. The probabilistic arguments are later used to show that secret sharing with shift derived from Shamir scheme allows to achieve a secret sharing which is uniformly perfect.

1 Introduction

Secret sharing introduced by Blakley [1] and Shamir [9] allows to hold a secret piece of information collectively by a group of participants. Each participant keeps a share of the secret. The secret can be reconstructed only if a big enough group of participants pool their shares together. Incorporated into cryptographic algorithms, secret sharing enables groups to perform collectively variety of cryptographic operations such as encryption, decryption, signature, etc.

Shares of the secret can be treated as atomic pieces of information which are given or withheld by their owners in their totality. As noted in [7], there is no reason why the shares should be treated as atomic entities because any secret sharing can be transformed into its equivalent in which participants are allowed to give pieces of their shares. This allows participants to increase the threshold of secret sharing just by agreement that they will donate smaller pieces of shares (also called subshares).

Karnin, Greene and Hellman [5] observed that Shamir secret sharing is a subclass of a broader class of linear secret sharing. We are going to investigate Shamir secret sharing and their relation to linear secret sharing. More precisely, we examine a Shamir secret sharing with atomic shares and a Shamir secret sharing with subshares and their conversion into their linear equivalent. We also define a notion of uniformly perfectness to characterise the ability of secret sharing to recover the secret from any big enough collection of subshares.

The paper is structured as follows. In Section 2, we present a motivation for this research and a rationale for a secret sharing design with shares divisible into independent pieces called subshares. Section 3 defines two classes of Shamir schemes which are studied in this paper. Linear secret sharing is introduced in Section 4 together with the notion of uniform perfectness. Probabilistic linear secret sharing is discussed in Section 5 and some probabilistic arguments are developed to evaluate uniform perfectness of the schemes obtained. Finally, Section 6 examines secret sharing with shift.

2 Motivation

Secret sharing allows a group of participants to hold collectively a secret. The collective ownership means that the decision about recovery of the secret is left to the group. Moreover, each participants is free to handle their shares – they can submit as little as the information that their shares are different from a specific value or belong to a specific set. On the other hand, participants can treat shares as atomic and donate them in their totality. Note that there is no way that the group can force a recalcitrant member who refuses to give her share atomically. On the contrary, the group may decide to pool their partial information about their shares. In some circumstances, the group may be strongly encouraged to pool subshares instead shares. These may include the following.

- Adjustment of the threshold which better reflects the current trust among the members of a group (this was considered in [7]).
- Ability to detect cheating by some participants. In this case, a share is divided into two pieces. One is used for secret recovery and the other is left for verification purposes.
- Protection of shares during a long period of time. This applies for secret sharing whose lifetime may extend for years or tens of years. Instead of storing a single share in one place, participants may divide their shares into subshares and store them in different places. The advantage of this is that even if some subshares have been corrupted or lost, the other can still be used without running typically very expensive share refreshment algorithm (for details for proactive secret sharing, see [4]).
- Necessity for access structure modification. Participants may modify the access structure via distribution of subshares (instead of shares). For instance, a participant who is no longer interested in secret sharing may evenly distribute her subshares among the other participants. On the other hand, a new participant can be given a share composed from single subshares donated by other participants.
- Hierarchical access structures are better supported by divisible shares.

Shamir secret sharing is by far the most popular and is easy to implement in any field. We study how the size of the field influences divisibility of shares.

3 Shamir Secret Sharing

Computations in Shamir scheme are done in a field $GF(q)$. The group of participants is $\mathcal{P} = \{P_1, \ldots, P_n\}$. Access structure Γ is a collection of all subsets of $\mathcal{P}$ allowed to recover jointly the secret. Secret sharing whose access structure is defined as

$$\Gamma = \{\mathcal{A} | \#\mathcal{A} \geq t\}$$

is called a t out of n threshold scheme or simply a (t, n) secret sharing ($\#\mathcal{A}$ stands for cardinality of the set $\mathcal{A}$).

Secret sharing is a collection of two algorithms: the dealer and the combiner. The dealer must be trusted. For a given secret, the dealer takes an access structure, a security parameter (typically, the size of the field in which the computations are performed), and computes shares for the participants. Shares are distributed to all participants via secure channels.

The combiner is activated by a group which wishes to recover the secret. Clearly, combiner must be trusted. Upon receiving shares from a group of participants, the combiner recovers the secret only if the group belongs to the access structure. Otherwise, the combiner fails to recover the secret with an overwhelming probability.

A (t, n) Shamir Scheme ([9]).

Dealer: For given parameters t and n, the dealer selects at random a polynomial $f(x) = a_0 + a_1 x + \ldots, x_{t-1} x^{t-1}$ by random choice of coefficients $a_i \in_R GF(q)$; $i = 0, \ldots, t-1$. The polynomial $f(x)$ is of degree at most $(t-1)$. Next, the dealer assigns an x_i co-ordinate to each participant P_i. These co-ordinates are public. Finally, the dealer computes shares $s_i = f(x_i)$; $i = 1, \ldots, n$ and communicates them secretly to corresponding participants. The secret is $s = f(0)$.

Combiner: It collects shares from a group of participants. Knowing t points $(x_{i_1}, s_{i_1}), \ldots, (x_{i_t}, s_{i_t})$, the combiner applies the Lagrange polynomial interpolation and recovers the polynomial $f(x)$ together with the secret $s = f(0)$.

A (t, n) Shamir Scheme with v Subshares ([7])

Dealer: The dealer assigns v co-ordinates $x_{i_1}, \ldots, x_{i_v}$ to each participant P_i; $i = 1, \ldots, n$. Co-ordinates are different from $0, 1, \ldots, v-1$. These co-ordinates are public. Next the dealer selects at random a polynomial $f(x) = a_0 + a_1 x + \ldots, a_{tv-1} x^{tv-1}$ where $a_i \in_R GF(q)$ for $i = 0, \ldots, tv - 1$. The polynomial $f(x)$ is of degree at most $tv - 1$. Finally, the dealer computes v subshares $s_{i_j} = f(x_{i_j})$; $j = 1, \ldots, v$ for the participant P_i; $i = 1, \ldots, n$. The subshares are distributed secretly to all corresponding participants. The secret is $s = (f(0), f(1), \ldots, f(v-1))$.

Combiner: Any group of participants who collectively pools tv subshares is able to successfully run the combiner algorithm. The combiner knowing tv

points, can use the Lagrange interpolation to recover the polynomial $f(x)$ and the secret $s = (f(0), f(1), \ldots, f(v-1))$.

To simplify the notation, we denote (t, n, v) Shamir scheme as the (t, n) Shamir scheme with v subshares.

4 Linear Secret Sharing

Karnin, Greene and Hellman [5] observed that Shamir schemes are, in fact, Reed-Solomon linear codes. Assume that we have two Shamir schemes. The first is designed in $GF(q^v)$ and it is a (t, n) threshold scheme. The second is built over $GF(q)$ and is a (t, n) threshold scheme with v subshares. We observe that the Shamir scheme in $GF(q^v)$ can always be converted into its equivalent with v subshares. To show this, note that shares in the scheme are $S_i = f(X_i)$ where $X_i \in GF(q^v)$ is public and $f(X)$ is the underlying polynomial of degree at most t. The relation between shares and the polynomial coefficients $(A_0, A_1, \ldots, A_{t-1})$ is captured by the following equation:

$$\begin{bmatrix} S_1 \\ S_2 \\ \vdots \\ S_n \end{bmatrix} = \mathcal{X} \times \begin{bmatrix} A_0 \\ A_1 \\ \vdots \\ A_{t-1} \end{bmatrix} = \begin{bmatrix} 1 & X_1 & \ldots & X_1^{t-1} \\ 1 & X_2 & \ldots & X_2^{t-1} \\ & & \vdots & \\ 1 & X_n & \ldots & X_n^{t-1} \end{bmatrix} \begin{bmatrix} A_0 \\ A_1 \\ \vdots \\ A_{t-1} \end{bmatrix} \tag{1}$$

The secret is $f(0) = A_0$. The public matrix $\mathcal{X}$ has n rows and t columns and any collection of t rows creates a nonsingular Vandermonde matrix. The arithmetics is done in $GF(q^v)$ modulo an irreducible polynomial $q(x)$ of degree v with coefficients in $GF(q)$. It is known [6] that any element from $GF(q^v)$ can be represented as a vector where the basis is $(1, \alpha, \alpha^2, \ldots, \alpha^{v-1})$ and α is the zero of the polynomial $q(x)$ or $q(\alpha) = 0$. Consider Equation (1). If the vector $(S_1, \ldots, S_n)$ is represented using the basis in $GF(q^v)$ then $S_i = (s_{i,1}, \ldots, s_{i,v})$ for $i = 1, \ldots, n$. If the same is done for the elements $A_i \in GF(q^v)$; $i = 0, \ldots, t-1$, then $A_i = (a_{i,1}, \ldots, a_{i,v})$ for $i = 0, \ldots, t-1$. The $n \times t$ matrix $\mathcal{X}$ is converted into a $nv \times tv$ matrix $\mathcal{Y}$ where its entries from $GF(q)$ can be found from Equation (1). Being more specific, assume that the vector $(S_1, \ldots, S_n)$ over $GF(q^v)$ is converted into its corresponding vector $(s_1, \ldots, s_{nv})$ over $GF(q)$. Similarly, the vector $(A_0, A_1, \ldots, A_{t-1})$ over $GF(q^v)$ is written as a vector $(a_0, \ldots, a_{tv-1})$ over $GF(q)$. Now we are looking for a $(nv \times tv)$ matrix $\mathcal{Y}$ such that

$$\begin{bmatrix} s_1 \\ s_2 \\ \vdots \\ s_{nv} \end{bmatrix} = \mathcal{Y} \times \begin{bmatrix} a_0 \\ a_1 \\ \vdots \\ a_{tv-1} \end{bmatrix} \tag{2}$$

The i-th column of $\mathcal{Y}$ can be identified by taking the vector $a^{(i)}$ which consists of all zeros except the i-th co-ordinate which is "1". The vector $a^{(i)}$ is now converted into its equivalent vector $A^{(i)} \in GF(q^v)$. The product $\mathcal{X} \times A^{(i)}$ gives a $(n \times 1)$ vector over $GF(q^v)$ which after conversion into $GF(q)$ gives the i-th column of $\mathcal{Y}$. So we have proved the following theorem.

Theorem 1. *A (t, n) threshold Shamir scheme built in $GF(q^v)$ can be converted into its linear equivalent over $GF(q)$ where each share splits into v subshares.*

Note that if participants have agreed to submit some number of their subshares, then the combiner may or may not be able to recover the secret knowing tv subshares. Ideally, one would expect that subshares should behave as independent pieces so every collection of tv subshares should allow to recover the secret.

4.1 Uniform Perfectness

Recall that secret sharing is *perfect* if the entropy of the secret $H(S)$ remains the same for all outsiders and participants in groups which do not belong to the access structure. This can be expressed as

$$H(S|\mathcal{A}) = \begin{cases} H(S) & \text{if } \mathcal{A} \notin \Gamma \\ 0 & \text{if } \mathcal{A} \in \Gamma \end{cases}$$

The notion of perfectness needs to be redefined if shares are non-atomic and participants can submit subshares.

Definition 1. *Given a (t, n) perfect threshold secret sharing scheme with shares divisible into v subshares. The scheme is called uniformly perfect if the following conditions hold*

$$H(S|\mathcal{A}) = \begin{cases} H(S) & \text{if } \#\mathcal{A} \leq (t-1)v \\ \frac{k}{v} H(S) & \text{if } \#\mathcal{A} = tv - k; \ k = 0, \ldots, v \end{cases}$$

where $\#\mathcal{A}$ stands for the number of subshares donated by the group $\mathcal{A}$ to the combiner.

Clearly, ramp schemes defined in [2] relate to uniformly perfect schemes. The relation between ramp schemes and thresholds schemes with divisible shares is discussed in [7].

Theorem 2. *Given a (t, n, v) Shamir secret sharing over $GF(q)$. Then the scheme is uniformly perfect.*

Proof. The entropy of the secret is $H(S) = \log_2 q^v$. Note that the subshares $(s_1, \ldots, s_{nv})$ are computed from

$$\begin{bmatrix} s_1 \\ s_2 \\ \vdots \\ s_{nv} \end{bmatrix} = \begin{bmatrix} 1 & x_1 & \ldots & x_1^{tv-1} \\ 1 & x_2 & \ldots & x_2^{tv-1} \\ & \vdots & \\ 1 & x_{nv} & \ldots & x_{nv}^{tv-1} \end{bmatrix} \times \begin{bmatrix} a_0 \\ a_1 \\ \vdots \\ a_{tv-1} \end{bmatrix}$$

For any collection of tv different subshares, the dealer creates the corresponding public matrix of powers of x co-ordinates. This is a Vandermonde matrix that is always nonsingular so the vector $(a_0, \ldots, a_{tv-1})$ is identified and secret recovered. For any smaller collection of $tv - k$ subshares, there are k free coefficients in the vector $(a_0, \ldots, a_{tv-1})$, that is the solution space consists of $GF(q)^k$ members. This also means that if $k \geq v$, the secret $(a_0, \ldots, a_{v-1})$ is completely undetermined and $H(S) = \log_2 q^v$. For $1 \leq k \leq v$, the space of possible solutions contains q^k equally probable members so the entropy $H(S|vt - k$ subshares $)$ is $\log_2 q^k = \frac{k}{v} \log_2 q^v$.

Theorem 3. *Given a (t, n) Shamir secret sharing over $GF(q^v)$. Then the scheme is not uniformly perfect.*

Proof. Recall that the matrix $\mathcal{X}$ of the scheme is

$$\mathcal{X} = \begin{bmatrix} 1 & X_1 & \ldots & X_1^{t-1} \\ 1 & X_2 & \ldots & X_2^{t-1} \\ & & \vdots & \\ 1 & X_n & \ldots & X_n^{t-1} \end{bmatrix}$$

where all entries are from $GF(q^v)$. Assume that arithmetics in $GF(q^v)$ is done modulo $q(x)$. Each element $A \in GF(q^v)$ can be represented as a vector $A = (a_1, a_2\alpha, \ldots, a_v\alpha^{v-1})$ where $a_i \in GF(q)$ for $i = 1, \ldots, v)$. The element α is the zero of the polynomial $q(x)$ or $q(\alpha) = 0$. Given three elements B, C, D from $GF(q^v)$ and $B = (b_1, \ldots, b_v)$, $C = (c_1, \ldots, c_v)$, $D = (d_1, \ldots, d_v)$. Let $B = CD$ and we take C as the multiplier. If we present B and D as vectors then the multiplier C must be a $v \times v$ matrix. The matrix representation of $C = (C_1, \ldots, C_v)$ where $C_i = (c_1, \ldots, c_v)\alpha^{i-1}$ for $i = 1, \ldots, v$. After conversion of entries from $GF(q^v)$ into their matrix representations over $GF(q)$, the matrix $\mathcal{X}$ becomes

$$\mathcal{X} = \begin{bmatrix} \begin{bmatrix} 1 & 0 & \ldots & 0 \\ & & \vdots & \\ 0 & 0 & \ldots & 1 \end{bmatrix} & [X_1] & \ldots & [X_1^{t-1}] \\ \begin{bmatrix} 1 & 0 & \ldots & 0 \\ & & \vdots & \\ 0 & 0 & \ldots & 1 \end{bmatrix} & [X_2] & \ldots & [X_2^{t-1}] \\ & & \vdots & \\ \begin{bmatrix} 1 & 0 & \ldots & 0 \\ & & \vdots & \\ 0 & 0 & \ldots & 1 \end{bmatrix} & [X_n] & \ldots & [X_n^{t-1}] \end{bmatrix}$$

where $[X_i^j]$ are $v \times v$ matrices for $i = 1, \ldots, n$ and $j = 1, \ldots, t-1$. Now we select first $tv + t$ rows of the matrix $\mathcal{X}$ and remove the following rows: $v, 2v, \ldots, tv$. The resulting matrix has tv rows and tv columns and the matrix is singular as the v-th column contains all zeros. This proves the theorem.

Due to the fact that first v columns of the matrix $\mathcal{X}$ are composed by $v \times v$ identity matrices, each participant has a collection of subshares which any single one depends on a single piece of the secret. Consequently, subshares are not interchangeable. There is an interesting question: *is it possible to convert a (t, n) threshold scheme in $GF(q^v)$ into its uniformly perfect equivalent over $GF(q)$?*

Consider a (t, n) threshold Shamir scheme over $GF(q^v)$. As shown previously, the scheme can be converted into a linear scheme over $GF(q)$ with the matrix $\mathcal{X}$ which describes the relation between the subshare vector and the vector A, i.e. $S = \mathcal{X} \times A$. Assume that we have a uniformly perfect (t, n) threshold Shamir scheme over $GF(q)$ with its matrix $\mathcal{X}_{Sh}$ so $S = \mathcal{X}_{Sh} \times A$. We can always convert the scheme defined by $\mathcal{X}$ into the scheme defined by $\mathcal{X}_{Sh}$ if we can find a linear transformation described by a $nv \times nv$ matrix T such that

$$\mathcal{X}_{Sh} = T \times \mathcal{X}$$

Note that matrices $\mathcal{X}_{Sh}$ and $\mathcal{X}$ are known and we are looking for the matrix T. This is a simple task of finding $(nv)^2$ unknowns having $nv \times tv$ linear equations. It is clear that we do not get a unique solution but instead we obtain space of solutions. Each one is a good one.

This conversion although possible has many drawbacks. The need of conversion may arise in two cases: (1) when the scheme has just been created and the dealer is still active, and (2) when the dealer is no longer active and the participants would like to make the conversion by themselves. In the first case, the conversion would amount to design a uniformly perfect Shamir scheme so the dealer may well redesign secret sharing using uniformly perfect Shamir scheme – in this case the conversion is not really an option. In the second case, however, participants may find the matrix T but to correct their subshares, they would need to collectively compute a new subshare vector S' such that $S' = T \times S$. Note that each participant knows their v co-ordinates of S and should know the same number of co-ordinates from S'. To do this without disclosing their subshares to other participants, they would need to engage themselves into a complex interaction with other participants such as those studied in [3]. This interaction may well exceed the effort needed to reissue a new threshold scheme (collectively or by the dealer).

The only practical solution for the conversion problem is to restrict the form of transformations defined by the matrix T so every participant can correct their subshares without interaction with others. In other words, we are looking for a nonsingular $nv \times nv$ matrix T such that

1. $T \times \mathcal{X}$ is a matrix of a uniformly perfect linear secret sharing,
2. the matrix T has the following form:

$$T = \begin{bmatrix} T_1 & 0 & \dots & 0 \\ 0 & T_2 & \dots & 0 \\ & & \vdots & \\ 0 & 0 & \dots & T_n \end{bmatrix} \tag{3}$$

where T_i are $v \times v$ matrices for $i = 1, \ldots, n$ and 0 are $v \times v$ zero matrices. Matrices T of this form allow participants to recalculate their new subshares from their old ones.

Corollary 1. *Given a (t, n) Shamir secret sharing over $GF(q^v)$ converted into its linear equivalent over $GF(q)$ defined by the matrix $\mathcal{X}$.*

- *For a large enough $GF(q)$, the (t, n) Shamir scheme can always be converted into uniformly perfect (t, n) Shamir scheme over $GF(q)$ but this conversion needs subshares to be corrected jointly by participants.*

- *The conversion is limited by the size of $GF(q)$ and can be made only if a uniformly perfect Shamir scheme can be designed over $GF(q)$. For instance, the conversion cannot be made for $GF(2^v)$ as the matrix $\mathcal{X}_{Sh}$ over $GF(2)$ cannot be constructed.*

- *The conversion determined by the matrix T of the form (3) allows each participant to correct their subshares but in general, it is impossible to convert the (t, n) Shamir scheme into its uniformly perfect Shamir equivalent over $GF(q)$.*

Consider an example. Given $GF(2^3)$ with arithmetics modulo $q(x) = x^3 + x + 1$. Let α be a root of $q(x)$, that is $q(\alpha) = 0$. A $(3, 5)$ threshold scheme is designed for five participants whose public co-ordinates are $X_1 = 1$, $X_2 = \alpha$, $X_3 = \alpha + 1$, $X_4 = \alpha^2$ and $X_5 = \alpha^2 + 1$. The scheme is described by the matrix $\mathcal{X}$ of the form

$$\mathcal{X} = \begin{bmatrix} 1 & X_1 & X_1^2 \\ 1 & X_2 & X_2^2 \\ 1 & X_3 & X_3^2 \\ 1 & X_4 & X_4^2 \\ 1 & X_5 & X_5^2 \end{bmatrix} = \begin{bmatrix} 1 & 1 & 1 \\ 1 & \alpha & \alpha^2 \\ 1 & \alpha+1 & \alpha^2+1 \\ 1 & \alpha^2 & \alpha^2+\alpha \\ 1 & \alpha^2+1 & \alpha^2+\alpha+1 \end{bmatrix}$$

The matrix $\mathcal{X}$ can be equivalently represented over $GF(2)$ as follows

$$
\mathcal{Y} = \begin{bmatrix}
1\,0\,0\,1\,0\,0\,1\,0\,0 \\
0\,1\,0\,0\,1\,0\,0\,1\,0 \\
0\,0\,1\,0\,0\,1\,0\,0\,1 \\
\\
1\,0\,0\,0\,0\,1\,0\,1\,0 \\
0\,1\,0\,1\,0\,1\,0\,1\,1 \\
0\,0\,1\,0\,1\,0\,1\,0\,1 \\
\\
1\,0\,0\,1\,0\,1\,1\,1\,0 \\
0\,1\,0\,1\,1\,1\,0\,0\,1 \\
0\,0\,1\,0\,1\,1\,1\,0\,0 \\
\\
1\,0\,0\,0\,1\,0\,0\,1\,1 \\
0\,1\,0\,0\,1\,1\,1\,1\,0 \\
0\,0\,1\,1\,0\,1\,1\,1\,1 \\
\\
1\,0\,0\,1\,1\,0\,1\,1\,1 \\
0\,1\,0\,0\,0\,1\,1\,0\,0 \\
0\,0\,1\,1\,0\,0\,1\,1\,0
\end{bmatrix}
$$

Note that each subshares depends on a subset of elements from the vector A. Zeroes in the corresponding row of the matrix $\mathcal{Y}$ indicate which elements do not influence the subshare. At the pooling time participants must submit tv subshares for which the corresponding part of matrix $\mathcal{Y}$ is nonsingular. Before submitting subshares participants must agree on which subshare to use so they can check whether the corresponding part of $\mathcal{Y}$ is nonsingular.

5   Probabilistic Linear Secret Sharing

Given a field $GF(q)$ and the parameters of a uniformly perfect (t, n) linear threshold scheme. To design the scheme it is enough to construct a $(nv \times tv)$ matrix $\mathcal{X}$ such that any collection of tv rows of the matrix $\mathcal{X}$ constitutes a nonsingular matrix. The idea is to select rows (vectors) of $\mathcal{X}$ at random and see what is the probability that the resulting matrix $\mathcal{X}$ gives a uniformly perfect scheme.

Lemma 1. *Given $GF(q)$ and an $k \times m$ matrix $\mathcal{X}$ with k rows $R_i = (r_{i,1}, \ldots, r_{i,m})$; $i = 1, \ldots, k$ whose elements are selected at random from all nonzero elements of $GF(q)$, that is $r_{i,j} \in_R GF(q) \setminus \{0\}$. Then the rows of the matrix $\mathcal{X}$ are linearly independent with the probability*

$$
P(k) = \prod_{i=1}^{k-1} \left(1 - \frac{1}{(q-1)^{m-i}} \right). \tag{4}
$$

for $k = 2, \ldots, m$.

Proof. By induction. Consider $k = 2$. Let R_1 and R_2 be the two chosen at random vectors. The vector R_2 is linearly dependent on R_1 if $R_2 = \alpha_1 R_1$ for some $\alpha_1 \in GF(q)$. If such α_1 exists, it can be found by solving the equation

$$r_{2,1} = \alpha_1 r_{1,1}.$$

Now α_1 is used to check whether other co-ordinates are related linearly so

$$r_{2,j} \stackrel{?}{=} \alpha_1 r_{1,j}$$

for $j = 2, \ldots, m$. Each check holds with probability $(q-1)^{-1}$. This means that two vectors are linearly dependent with probability $(q-1)^{-m+1}$. In other words, two random vectors are linearly independent with the probability $(1-(q-1)^{1-m})$.

The induction step. Assume that the lemma is true for $k-1$ and let $2 < k \leq m - 1$. The collection of vectors $\{R_1, \ldots, R_k\}$ is linearly dependent if there are $\alpha_1, \ldots, \alpha_{k-1}$ such that

$$R_k = \alpha_1 R_1 + \ldots + \alpha_{k-1} R_{k-1}$$

The numbers $\alpha_1, \ldots, \alpha_{k-1} \in GF(q)$ can be determined by solving the following system of linear equations

$$r_{k,1} = \alpha_1 r_{1,1} + \ldots + \alpha_{k-1} r_{k-1,1}$$
$$r_{k,2} = \alpha_1 r_{1,2} + \ldots + \alpha_{k-1} r_{k-1,2}$$
$$\vdots$$
$$r_{k,k-1} = \alpha_1 r_{1,k-1} + \ldots + \alpha_{k-1} r_{k-1,k-1}$$

The system has a unique solution if the determinant is different from zero. This occurs when $\{R_1, \ldots, R_{k-1}\}$ are linearly independent. To test for linear dependency, we check

$$r_{k,j} \stackrel{?}{=} \alpha_1 r_{1,j} + \ldots + \alpha_{k-1} r_{k-1,j}$$

for $j = k, \ldots, m$. The test is successful with the probability q^{k-m-1}. Or in other words, the collection of k rows is linearly independent with the probability $(1 - (q-1)^{k-m-1})$ provided $(R_1, \ldots, R_{k-1})$ are linearly independent. Denote $P_R(k)$ to be the probability that vectors $(R_1, \ldots, R_k)$ are linearly independent. Let $P_R(k \mid k-1)$ be the probability that vector R_k is not a linear combination of $(R_1, \ldots, R_{(k-1)})$ provided $(R_1, \ldots, R_{k-1})$ are independent. So

$$P_R(k) = P_R(k \mid k-1) \times P_R(k-1).$$

Using our induction assumption, we obtain the final expression.

Definition 2. *Given a collection $\{v_1, \ldots, v_k\}$ where each vector $v_i \in (GF(q))^k$ and $k > m$. The collection is m-wise independent if any subset of m vectors from it, is linearly independent.*

Lemma 2. *Given $GF(q)$ and an $(k \times m)$ matrix X with k rows $R_i = (r_{i,1}, \ldots, r_{i,m})$; $i = 1, \ldots, k$, whose elements are selected at random from all nonzero elements of $GF(q)$, that is $r_{i,j} \in_R GF(q) \setminus \{0\}$. Then the matrix contains rows which are m-wise independent with probability*

$$P(k|m) \geq \prod_{i=1}^{k} \left(1 - \binom{m+i}{m}\right)(q-1)^{-k+m} P(m)$$

where $P(m)$ is the probability given by the expression (4) and $k = m+1, \ldots$.

Proof. Denote $P_m(k)$ to be the probability that $(R_1, \ldots, R_k)$ are m-wise independent. $P_m(k \mid k-1)$ means the probability that all m-subsets of rows with the row R_k are independent provided that the rows $(R_1, \ldots, R_{k-1})$ are m-wise independent.

By induction. We start from the case when the number of rows is $k = m+1$. It is clear that the collection $(R_1, \ldots, R_{m+1})$ is m-wise independent if all subsets of m rows are independent so

$$P_m(m+1) = P_m(m+1 \mid m) \times P_m(m).$$

The collection of m-subsets with R_{m+1} are

$$
\begin{array}{ccccc}
0 & R_2 & \ldots R_m & R_{m+1} & \to (\alpha_1^{(1)}, \ldots, \alpha_{m-1}^{(1)}) \\
R_1 & 0 & \ldots R_m & R_{m+1} & \to (\alpha_1^{(2)}, \ldots, \alpha_{m-1}^{(2)}) \\
& & \vdots & & \\
R_1 & R_2 & \ldots \quad 0 & R_{m+1} & \to (\alpha_1^{(m)}, \ldots, \alpha_{m-1}^{(m)})
\end{array}
$$

where $(\alpha_1^{(i)}, \ldots, \alpha_{m-1}^{(i)})$ is the vector which establishes a linear dependency among the i-th collection of vectors $\{R_1, \ldots, R_{m+1}\} \setminus \{R_i\}$ using first $m-1$ co-ordinates of the vectors (see Lemma (1)). The vector $\alpha_1^{(i)}, \ldots, \alpha_{m-1}^{(i)}$ is next used to produce the values

$$u_i = \alpha_1^{(i)} r_{1,m} + \ldots + \alpha_{m-1}^{(i)} r_{m-1,m}$$

for $i = 1, \ldots, m$. The random vector R_{m+1} is linearly independent if its last co-ordinate $r_{m+1,m}$ is different from all u_i; $i = 1, \ldots, m$. As the set $\mathcal{U} = \{u_1, \ldots, u_m\}$ may have repetitions, the probability that all m-subsets with R_{m+1} are all independent is bigger or equal to

$$\left(1 - \frac{m}{q-1}\right).$$

Thus,

$$P_m(m+1) \geq \left(1 - \frac{m}{q-1}\right) \times P(m)$$

In the induction step we assume that the lemma is true for $(k-1)$. To conclude the proof, it is enough to notice that for arbitrary $k > m$, the size of the set $\mathcal{U}$ contains $\binom{k}{m}$ elements so after using the induction assumption, the final bound on the probability follows.

The last lemma can be rephrased in the form of the theorem.

Theorem 4. *Given $GF(q)$ and a (t, n) threshold linear scheme with v subshares with a $(nv \times tv)$ matrix $\mathcal{X}$ whose rows are selected at random from all nonzero elements of $GF(q)$. Then the resulting secret sharing is uniformly perfect with the probability*

$$P_v \geq \prod_{i=1}^{nv} \left(1 - \binom{tv+i}{tv}\right)(q-1)^{-nv+tv} P(tv)$$

where $P(tv)$ is the probability given by the expression (4).

We make the following observations.

- The probabilistic construction does not work if the set $\mathcal{U}$ defined in the proof of Lemma (2) is getting too big, i.e. $q \approx \binom{nv}{tv}$. In other words, the probabilistic approach is practical only if

$$q \gg \binom{nv}{tv}.$$

- A probabilistic Shamir scheme over $GF(q)$ generates a $(nv \times tv)$ matrix $\mathcal{X}_{Sh}$ whose tv-subsets of rows are independent. This construction exists only if $nv < q$. If the parameter nv is bigger than q, a Shamir secret sharing over $GF(q^v)$ can be constructed. This time the number of different rows in the underlying array $\mathcal{X}$ with entries from $GF(q^v)$ is upper bounded by $q^v - 1$ (or equivalently by $v(q^v - 1)$ if rows are considered over $GF(q)$). Clearly, the resulting scheme is not uniformly perfect.

- After a probabilistic linear secret sharing is constructed, it is going to be used only once for a very specific collection of subshares. It can be argued that probabilistic construction can be an option even if $q \approx \binom{nv}{tv}$ as the combiner is going to use one single tv-subset of the matrix $\mathcal{X}$ which is non-singular with the probability $\approx q^{-1}$.

6 Secret Sharing with Linear Shift

In this section we address the problem of conversion of Shamir secret sharing designed over $GF(q^v)$ into its linear equivalent in such a way that for any tv subshares, the dealer can always recover the secret. We noted that (t, n) Shamir secret sharing over $GF(q^v)$ is not uniformly perfect. We also observed that participants can convert the underlying matrix $\mathcal{X}$ using the matrix T of the form (3). The conversion can be done without interaction among them.

Given a (t, n) Shamir secret sharing over $GF(q^v)$ defined by the matrix $\mathcal{X}$ of the form

$$\mathcal{X} = \begin{bmatrix} 1 & X_1 & \dots & X_1^{t-1} \\ 1 & X_2 & \dots & X_2^{t-1} \\ & & \vdots & \\ 1 & X_n & \dots & X_n^{t-1} \end{bmatrix}$$

where $X_i \in GF(q^v)$ is a public value assigned to participant P_i. Participants are going to convert $\mathcal{X}$ using a non-singular transformation of the form

$$T = T(X_1, \ldots, X_n) = \begin{bmatrix} X_1 & 0 & \ldots & 0 \\ 0 & X_2 & \ldots & 0 \\ & & \vdots & \\ 0 & 0 & \ldots & X_n \end{bmatrix} \tag{5}$$

Note that T describes a nonsingular transformation as the inverse matrix is $T^{-1} = T(X_1^{-1}, \ldots, X_n^{-1})$. Next each participant converts the secret sharing by finding the new matrix

$$\mathcal{X}^{(1)} = T \times \mathcal{X}$$

and corrects their share $S_i' = S_i \times X_i$. Now assume that participants would like to pool their subshares. Before sending subshares, participants communicate the combiner the positions of subshares so the combiner computes the corresponding submatrix $\mathcal{X}_A$ and checks whether or not, it is nonsingular. If it is nonsingular, combiner asks participants for their subshares and recovers the secret. Otherwise, the combiner computes $T^i \times \mathcal{X}_A$ for some small i until the resulting matrix is non-singular. If such non-singular matrix is found, the combiner sends the exponent i to all active participants who correct their shares and send suitable subshares back to the combiner. The combiner recovers the secret and distributes it to the active participants.

It is not difficult to note that each row $(1, X_i, \ldots, X_i^{t-1})$ after j consecutive transformation T, becomes $(X_i^j, X_i^{j+1}, \ldots, X_i^{j+t-1})$. In other words, rows are part of a cycle generated by the element X_i. It is also obvious that rows can also be seen as parts of the sequence generated by a linear feedback shift register (LFSR) - see [8]. LFSRs are used commonly as a source of pseudorandom number generators. Their randomness specifically relates to the length of the cycle. The longer the cycle the better pseudorandom behaviour. The longest cycle is obtained when X_i is a primitive element of $GF(q)$.

If we assume that every conversion using the matrix T is equivalent to a random selection of a $tv \times tv$ matrix, then the probability of the matrix to be nonsingular is $\approx (1 - q^{-1})$. So for a big enough q, the first transformation is enough to guarantee with a very high probability that the intended collection of subshares will enable the combiner to recover the secret.

(t, n) Secret Sharing with Shift

Dealer: For given $GF(q^v)$, the dealer selects public elements $X_i \in GF(q^v)$; $i = 1, \ldots, n$. X_i are primitive elements. Each participant is assigned a single X_i. This assignment is public. Next the dealer generates the underlying polynomial $f(X) = A_0 + A_1 X + \ldots, A_{t-1} X^{t-1}$ by random selection of elements $A_i \in_R GF(q^v)$. A share of participant P_i is $S_i = f(X_i)$. The secret is $f(0) = A_0$. Shares are distributed to participants via secure channels.

Participants: Every participant creates the defining matrix $\mathcal{X}$ whose entries are from $GF(q)$ and calculates the matrix $\mathcal{X}^{(1)} = T \times \mathcal{X}$ where the matrix

T contains X_i on its main diagonal having other entries equal to zero. Also participants correct their shares by $S_i' = X_i S_i$.

Combiner: It works in two stages. First the combiner asks participants to identify rows of the matrix $\mathcal{X}^{(1)}$ which will be used later for key recovery. Knowing the rows, the combiner creates $(tv \times tv)$ matrix $\mathcal{X}_A^{(1)}$. If $det(\mathcal{X}_A^{(1)}) = 0$ then the combiner finds a smallest j for which the matrix $T^j \times \mathcal{X}^{(1)}$ produces a nonsingular matrix on the positions indicated by participants. The combiner sends j back to participants.

Participants: Having j from the combiner, participants correct their shares $\bar{S}_i = S_i' \times X_i^j$ and submit their subshares.

Combiner: It computes the secret and distributes it to all active participants via secure channels.

Consider an example. Given $GF(97^3)$ with the irreducible polynomial $q(x) = x^3 + 75x^2 + 30x + 75$ modulo 97 and $q(\alpha) = 0$. We design a $(3,9)$ threshold scheme with 3 subshares (i.e. $t = 3$, $n = 9$ and $v = 3$). The collection of primitive elements is

$$X_1 = 44 + 33\alpha + 5\alpha^2$$
$$X_2 = 40 + 82\alpha + 32\alpha^2$$
$$X_3 = 26 + 47\alpha + 80\alpha^2$$
$$X_4 = 34 + 18\alpha + 38\alpha^2$$
$$X_5 = 75 + 79\alpha + 5\alpha^2$$
$$X_6 = 69 + 2\alpha + 88\alpha^2$$
$$X_7 = 5 + 85\alpha + 37\alpha^2$$
$$X_8 = 85 + 41\alpha + 30\alpha^2$$
$$X_9 = 7 + 30\alpha + 12\alpha^2$$

The matrix $\mathcal{X}^{(1)}$ has a form

$$\begin{bmatrix} X_1 & X_1^2 & X_1^3 \\ X_2 & X_2^2 & X_2^3 \\ & \vdots & \\ X_9 & X_9^2 & X_9^3 \end{bmatrix}$$

which can be converted into a 27×9 equivalent

$$
\mathcal{X}^{(1)} =
\begin{bmatrix}
44 & 13 & 42 & 53 & 59 & 13 & 76 & 77 & 87 \\
33 & 88 & 88 & 43 & 96 & 6 & 46 & 68 & 73 \\
5 & 46 & 33 & 60 & 5 & 12 & 52 & 26 & 58 \\
\\
40 & 25 & 26 & 20 & 85 & 21 & 3 & 13 & 81 \\
82 & 50 & 16 & 35 & 54 & 74 & 48 & 47 & 26 \\
32 & 10 & 76 & 70 & 23 & 75 & 5 & 61 & 31 \\
\\
26 & 14 & 81 & 54 & 5 & 81 & 68 & 35 & 69 \\
47 & 51 & 27 & 56 & 56 & 18 & 96 & 82 & 82 \\
80 & 61 & 35 & 84 & 61 & 40 & 6 & 34 & 54 \\
\\
34 & 60 & 67 & 29 & 76 & 50 & 83 & 57 & 7 \\
18 & 58 & 48 & 85 & 40 & 96 & 80 & 67 & 21 \\
38 & 78 & 28 & 74 & 64 & 90 & 7 & 40 & 74 \\
\\
75 & 13 & 84 & 88 & 65 & 57 & 89 & 58 & 70 \\
79 & 22 & 66 & 39 & 17 & 49 & 51 & 54 & 86 \\
5 & 92 & 9 & 25 & 7 & 74 & 82 & 12 & 27 \\
\\
69 & 93 & 53 & 8 & 8 & 56 & 64 & 46 & 29 \\
2 & 48 & 56 & 21 & 50 & 11 & 92 & 63 & 77 \\
88 & 95 & 4 & 18 & 29 & 9 & 55 & 41 & 92 \\
\\
5 & 38 & 87 & 72 & 38 & 73 & 83 & 86 & 75 \\
85 & 59 & 34 & 5 & 29 & 9 & 10 & 1 & 19 \\
37 & 26 & 49 & 37 & 43 & 5 & 48 & 96 & 76 \\
\\
85 & 78 & 96 & 14 & 74 & 52 & 61 & 29 & 40 \\
41 & 58 & 0 & 43 & 63 & 56 & 61 & 92 & 45 \\
30 & 22 & 57 & 21 & 20 & 18 & 41 & 90 & 35 \\
\\
7 & 70 & 66 & 31 & 34 & 5 & 36 & 77 & 28 \\
30 & 35 & 77 & 50 & 64 & 36 & 83 & 28 & 30 \\
12 & 3 & 4 & 28 & 84 & 69 & 52 & 63 & 56
\end{bmatrix}
$$

When we select 9 rows from the matrix at random, then the probability of choosing a non-singular 9×9 submatrix is greater than $(1 - 97^{-1})$. We now created all 9×9 submatrices with single rows taken from those assigned to a single participant. It is easy to compute that there are $3^9 = 19683$ such submatrices. From those only 195 are singular. In other words, random selection of rows from the matrix $\mathcal{X}^{(1)}$ gives with probability $1 - \frac{195}{19683} \approx 0.99$ nonsingular submatrix. This is a bit better than expected from the lower bound $(1 - 97^{-1}) \approx 0.989$. The very first singular submatrix found consists of rows $\{1, 4, 7, 10, 15, 17, 21, 24, 27\}$. Now if we create the matrix $\mathcal{X}^{(2)}$, then these specific collection of rows becomes non-singular with the determinant equal to 84.

Secret sharing with linear shift exhibits the following remarkable properties:

1. It provides a practical uniformly perfect secret sharing when the (t, n, v) Shamir scheme fails because the number of participants n is bigger than $q - 1$ (the matrix $\mathcal{X}$ cannot be constructed).

2. It can be applied for any already existing (t, n) Shamir secret sharing, demonstrating how any Shamir secret sharing over $GF(q^v)$ can be converted into a practical uniformly perfect scheme. More precise, uniform perfectness is probabilistic in nature as to achieve it the underlying matrix $\mathcal{X}^{(1)}$ may need to be subject to a number of transformations. The average number of transformations is $\approx q^{-1}$.

86

3. The security of the scheme is the same as the security of Shamir schemes and the interaction between the combiner and participants can be conducted via public (although authenticated) discussion. Clearly, subshares must be sent privately.

4. The interaction between participants can be eliminated if the size q of the field is very large. This of course means that the secret recovery will fail with the probability $\approx q^{-1}$ which is negligible for a large q.

Acknowledgement

The author wishes to thank anonymous referees for their helpful comments and suggestions.

References

1. G. R. Blakley. Safeguarding cryptographic keys. In *Proc. AFIPS 1979 National Computer Conference*, pages 313–317. AFIPS, 1979.
2. G.R. Blakley and Catherine Meadows. Security of ramp schemes. In G. R. Blakley and D. C. Chaum, editors, *Advances in Cryptology - CRYPTO'84*, pages 242–268. Springer, 1985. Lecture Notes in Computer Science No. 196.
3. R. Gennaro, S. Jarecki, H. Krawczyk, and T. Rabin. Robust threshold DSS signatures. In U. Maurer, editor, *Advances in Cryptology - EUROCRYPT'96*, pages 354–371. Springer, 1996. Lecture Notes in Computer Science No. 1070.
4. A. Herzberg, S. Jarecki, H. Krawczyk, and M. Yung. Proactive secret sharing or: how to cope with perpetual leakage. In D. Coppersmith, editor, *Advances in Cryptology - CRYPTO'95*, pages 339–352. Springer, 1995. Lecture Notes in Computer Science No. 963.
5. E.D. Karnin, J.W. Greene, and M.E. Hellman. On secret sharing systems. *IEEE Transactions on Information Theory*, IT-29:35–41, 1983.
6. Rudolf Lidl and Harald Niederreiter. *Finite Fields*. Addison-Wesley, Reading, MA, 1983.
7. K. Martin, J. Pieprzyk, R. Safavi-Naini, and H. Wang. Changing thresholds in the absence of secure channels. In *Proceedings of the Fourth Australasian Conference on Information Security and Privacy (ACISP99)*, volume 1587, pages 177–191. Springer-Verlag, 1999.
8. A. Menezes, P. van Oorschot, and S. Vanstone. *Handbook of Applied Cryptography*. CRC Press, Boca Raton, 1997.
9. A. Shamir. How to share a secret. *Communications of the ACM*, 22:612–613, November 1979.

Efficient Publicly Verifiable Secret Sharing Schemes with Fast or Delayed Recovery

Fabrice Boudot and Jacques Traoré

France Telecom - CNET
42, rue des Coutures
BP 6243, F-14066 Caen Cedex 4, France
fabrice.boudot@cnet.francetelecom.fr
jacques.traore@cnet.francetelecom.fr

Abstract. A publicly verifiable secret sharing scheme is a secret sharing scheme in which everyone, not only the shareholders, can verify that the secret shares are correctly distributed. We present new such schemes and use them to share discrete logarithms and integer factorizations. The shareholders will be able to recover their shares quickly (fast recovery) or after a predetermined amount of computations (delayed recovery) to prevent the recovery of all the secrets by un-trustworthy shareholders (e.g. if these schemes are used for escrowing secret keys). The main contribution of this paper is that all the schemes we present need much less computations and communicated bits than previous ones [BGo, FOk, Mao, Sta, YYu]. By the way, we introduce in this paper several tools which are of independent interest: a proof of equality of two discrete logarithms modulo two different numbers, an efficient proof of equality of a discrete logarithm and a third root, and an efficient proof of knowledge of the factorization of any composite number n, where it is not necessary to prove previously that n is the product of two prime factors.

Keywords: Publicly Verifiable Secret Sharing, Key Escrow Systems, Zero Knowledge Proofs.

1 Introduction

In a secret sharing scheme, a *dealer* wants to split his secret s between l *shareholders*. Each shareholder receives through a secure channel his share s_i. If the dealer is honest, the collaboration of at least k shareholders allows to recover the secret s, while the collaboration of only $k - 1$ shareholders does not reveal any extra-information about s. The first secret sharing scheme has been presented by [Sha].

Unfortunately, if the dealer is dishonest, the collaboration of the shareholders fails to retrieve the secret s. To overcome this problem, Chor et al. proposed a verifiable secret sharing in [CGMA], where the shareholders can verify the validity of their shares (i.e. that at least k of them can recover the secret s).

Another property, called *public verifiability*, was introduced by Stadler in [Sta]. In a publicly verifiable secret sharing (PVSS) scheme, everyone, not only the shareholders, can verify the validity of the shares.

This is an informal description of a PVSS scheme: Each shareholder SH_i has a public-key encryption function $\mathcal{E}_i$. The dealer wants to share a secret s, which is a discrete logarithm or the factorization of a large number. The dealer splits his secret into l shares $s_1, \ldots, s_l$, computes for all i $E_i = \mathcal{E}_i(s_i)$ and a *certificate of recoverability* $Cert$ which proves that the E_i are correctly computed. Then, he sends $(E_1, \ldots, E_l, Cert)$ to a verifier who checks the validity of the certificate of recoverability. When the shareholders want to recover the dealer's secret, the verifier sends E_i to each shareholder SH_i who can recover s_i. Finally, the collaboration of at least k shareholders allows to recover the dealer's secret s.

PVSS schemes can be used to build a key escrow system [Mic]. In such systems, the user is allowed to use cryptologic products by his government only if he discloses his secret key to a *trusted third party*. This user plays the role of the dealer who shares his secret key, the certification authority plays the role of the verifier and certifies the user's public key only if the certificate of recoverability is valid, and the escrow authorities play the role of the shareholders who recover the user's key on court order.

In some applications, it seems desirable that the shareholders need a certain amount of work to recover the secret key. For example, in a key escrow system, this avoids that an inquisitive government be able to recover a lot of keys in short time. We will call such a system a PVSS scheme with delayed recovery.

Our work: We present quite efficient PVSS schemes to share discrete logarithms and integer factorizations, the recovery of which can be made fast or delayed. In particular, these schemes can be used for sharing RSA keys.

Organization of the paper: In Section 2, we give some definitions and recall previous results about these topics. Then, in Section 3, we present the basic tools we will use in our schemes. We describe these schemes in Section 4 and compare their efficiency. Finally, we conclude in Section 5.

2 Background

The following definition is derived from [YYu].

Definition 1: A $(k, l)-Publicly\ Verifiable\ Secret\ Sharing$ (PVSS) scheme is a triple (GEN, VER, REC) such that:

1. GEN is a poly-time probabilistic Turing machine, run by the dealer, that takes no input and generates $(K_P, K_S, (E_1, \ldots, E_l), Cert)$. K_P is the public key of the dealer, K_S his private key (or his secret), $(E_1, \ldots, E_l)$ the messages dedicated to the shareholders and $Cert$ the certificate of recoverability.
2. VER is a poly-time probabilistic Turing machine, run by the verifier, that takes $(K_P, (E_1, \ldots, E_l), Cert)$ as its input and returns a boolean value. With very high probability, VER returns true if and only if $(E_1, \ldots, E_l)$ can allow at least k shareholders to recover K_S.

3. *REC* is a set of probabilistic Turing machines $REC_1, \ldots, REC_l, END$. Each REC_i is a Turing machine with private input and belongs to the shareholder SH_i. If VER returns "true", then a) each REC_i, on input E_i, outputs a share s_i of K_S, b) on any set of k different elements from $\{s_1, \ldots, s_l\}$, END outputs K_S, c) the knowledge of less than k different elements from $\{s_1, \ldots, s_l\}$ gives no information about K_S.

4. It is intractable to recover K_S given $(K_P, (E_1, \ldots, E_l), Cert)$.

Definition 2:

1. A PVSS is a *PVSS with fast recovery* if the algorithms REC_i and END run rapidly.
2. A PVSS is a *PVSS with delayed recovery* if any set of k REC_i and/or END needs a non-trivial amount of work to output correct values.

In PVSS with delayed recovery, the shareholders are not able to recover a large number of private keys. This property was first discussed by Bellare and Goldwasser in [BGo]. In their schemes, the full secret is not shared, but only a part of it, and the shareholders are obliged to find the missing bits by exhaustive research to recover the full secret. Unfortunately, other schemes with delayed recovery may suffer from *early recovery*, i.e. one of the shareholders can recover by himself the missing bits. So, if the PVSS scheme is used for escrowing the users' private keys, and if one of the escrow authorities (shareholders) belongs to a political group with totalitarian projects, he can recover the missing bits of all users before seizing power and getting all the users' keys just after the coup, when the other escrow authorities are under his control. We propose another way to achieve delayed recovery: we will use an encryption scheme such that the decryption is very slow. Then each shareholder needs a lot of work to recover one share, and if a malicious shareholder recovers all its shares, this does not help him to recover faster the dealer's secret, as the time needed to recover a secret is the maximum of the times for computing each share (the shares are computed by each shareholder in parallel). Note that our definition of delayed recovery is less restrictive than [BGo]'s one: they claim that a PVSS does not suffer from early recovery only if the shareholders need a non-trivial amount of work to recover the secret *after* the cooperation of the shareholders (i.e. only if END runs slowly). In our definition, we also include the case when all the shares are recovered slowly.

Note that other works about delayed recovery have been done by [RSW].

Related results: An efficient PVSS scheme was presented by Stadler in [Sta], and allows to share a discrete logarithm. A similar result was presented by Young and Yung in [YYu]. Mao presented a PVSS scheme for factorization in [Mao]. All these results were improved by Fujisaki and Okamoto in [FOk]. A PVSS scheme with delayed recovery was presented by Bellare and Goldwasser in [BGo] for discrete logarithm. PVSS schemes with delayed recovery were presented by Fujisaki and Okamoto in [FOk], but the delayed recovery property is obtained only if the number of shareholders is about 48. We present in this paper two PVSS with fast recovery, one for discrete logarithm and the other for factorization. Our PVSS schemes are more efficient than the previous ones. We also present for both problems two PVSS with delayed recovery. They are more efficient than previous ones and allow any number of shareholders, including one only. Our schemes are at least 3 times more efficient in terms of bits communicated and 10 times more efficient in terms of computations needed than all the previous ones.

Notations: Throughout this paper, $\mathbb{Z}_n$ denotes the residue class ring modulo n, and $\mathbb{Z}_n^*$ denotes the multiplicative group of invertible elements in $\mathbb{Z}_n$. Let $\varphi(n)$ be the cardinality of $\mathbb{Z}_n^*$ (the Eulerian number of n) and $\lambda(n)$ be the maximal order of an element in $\mathbb{Z}_n^*$ (the Carmichael number of n). $|\cdot|$ denotes binary length, $\mathrm{abs}(\cdot)$ denotes the absolute value, $a \parallel b$ is the concatenation of the strings a and b. We denote $\log_g(a)[\mathrm{mod}\ n]$ the discrete logarithm of a in base g modulo n which belongs to $\{-\mathrm{ord}(g)/2,\ldots,\mathrm{ord}(g)/2-1\}$. We note $\mathrm{Proof}(\mathcal{R})$ a zero-knowledge proof of the assertion $\mathcal{R}$, and $\mathrm{Proof}(x, \mathcal{R}(x))$ a zero-knowledge proof of knowledge of x such that $\mathcal{R}(x)$ is true.

3 Basic Tools

In this section, we present all the basic tools we will use in Section 4. Some of them are already known: a proof of knowledge of a discrete logarithm modulo a composite number by [Gir] in Section 3.1, that we extend to a proof that this logarithm is range-bounded, and a verifiable proof of secret sharing by [Fel] in Section 3.3.

Other ones are variants or extensions of existing schemes: We present a proof of equality of two discrete logarithms with respect to different moduli, in Section 3.2. This proof is related to the protocol due to [CPe] (but in their protocols the moduli are the same) and also to a protocol due to [Bao] (where the order of the bases are different, one known and the other unknown). We also present a proof of equality of a discrete logarithm and a third root related to [FOk] in Section 3.6. We will introduce a cryptosystem with arbitrarily slow decryption in Section 3.7. Such a cryptosystem is used by [MYa] for other purposes.

The remaining tools are (to our knowledge) new: we will introduce in Section 3.4 a new problem equivalent to the factorization problem and use it in Section 3.5 to prove that the knowledge of a particular discrete logarithm allows to factor a given integer. This is an original method for sharing the factorization of an integer. The main advantage of our method is, unlike [Mao] and [FOk] schemes, that we do not have to prove that this integer is the product of exactly two primes. Indeed, the protocols that can be used to prove that an integer is the product of exactly two primes [vGP, Mic] are relatively inefficient.

All the proofs described in this section are made non-interactive using the Fiat-Shamir [FSh] heuristic.

3.1 Proof of knowledge of a range-bounded discrete logarithm modulo a composite number

The following protocol allows Alice to prove that she knows a discrete logarithm and that this discrete logarithm is range-bounded. It is an extension (related e.g. to [CFT]) of a previous protocol by Girault [Gir], proven secure in the random oracle model by Poupard and Stern [PSt].

Let n be a composite number whose factorization is unknown by Alice, and $g \in \mathbb{Z}_n^*$ an element of maximal order. Let $h : \{0,1\}^* \longrightarrow \{0, \dots, k-1\}$ be a hash function.

Let b, B, t and k be four integers such that bt/B and $1/k$ are negligible. Assume that Alice knows an integer $x \in \{0, \dots, b\}$ such that $y = g^x \bmod n$.

The following protocol is a non-interactive zero-knowledge proof that she knows a discrete logarithm of y in base g modulo n and that this common discrete logarithm is in $\{-B, \dots, B\}$ (in default of proving that it belongs to $\{0, \dots, b\}$).

Protocol: $\text{Proof1}(x, y = g^x \bmod n \wedge \text{abs}(x) < B)$

1. Alice chooses $r \in \{0, \dots, tkb - 1\}$ and computes $W = g^r \bmod n$.
2. Alice computes $c = h(W)$.
3. Alice computes $D = r + cx$ (in $\mathbb{Z}$) and sends the proof (W, c, D) if $D \in \{cb, \dots, tkb - 1\}$, otherwise she starts the protocol again.
4. The verifier checks that $W = g^D y^{-c} \bmod n$ and that $D \in \{cb, \dots, tkb - 1\}$.

3.2 Proof of equality of two discrete logarithms

We design in this paragraph a protocol which allows Alice to prove that she knows a discrete logarithm common to two (or more) values and that this common discrete logarithm is range-bounded.

Let n be a composite number whose factorization is unknown by Alice, and $g \in \mathbb{Z}_n^*$ an element of maximal order. Let n' be an integer, prime or composite, and $g' \in \mathbb{Z}_{n'}^*$ an element of maximal order. We assume that n and n' are equal or relatively prime. Let $h : \{0,1\}^* \longrightarrow \{0, \dots, k-1\}$ be a hash function.

Let b, B, t and k be four integers such that bt/B and $1/k$ are negligible. Assume that Alice knows an integer $x \in \{0, \dots, b\}$ such that $y = g^x \bmod n$ and $y' = g'^x \bmod n'$.

The following protocol is a non-interactive zero-knowledge proof that she knows a discrete logarithm common to y in base g modulo n and to y' in base g' modulo n' and that this common discrete logarithm is in $\{-B, \dots, B\}$ (in default of proving that it belongs to $\{0, \dots, b\}$).

The first proof of equality of two discrete logarithms is due to [CEG], and has been improved by [CPe] and [Bao] (which prove the equality of two discrete logarithms modulo the same number).

Protocol: $\text{Proof2}_b(x, y = g^x \bmod n \wedge y' = g'^x \bmod n' \wedge \text{abs}(x) < B)$

1. Alice chooses $r \in_R \{1, \ldots, tkb - 1\}$ and computes $W = g^r \bmod n$ and $W' = g'^r \bmod n'$.
2. Alice computes $c = h(W, W')$.
3. Alice computes $D = r + cx$ (in $\mathbb{Z}$) and sends the proof (W, W', c, D) if $D \in \{cb, \ldots, tkb - 1\}$, otherwise she starts the protocol again.
4. The verifier checks that $D \in \{cb, \ldots, tkb - 1\}$, $W = g^D y^{-c} \bmod n$ and $W' = g'^D y'^{-c} \bmod n'$.

Lemma 1: The above protocol is a non-interactive zero-knowledge proof system, i.e.:

1. *Completeness*: If Alice behaves honestly and if $x \in \{0, \ldots, b\}$, then the protocol succeeds with probability greater than $1 - 1/t$.
2. *Soundness*: If Alice does not know any common discrete logarithm, then the protocol succeeds with probability less than $2/k$. Moreover, if Alice succeeds, then $\text{abs}(x) < B$ with probability greater than $1 - bt/B$.
3. *Non-Interactive Zero-Knowledge*: There exists a simulator in the random oracle model which outputs conversations indistinguishable from the true ones. In the random oracle model, the simulator is allowed to simulate the random oracle, i.e. the hash function h [BRo].

How to use Proof2 when $n \neq n'$: If the verifier is convinced by Proof2 that there exists a common discrete logarithm of y and of y', then he knows that someone who is able to compute $\log_{g'}(y') [\bmod n']$ is able to compute $\log_g(y) [\bmod n]$. Assume that $|\lambda(n')| = |n| + \beta$, where $\beta \geq 91$. Let $b = n$ and $B = \lambda(n')/2$. For $|t| = 10$, the protocol proves that $\text{abs}(x) > \lambda(n')/2$ with probability less than $2tn/\lambda(n')$, and so less than 2^{-80} which is negligible. Let $\tilde{x} = \log_{g'}(y') [\bmod n']$, Proof2 convinces the verifier that $\tilde{x} = x \bmod \lambda(n')$ and that $\text{abs}(x) < \lambda(n')$. Hence, $\tilde{x} = x$ in $\mathbb{Z}$ and then someone who is able to compute $\tilde{x}$ is able to compute a discrete logarithm of y in base g modulo n.

In the rest of this paper, we will assume that $t = 2^{10}$. Then, if $b/B < 2^{-91}$, Alice can prove that $\text{abs}(x) < B$ when she uses a secret $x \in \{0, \ldots, b\}$.

Note: Such a protocol can be used to prove the equality of more than two discrete logarithms.

Correctness of lemma 1:

Completeness: If Alice behaves honestly, the protocol succeeds if and only if $D \in \{cb, \ldots, tkb - 1\}$, therefore if $r \in \{cb - cx, \ldots, tkb - 1 - cx\}$. As r is taken randomly and uniformly over $\{1, \ldots, tkb - 1\}$ and x belongs to $\{0, \ldots, b\}$, the probability that the protocol succeeds is greater than $(tkb - cb)/(tkb - 1) > (tk - c)/tk > 1 - 1/t$

Soundness: We first show that the discrete logarithms are equal with probability greater than $1 - 2/k$, using [PSt] lemma : if Alice knows $(g, y, c_1, c_2, D_1, D_2)$ such that $g^{D_1} y^{-c_1} \equiv g^{D_2} y^{-c_2} \pmod{n}$, then she knows a discrete logarithm of y in base g unless she can factor n.

Case 1: $n = n'$

Assume that Alice can, given W and W', compute correct proofs (W, W', c_1, D_1) and (W, W', c_2, D_2) with $c_1 \neq c_2$. Then, we have $W = g^{D_1} y^{-c_1} \bmod n =$

$g^{D_2}y^{-c_2} \bmod n$ and $W' = g'^{D_1}y'^{-c_1} \bmod n = g'^{D_2}y'^{-c_2} \bmod n$. Using [PSt] lemma, such equalities imply that Alice knows $x = \log_g(y) = \log_{g'}(y')$.

Case 2: n and n' are relatively prime.

Let ϕ be the Chinese Remainder Theorem isomorphism :

$$\phi : \mathbb{Z}^*_{nn'} \longrightarrow \mathbb{Z}^*_n \times \mathbb{Z}^*_{n'}$$
$$x \longmapsto (x \bmod n, x \bmod n')$$

We note $\bar{W} = \phi^{-1}(W, W')$, $\bar{g} = \phi^{-1}(g, g')$ and $\bar{y} = \phi^{-1}(y, y')$. If the proof succeeds, then $\bar{W} = \bar{g}^D \bar{y}^{-c} \bmod nn'$.

Assume that Alice can, given W and W', compute correct proofs (W, W', c_1, D_1) and (W, W', c_2, D_2). Then we have : $\bar{W} = \bar{g}^{D_1}\bar{y}^{-c_1} \bmod nn' = \bar{g}^{D_2}\bar{y}^{-c_2} \bmod nn'$. Using [PSt] lemma, this equality implies that Alice knows $x = \log_{\bar{g}}(\bar{y})[\bmod nn']$ unless she can factor n. So she knows the discrete logarithm $x = \log_g(y)[\bmod n] = \log_{g'}(y')[\bmod n']$.

To prove that $\mathrm{abs}(x) < B$ with probability less than tb/B, it is sufficient to prove that for all u, $\Pr[success/\mathrm{abs}(x) > tub] < 1/u$. Let $\bar{r} = \log_{\bar{g}}(\bar{W})[\bmod nn']$, then we have $D = \bar{r} + xc$. Alice succeeds only if $D \in \{cb, \ldots, tkb - 1\}$, i.e. c must be in the set $I = \{\lceil \bar{r}/x \rceil, \ldots, \lfloor (\bar{r} - tkb - 1)/x \rfloor\}$. If $\mathrm{abs}(x) > tub$, her probability of success is $\Pr(success/\mathrm{abs}(x) > tub) = (\sharp I)/k < (k/u)/k = 1/u$.

Non-Interactive Zero-Knowledge : The simulator picks random $c \in \{0, 1\}^k$ and $D \in \{cb, \ldots, tkb - 1\}$ and sets $W = g^D y^{-c} \bmod n$, $W' = g'^D y'^{-c} \bmod n'$ and $h(W, W') = c$.

3.3 Verifiable proof of secret sharing

In [Sha], Shamir presents a (k, l)-threshold secret sharing in which a secret s is split into l shares such that the secret can be computes if k shares are known, but no information about s can be computed if only $k - 1$ shares are known.

Let p and q be two large prime numbers such that $p = 2q + 1$, g be an element of order q in $\mathbb{Z}^*_p$ and $s \in \mathbb{Z}_q$ the secret a dealer wants to share between l shareholders. The following protocol shares s into l shares $s_1, \ldots, s_l$.

Protocol: $Share(s) = ((s_1, \ldots, s_l), (a_1, \ldots, a_{k-1}))$

1. For $j = 1, \ldots, k - 1$, pick random $a_j \in \mathbb{Z}_q$.
2. Set $P(X) = s + \sum_{j=1}^{k-1} a_j X^j \bmod q$.
3. For $i = 1, \ldots, l$ compute $s_i = P(i)$.

If k shares are known, then s can be recovered using Lagrange's interpolation.

Let $S = g^s \bmod p$, $S_i = g^{s_i} \bmod p$ for $i = 1, \ldots, l$ and $A_j = g^{a_j} \bmod p$ for $j = 1, \ldots, k - 1$. The following protocol due to [CPS] and [Fel] proves that $\log_g(S_i)$ are the shares.

Protocol : $Proof3(Share(\log_g(S)) = ((\log_g(S_1), \ldots, \log_g(S_l)), (\log_g(A_1), \ldots, \log_g(A_k - 1)))$

1. The dealer sends $(S, S_1, \ldots, S_l, A_1, \ldots, A_k - 1)$ to the verifier.
2. The verifier checks that for all $i = 1, \ldots, l$, $S_i = S \cdot \prod_{j=1}^{k-1}(A_j)^{i^j} \bmod p$.

3.4 A problem equivalent to factorization

Let n be a composite number, e be coprime with $\lambda(n)$ and d be such that $de \equiv 1 \bmod \lambda(n)$. We prove in this subsection the equivalence between the factorization and the knowledge of the common discrete logarithm of g_i in base g_i^e for $i = 1, 2$, where g_i is computed (but not freely chosen) by the dealer who knows the factorization of n. We will prove that this common discrete logarithm allows to factor n.

Lemma: Assume that a time-bounded dealer has chosen a composite number n and has computed $g_i = h(n \parallel i) \bmod n$ for $i = 1, 2$, where h is a hash function which outputs $|n|$-bit values. Let e be coprime with $\lambda(n)$. Assume that this dealer has disclosed to a party an integer α, a common discrete logarithm of g_i in base g_i^e for $i = 1, 2$. Then this party, on input α, can factor n. In other words, the dealer cannot compute in polynomial time a composite number n such that a common discrete logarithm of g_i in base g_i^e for $i = 1, 2$ does not reveal the prime factors of n.

Sketch of proof: First note that the dealer can compute such α if and only if e is coprime with $\lambda(n)$: d such that $de \equiv 1 \pmod{\lambda(n)}$ satisfies the required properties.

We first state this result (that we do not prove for lack of space): let n be a composite number and g be an element randomly selected over $\mathbb{Z}_n^*$. If β divides $\lambda(n)$, the probability that $\lambda(n)/\mathrm{ord}(g)$ is equal to β or a multiple of β is less than or equal to $1/\beta$.

Once the dealer has chosen the composite number n, he has no control on the values of g_1 and g_2 (due to the use of the hash function h). So, given β, his only chance to get (n, g_1, g_2) such that both $\lambda(n)/\mathrm{ord}(g_1)$ and $\lambda(n)/\mathrm{ord}(g_2)$ are equal to β or a multiple of β is to try different values of n until $g_i^{\lambda(n)/\beta} = 1$ for $i = 1, 2$, which happens after more than β^2 tests on average: as the probability that $\mathrm{ord}(g_i)$ is equal to β or a multiple of β is less than $1/\beta$ for $i = 1, 2$ and that these probabilities are independent, the probability that both $\mathrm{ord}(g_1)$ and $\mathrm{ord}(g_2)$ are equal to $\lambda(n)/\beta$ or a divisor of $\lambda(n)/\beta$ is less than $1/\beta^2$.

Remark : The cheating dealer can also take several values β_j $(j = 1..l)$ which divide $\lambda(n)$ (e.g. all the divisors of $\lambda(n)$ less than $\lambda(n)/2^{40}$) and test for all β_j if $g_i^{\lambda(n)/\beta_j} = 1$ for $i = 1, 2$. Then, his probability of success is $\sum_j 1/\beta_j^2$ which is less than $l/(\min(\beta)^2)$, but the number of tests "$g_i^{\lambda(n)/\beta} = 1$" is multiplied by l. Thus, this strategy is worse than the previous one.

So, as the dealer is time-bounded, we assume that it is infeasible for the dealer to try more than β^2 different values of n (for example, $\beta > 2^{40}$).

Let $L = \mathrm{lcm}(\mathrm{ord}(g_1); \mathrm{ord}(g_2))$ and $M = \lambda(n)/L$. As the dealer is time-bounded, we can assume that he is unable to find n such that M is a multiple of β. As α is a discrete logarithm of g_i^e in base g_i for $i = 1, 2$, we have $g_i^{e\alpha} = g_i \bmod n$ and thus $e\alpha = 1 \bmod \mathrm{ord}(g_i)$ for $i = 1, 2$. Consequently, $e\alpha = 1 \bmod L$; so there exists $k \in \mathbb{Z}$ such that $e\alpha - 1 = kL$ (in $\mathbb{Z}$) and hence $M(e\alpha - 1) \equiv 0 \bmod \lambda(n)$.

To factor n, we first choose an element y in $\mathbb{Z}_n^*$, we compute M', a discrete logarithm of 1 in base $y^{e\alpha-1}$ (which is M or a divisor of M). This discrete logarithm is less than β, so it can be computed using $2\sqrt{\beta}$ modular multiplications using Shanks [Sch] algorithm. Then, $M'(e\alpha - 1) \equiv 0 \mod \text{ord}(y)$. So we get $V_1 = M'(e\alpha - 1)$, a multiple of $\text{ord}(y)$.

Then, we choose another element z and we compute M'', the discrete logarithm of 1 in base z^{V_1}. So we get $V_2 = M'M''(e\alpha - 1)$, a multiple of $\text{ord}(y)$ and of $\text{ord}(z)$. Note that as y was randomly selected over $\mathbb{Z}_n^*$, its order α is such that $\lambda(n)/\alpha$ is very small. So M'', which divides $\lambda(n)/\alpha$, is very small too and can be computed by exhaustive search.

We repeat this process until we are convinced that there exists no element whose order does not divides V_j (i.e. even the elements of maximal order). This is the case after 80 rounds, as the probability that K elements do not generate the whole group $\mathbb{Z}_n^*$ is less than $\zeta(K) = \sum_{i=2..\infty} 1/i^K \approx 2^{-K}$ (according to the result stated at the beginning of this proof). Then, V_j is a multiple of $\lambda(n)$ and the party can factor n by Miller's lemma [Mil] (it is well known that given n and a multiple of $\lambda(n)$ we can factor n even if n has more than two prime factors).

Efficiency of the computation of a multiple of the order of y with respect to the computational effort of a cheating dealer: We recall that a cheating dealer can output a composite n such that $\text{lcm}(\text{ord}(g_1);\text{ord}(g_2))$ is greater than β only after trying and testing β^2 moduli n, and that, in this case, we need $2\sqrt{\beta}$ modular multiplications using Shanks algorithm to be able to compute a multiple of the order of y. For $\beta = 2^{40}$, the cheating dealer needs 2^{80} tries (which is infeasible nowadays) and Shanks algorithm requires 2^{21} modular multiplications (which can be done in a few seconds). This fact will discourage any dealer from attempting to cheat. And, if the dealer does not cheat, M is very small (in 99% of the cases, M is less than 10) and can be found by exhaustive search.

3.5 Proof that a discrete logarithm allows to factor n

In this subsection, we give an efficient proof of the knowledge of a discrete logarithm and that this discrete logarithm allows to factor n. In this proof, it is unnecessary to prove that n is the product of only two prime factors, unlike [Mao] and [FOk] schemes. This avoids to use inefficient protocols to prove this fact, like [Mic] or [vGP]. Note that, for this reason, this proof is the more efficient proof of knowledge of the factorization of a composite number n.

Let p be a large prime number, and g an element of maximal order in $\mathbb{Z}_p^*$, N be a large composite number whose factorization is unknown by all the participants (the dealer, the verifier and the shareholders) which is generated by a trusted third party, and G an element in $\mathbb{Z}_N^*$ (randomly selected by the shareholders). Assume that Alice knows the factorization of a large composite number n. Let d be such that $de \equiv 1 \pmod{\lambda(n)}$, and $Y = g^d \mod p$. If $n/\text{ord}(g) < 2^{-91}$, the following protocol allows Alice to prove that $\log_g(Y)[\mod p]$ allows to factor n.

Protocol: Proof5($d, Y = g^d \mod p \wedge d$ allows to factor n)

1. Alice computes $g_1 = h(n \parallel 1) \mod n$ and $g_2 = h(n \parallel 2) \mod n$.

2. Alice sends $W = G^d \bmod N$.
3. Alice executes $\text{Proof2}_n(d, Y = g^d \bmod p \wedge g_1 = (g_1^e)^d \bmod n \wedge g_2 = (g_2^e)^d \bmod n \wedge W = G^d \bmod N \wedge \text{abs}(d) < \text{ord}(g)/2)$.

Correctness : First note that Proof2 (which is sound only if the factorization of one of the moduli is unknown by Alice) can be used because the factorization of N is unknown by Alice (this is the only utility of the number W). Let α be the common discrete logarithm of Y in base g modulo n, g_1 in base $g_1^e \bmod n$ and g_2 in base $g_2^e \bmod n$. Using the method described in Section 3.4., it is possible, knowing α, to factor n.

3.6 Proof of equality of a discrete logarithm and a third root

Let p be a large prime number and g an element of maximal order in Z_p^*. Let n_1 be a large composite number such that its factorization is unknown by Alice. We assume that $p/n_1 < 2^{-100}$.

Given G and E, the following protocol proves that there exists x such that $G = g^x \bmod p$, $E = x^3 \bmod n_1$ and that $\text{abs}(x) < (n_1 - 1)/2$. This protocol is related to [FOk].

Let N be a large composite number, whose factorization is unknown by all the participants, such that $N < n_1$ and that $p/\lambda(N) < 2^{-91}$, and $\bar{g}$ an element in Z_N^*. The protocol runs as follows:

Protocol: $\text{Proof6}(x, G = g^x \bmod p \wedge E = x^3 \bmod n_1 \wedge \text{abs}(x) < (n_1 - 1)/2)$

1. Alice computes $\alpha = \frac{E - x^3}{n_1}$, $G_1 = \bar{g}^x \bmod N$, $G_2 = \bar{g}^{x^2} \bmod N$, $G_3 = \bar{g}^{x^3} \bmod N$ and $Z = \bar{g}^{\alpha n_1} \bmod N$.
2. Alice computes $\text{Proof2}_p(x, G = g^x \bmod p \wedge G_1 = \bar{g}^x \bmod N \wedge G_2 = G_1^x \bmod N \wedge G_3 = G_2^x \bmod N \wedge \text{abs}(x) < \lambda(N)/2)$ and $\text{Proof1}(\alpha, Z = (\bar{g}^{n_1})^\alpha \bmod N)$.
3. The verifier checks Proof2 and Proof1, computes $T = \bar{g}^E \bmod N$ and checks that $G_3 = T/Z \bmod N$.

Correctness: If all the proofs succeed, then Alice knows x such that $G = g^x \bmod p$, $G_1 = \bar{g}^x \bmod N$, $G_3 = \bar{g}^{x^3} \bmod N$ and such that $\text{abs}(x) < \lambda(N)$, and α such that $Z = \bar{g}^{\alpha n_1} \bmod N$. Hence, $\bar{g}^{x^3} = T/\bar{g}^{\alpha n_1} \bmod N$, and this implies $x^3 \equiv E - \alpha n_1 \pmod{\lambda(N)}$. So $x^3 = E - \alpha n_1$ (if not, Alice would be able to factor N). It comes $E \equiv x^3 \pmod{n_1}$.

How to use Proof6: If the verifier is convinced by Proof6 that there exists x such that $G = g^x \bmod p$ and $E = x^3 \bmod n_1$ and $\text{abs}(x) < (n_1 - 1)/2$, then he knows that someone who is able to computes the third root of E is able to compute the discrete logarithm of G in base g modulo p.

Let $\tilde{x}$ be the third root of E, which is congruent to x modulo n_1. As it is proven that $\text{abs}(x) < \lambda(N) < n_1/2$, this third root (which theoretically belongs to $\{(-n_1 + 1)/2, \ldots, (n_1 - 1)/2\}$) is equal (in $\mathbb{Z}$) to the value x used by Alice in this protocol. So this value is the discrete logarithm of G_1 in base $\bar{g}$ modulo N (which belongs to $\{-\lambda(N)/2, \ldots, \lambda(N)/2\}$). Then, this value is also a discrete logarithm of G in base g modulo p, and it is easy to compute the discrete logarithm of G in base g modulo p, which is $x \bmod \mathrm{ord}(g)$.

Secrecy of x: Although x has only $|p| = 1024$ bits and n has 1500 bits, to compute x from $E = x^3 \bmod n_1$ is infeasible.

Thus, the third root of E is a discrete logarithm of G in base g modulo p.

3.7 A cryptosystem with arbitrarily slow decryption

We characterize composite numbers n for which solving the discrete logarithm modulo n is hard, but becomes easier when its factorization is known. Such numbers are used by Maurer and Yacobi [MYa] to provide an identity-based cryptosystem.

Let $n = pq$ such that $p - 1 = 2p_1 \ldots p_r$ and $q - 1 = 2q_1 \ldots q_{r'}$. We denote $m = \min(\max(p_i); \max(q_i))$ and $M = \max(\max(p_i); \max(q_i))$. Let g be an element of maximal order in $\mathbb{Z}_n^*$. In practice, $|n| = 1500$ and $|m| \simeq |M| \simeq 70$.

Solving a discrete logarithm modulo n is hard: This problem is harder than factoring n. As $p - 1$ and $q - 1$ have relatively small prime factors, some factorization algorithms become more efficient, like Pollard's algorithm [Pol]. Such algorithms run in time $O(m)$, so our choice of $|m|$ makes the factorization infeasible.

When the factorization is known, solving the problem is feasible: The Pohlig-Hellman's algorithm [PHe] consists of computing the discrete logarithm modulo each p_i and each q_i using Shanks [Shn] algorithm. Shanks algorithm is used $r + r' \simeq |n|/|M|$ times and needs $2\sqrt{M}$ modular exponentiations, each of them costs $3|M|/2$ modular multiplications. So, it finds the discrete logarithm in about $3|n|\sqrt{M}$. If we choose $|M| = 70$ and $|n| = 1500$, the algorithm needs 2^{47} modular multiplications.

Then, Alice can encrypt x setting $E = g^x \bmod N$ and Bob, who knows the factorization of n, can compute $x = \log_g(E)[\bmod n]$ using Pohlig-Hellman algorithm.

In the rest of this paper, we will call such number a Maurer-Yacobi (MY) number.

4 Our Schemes

4.1 How to share a discrete logarithm with fast recovery

Let $p = 2q + 1$ be a prime number, g be an element of order q in $\mathbb{Z}_p^*$, N be a composite number whose factorization is unknown by Alice, G be an element of maximal order in $\mathbb{Z}_N^*$ and, for $i = 1, \ldots, l$, n_i be a composite number whose factorization is only known by the shareholder SH_i. Typically, $|p| = 1024$, and $|n_i| = 1500$. Let s be Alice's secret and $S = g^s \bmod p$ be a public number.

To compute the shares and the certificate:

1. *Share:* Alice runs $Share(s) = ((s_1, \ldots, s_l), (a_1, \ldots, a_{k-1}))$, computes $S_i = g^{s_i} \bmod p$ for $i = 1, \ldots, l$ and $A_j = g^{a_j} \bmod p$ for $j = 1, \ldots, k - 1$. Then

she executes $\text{Proof3}(Share(\log_g(S)) = ((\log_g(S_1), \ldots, \log_g(S_l)), (\log_g(A_1), \ldots, \log_g(A_{k-1})))$.

2. *Encrypt each share:* For $i = 1, \ldots, l$, Alice computes $E_i = s_i^3 \bmod n_i$ and executes $\text{Proof6}(s_i, S_i = g^{s_i} \bmod p \wedge E_i = s_i^3 \bmod n_i \wedge \text{abs}(s_i) < (n_i - 1)/2)$.

To check the validity of the certificate:
The verifier checks that all proofs succeed.

To recover the secret:
At least k shareholders decrypt E_i and recover s_i, then they can recover s using Lagrange's interpolation.

4.2 How to share a factorization with fast recovery

Let $P = 2Q + 1$ be a prime number, g be an element of order Q in $\mathbb{Z}_P^*$, and, for $i = 1, \ldots, l$, n_i be a composite number whose factorization is only known by the shareholder SH_i. Let p, q be Alice's secret and $n = pq$ be a public number. Let d such that $de \equiv 1 \pmod{\lambda(n)}$. Typically, $|n| = 1024$, $|P| = 1400$ and $|n_i| = 1500$.

To compute the shares and the certificate:

1. *Transform the secret into a discrete logarithm :* Alice sets $S = g^d \bmod P$ and executes $\text{Proof5}(d, S = g^d \bmod P \wedge d$ allows to factor $n)$.
2. *Share:* Alice runs $Share(d) = ((s_1, \ldots, s_l), (a_1, \ldots, a_{k-1}))$, computes $S_i = g^{s_i} \bmod P$ for $i = 1, \ldots, l$ and $A_j = g^{a_j} \bmod P$ for $j = 1, \ldots, k - 1$. Then she executes $\text{Proof3}(Share(\log_g(S)) = ((\log_g(S_1), \ldots, \log_g(S_l)), (\log_g(A_1), \ldots, \log_g(A_{k-1})))$.
3. *Encrypt each share:* For $i = 1, \ldots, l$, Alice computes $E_i = s_i^3 \bmod n_i$ and executes $\text{Proof6}(s_i, S_i = g^{s_i} \bmod P \wedge E_i = s_i^3 \bmod n_i \wedge \text{abs}(s_i) < (n_i - 1)/2)$.

To check the validity of the certificate:
The verifier checks that all proofs succeed.

To recover the secret:
At least k shareholders decrypt E_i and recover s_i, then they can recover, using Lagrange's interpolation, a number which allows to factor n.

4.3 How to share a discrete logarithm with delayed recovery

Let $p = 2q + 1$ be a prime number, g be an element of order q in $\mathbb{Z}_p^*$, for $i = 1, \ldots, l$, n_i be a MY-number whose factorization is only known by the shareholder SH_i, and g_i be an element of maximal order in $\mathbb{Z}_{n_i}^*$. Typically, $|p| = 1024$ and $|n_i| = 1500$. Let s be Alice's secret and $S = g^s \bmod p$ be a public number.

To compute the shares and the certificate:

1. *Share:* Alice runs $Share(s) = ((s_1, \ldots, s_l), (a_1, \ldots, a_{k-1}))$, computes $S_i = g^{s_i} \bmod p$ for $i = 1, \ldots, l$ and $A_j = g^{a_j} \bmod p$ for $j = 1, \ldots, k - 1$. Then she executes $\text{Proof3}(Share(\log_g(S)) = ((\log_g(S_1), \ldots, \log_g(S_l)), (\log_g(A_1), \ldots, \log_g(A_{k-1})))$.

2. *Encrypt each share:* For $i = 1, \ldots, l$, Alice computes $E_i = g_i^{s_i} \bmod n_i$ and executes $\text{Proof2}_p(s_i, S_i = g^{s_i} \bmod p \wedge E_i = g_i^{s_i} \bmod n_i \wedge \text{abs}(s_i) < \lambda(n_i)/2)$.

To check the validity of the certificate:
The verifier checks that all proofs succeed.

To recover the secret:
At least k shareholders decrypt E_i and recover s_i using Pohlig-Hellman algorithm, then they can recover s using Lagrange's interpolation.

4.4 How to share a factorization with delayed recovery

Let $P = 2Q + 1$ be a prime number, g be an element of order Q in $\mathbb{Z}_P^*$, for $i = 1, \ldots, l$, n_i be a MY-number whose factorization is only known by the shareholder SH_i, and g_i be an element of maximal order in $\mathbb{Z}_{n_i}^*$. Let p, q be Alice's secret and $n = pq$ be a public number. Let d such that $de \equiv 1 \pmod{\lambda(n)}$. Typically, $|n| = 1024$, $|P| = 1400$ and $|N| = |n_i| = 1500$.

To compute the shares and the certificate:

1. *Transform the secret into a discrete logarithm:* Alice sets $S = g^d \bmod P$ and executes $\text{Proof5}(d, S = g^d \bmod P \wedge d$ allows to factor $n)$.
2. *Share:* Alice runs $Share(d) = ((s_1, \ldots, s_l), (a_1, \ldots, a_{k-1}))$, computes $S_i = g^{s_i} \bmod P$ for $i = 1, \ldots, l$ and $A_j = g^{a_j} \bmod P$ for $j = 1, \ldots, k-1$. Then she executes $\text{Proof3}(Share(\log_g(S)) = ((\log_g(S_1), \ldots, \log_g(S_l)), (\log_g(A_1), \ldots, \log_g(A_{k-1})))$.
3. *Encrypt each share:* For $i = 1, \ldots, l$, Alice computes $E_i = g_i^{s_i} \bmod n_i$ and executes $\text{Proof2}_P(s_i, S_i = g^{s_i} \bmod P \wedge E_i = g_i^{s_i} \bmod n_i \wedge \text{abs}(s_i) < \lambda(n_i)/2)$.

To check the validity of the certificate:
The verifier checks that all proofs succeed.

To recover the secret:
At least k shareholders decrypt E_i and recover s_i using Pohlig-Hellman algorithm, then they can recover, using Lagrange's interpolation, a number which allows to factor n.

4.5 Efficiency of our schemes

We evaluate the efficiency of each scheme and compare it to the efficiency of [FOk]'s schemes. The values are given for $k = 5$ and $l = 3$ (5 shareholders, such that at least 3 can recover the secret) as an example, but the savings are of the same order for any values of k and l.

This first table gives the number of 1024-bit multiplications the dealer must perform to compute its certificate.

Property of the scheme	Efficiency of [FOk] scheme (mult.)	Efficiency of our scheme (mult.)	% comp. saved
DL, fast recovery	144.10^3	16.10^3	88.9%
Fact, fast recovery	185.10^3	19.10^3	89.7%
DL, delayed recovery	144.10^3	11.10^3	92.4%
Fact, delayed recovery	185.10^3	14.10^3	92.4%

This second table gives the length of the message the dealer must send to the verifier.

Property of the scheme	Efficiency of [FOk] scheme (kBytes)	Efficiency of our scheme (kBytes)	% bits saved
DL, fast recovery	30.8	11.7	62.0%
Fact, fast recovery	40.8	12.3	69.9%
DL, delayed recovery	30.8	3.81	87.6%
Fact, delayed recovery	40.8	4.43	89.1%

DL = sharing of a discrete logarithm
RSA = sharing of the factorization

5   Conclusion

We have presented publicly verifiable secret sharing schemes for both discrete logarithm and integer factorization problems, and with fast or delayed recovery. Our schemes allow to save about 90% of computations with respect to the previous ones, and more than 62% of bits communicated. Moreover, there are no restricted conditions (such that 48 shareholders for schemes with delayed recovery) in our schemes, unlike [FOk] schemes.

Acknowledgements

We would like to thank Marc Girault for helpful discussions and comments.

References

[BGo]   M. Bellare and S. Goldwasser, Verifiable Partial Key Escrow, *Proceedings of the Fourth Annual Conference on Computer and Communications Security, pp. 78-91, 1997.*

[Bao]   F. Bao, An Efficient Verifiable Encryption Scheme for Encryption of Discrete Logarithms, *to be published in the proceedings of CARDIS'98, 1998.*

[BRo]   M. Bellare and P. Rogaway, Random Oracles are Practical: a Paradigm for Designing Efficient Protocols, *Proceedings of the First Annual Conference and Communications Security, pp. 62-73, 1993.*

[CEG]   D. Chaum, J.-H. Evertse and J. van de Graaf, An Improved Protocol for Demonstrating Possesion of Discrete Logarithm and Some Generalizations, *Proceedings of EUROCRYPT'87, LNCS 304, pp. 127-141*, 1988

[CFT]   A. Chan, Y. Frankel and Y. Tsiounis, Easy Come - Easy Go Divisible Cash, *Proceedings of EUROCRYPT'98, LNCS 1403, pp. 561-575*, 1998.

[CGMA]   B. Chor, S. Goldwasser, S. Micali and B. Awerbuch, Verifiable Secret Sharing and Achieving Simultaneity in the Presence of Faults, *Proceedings of FOCS, pp. 383-395*, 1985.

[CPe]   D. Chaum and T.P. Pedersen, Wallet Databases with Observers, *Proceedings of CRYPTO'92, LNCS 740, pp. 89-105*, 1992.

[CPS]   C. Charnes, J. Pieprzyk and R. Safavi-Naini, Conditionally Secure Secret Sharing Scheme with Disenrolment Capability, *Second ACM Conference on Computer and Communication Security, pp. 89-95*, 1994.

[Fel]   P. Feldman, A Practical Scheme for Non-Interactive Verifiable Secret Sharing, *Proceedings of the 28th IEEE Symposium on FOCS, pp. 427-437*, 1987.

[FOk]   E. Fujisaki and T. Okamoto, A Practical and Provably Secure Scheme for Publicly Verifiable Secret Sharing and Its Applications, *Proceedings of EUROCRYPT'98, LNCS 1403, pp. 32-46*, 1998.

[FSh]   A. Fiat and A. Shamir, How to Prove Yourself: Practical Solutions to Identification and Signature Problems, *Proceedings of CRYPTO'86, LNCS 263, pp. 186-194*, 1986.

[Gir]   M. Girault, Self-Certified Public Keys, *Proceedings of EUROCRYPT'91, LNCS 547, pp. 490-497*, 1991.

[vGP]   J. van de Graaf and R. Peralta, A Simple and Secure Way to Show the Validity of Your Public Key, *Proceedings of CRYPTO'87, LNCS 293, pp. 128-134*, 1987.

[Mao]   W. Mao, Guaranteed Correct Sharing of Integer Factorization with Offline Shareholders, *Proceedings of Public Key Cryptography 98, pp. 27-42*, 1998.

[Mic]   S. Micali, Fair Public-Key Cryptosystems, *Proceedings of CRYPTO'92, LNCS 740, pp. 113-138*, 1992.

[Mil]   G. Miller, Riemann's hypothesis and Tests for Primality, *Journal of Computer and System Sciences (13), pp. 300-317*, 1976.

[MYa]   U. Maurer and Y.Yacobi, Non-Interactive Public-Key Cryptography, *Proceedings of EUROCRYPT'91, LNCS 547, pp. 498-507*, 1991.

[PHe]   S.C. Pohlig and M. Hellman, An Improved Algorithm for Computing Logarithms over $GF(p)$ and its Cryptographic Significance, *Proceedings of IEEE Transactions on Information Theory, Vol. IT-24, pp. 106-110*, 1978.

[Pol]   J.M. Pollard, Theorems on Factorization and Primality Testing, *Proceedings of Cambridge Philos. Society, vol. 76, pp. 521-528*, 1974.

[PSt]   G. Poupard and J. Stern, Security Analysis of a Practical "on the fly" Authentication and Signature Generation, *Proceedings of EUROCRYPT'98, LNCS 1403, pp. 422-436*, 1998.

[RSW]   R. Rivest, A. Shamir and D. Wagner, Time-Lock Puzzles and Time-Release Crypto, *this paper is available at* http://theory.lcs.mit.edu/~rivest/publications.html, to appear.

[Sha]   A. Shamir, How to Share a Secret, *CACM, vol. 22, No. 11, pp. 612-613*, 1979.

[Shn]   D. Shanks, Five Number-Theoric Algorithms, *Proceedings of the 2nd Manitoba conference on numerical mathematics, pp. 51-70*, 1972.

[Sta]   M. Stadler, Publicly Verifiable Secret Sharing, *Proceedings of EURO-CRYPT'96, LNCS 1070, pp. 190-199*, 1996.

[YYu]   A. Young and M. Yung, Auto-Recoverable Auto-Certifiable Cryptosystems, *Proceedings of EUROCRYPT'98, LNCS 1403, pp. 17-31*, 1998.

Zero-Knowledge Proofs of Possession of Digital Signatures and Its Applications

Khanh Quoc Nguyen[1], Feng Bao[2], Yi Mu[1], and Vijay Varadharajan[1]

[1] School of Computing & Information Technology
University of Western Sydney, Nepean
PO Box 10, Kingswood, NSW 2747, Australia
{qnguyen,yimu,vijay}@cit.nepean.uws.edu.au
[2] Kent Ridge Digital Labs
21 Heng Mui Keng Terrace, Singapore 119597
baofeng@krdl.org.sg

Abstract. Demonstrating in zero-knowledge the possession of digital signatures has many cryptographic applications such as anonymous authentication, identity escrow, publicly verifiable secret sharing and group signature. This paper presents a general construction of zero-knowledge proof of possession of digital signatures. An implementation is shown for discrete logarithm settings. It includes protocols of proving exponentiation and modulo operators, which are the most interesting operators in digital signatures. The proposed construction is applicable for ElGamal signature scheme and its variations. The construction also works for the RSA signature scheme. In discrete logarithm settings, our technique is $O(l)$ times more efficient than previously known methods.

1 Introduction

1.1 The problem

Demonstrating the knowledge of secrets, which satisfy some specific relations while revealing no useful information about the secrets, is a major cryptographic primitive. It is used to realize many practical cryptographic protocols. In Camenisch-Stadler group signature[5] a group signature on the message m is a non-interactive proof of the signer's group membership certificate which is a RSA signature signed by the group manager. In identity escrow[17], users have to prove the possession of valid identity certificates which are digital signatures on some known messages. Some other protocols to deal with publicly verifiable secret sharing [25] and fair exchange of digital signatures[1] could also be considered as this type of problems. This paper focuses on zero-knowledge proof of secrets that form a valid digital signature.

1.2 Related Works

A digital signature scheme is referred to as a specific arithmetic relation. Theoretically, protocols to demonstrate any arithmetic relation can be derived from zero-knowledge proof for NP-language[16] and can be converted to proof of knowledge[14]. Those protocols are very inefficient because they require to encode the relations into a Boolean circuit and apply the zero-knowledge proof for each gate. The common methodology is to commit the secrets in some commitments, form an arithmetic circuit corresponding to the relation and then prove that the committed secrets satisfy the desired relations using the circuit.

There exist general constructions for zero-knowledge protocols to prove linear relations connected by Boolean operators[3] and polynomial relations[9,15]. However, proving such relations is relatively easy as all commitments are computed in the same finite field. For digital signature relations, proving knowledge possession is not an easy task. This is because commitments of signatures are often computed in different finite fields. If there is an exponentiation in digital signatures of which both the base and exponent are the secrets to be demonstrated, the proof becomes much more difficult.

1.3 Our contribution

This paper presents a general construction of zero-knowledge proof of possession of digital signatures. The construction yields zero-knowledge proofs for many digital signature schemes, including ElGamal, DSA. Nyberg-Rueppel and RSA signatures.

Our construction can be briefly described as follows. We first break the relation into a corresponding arithmetic circuit, that consists of gates, each is only a standard arithmetic relation. We then compute a commitment for all the secrets and the output of every gate. The protocol to prove the digital signature relation is a set of procedures proving that the committed inputs yield the committed output in all gates.

Thus our first task is to form a language $\mathcal{L}$ of all basic arithmetic relations required in digital signatures and show how to create an arithmetic circuit for any relation of $\mathcal{L}$. The remaining task is to construct building blocks of proving all those standard arithmetic relations that include addition, multiplication, exponentiation, modulo and equality. The building blocks are dependent on the underlying setting. We give a realization of these building blocks in discrete logarithm settings. Our general protocol for discrete logarithms is $O(l)$ times more efficient than current methods. Other modes of proofs and settings can be achieved with some slight modifications.

The general construction is not only useful for digital signature relations but can also be used to prove many other arithmetic relations. For instance, linear and polynomial relations between secrets committed in the same finite field can be demonstrated with addition and multiplication relation protocols. The equality relation protocol can turn the problem of showing the relations between secrets committed in different finite fields into the less difficult problem of proving the relations between secrets committed in the same finite field.

The outline of the paper: The construction of language $\mathcal{L}$ and the general zero-knowledge protocol to prove any arithmetic relation in $\mathcal{L}$ are given in section 2.

Section 3 presents the realization of the general protocol for discrete logarithm settings. It also discusses zero-knowledge proofs in other settings and models. The zero-knowledge protocols to prove possession of digital signatures are discussed in section 4.

2 General Construction

This section gives the description of our setting and the general zero-knowledge protocol to prove arithmetic relations. The construction is independent of any particular implementation.

2.1 Notations and Settings

For any number m, let Z_m denote the finite field of $0, ..., m-1$. Let $\mathcal{L} = \{F_i(x_1, ..., x_n)|i = 0, ..\}$ be the language of all functions $F_i(x_1, ..., x_n)$ computed over the finite field Z_{p_i}, parsed according to the following BNF rules:

$$F_j ::= c\,|\,x_i\,|\,F_j + F_j\,|\,F_j * F_j\,|\,F_j^{F_k}\,|\,F_j \bmod p_i$$

Here c is some constant, F_j represents an expression of $\mathcal{L}$ computed in Z_{p_j}, F_k in Z_{p_k} for any j, k. Using these parsing rules, one can form an arithmetic circuit corresponding to any member $F_i()$ of $\mathcal{L}$. For convenience, let the arithmetic circuit of function $F_i()$ be *the circuit* $F_i()$. Let each node in the circuit $F_i()$ corresponding to $F_j = (F_j + F_j)$, $(F_j * F_j)$, $(F_j^{F_k})$ and $(F_j \bmod p_i)$ be addition, multiplication, exponentiation and modulo gate respectively. The node corresponds to $F_j = x_i$ is referred to as a leaf. Note that gates are computed over many different finite fields Z_{p_i} $(i = 0, ..)$ and there might be several exponentiation gates in the circuit $F_i()$ computed over the *same* group Z_{p_j} but with a different order p_k.

2.2 Protocol Description

We now give a general construction for zero-knowledge proof of possession of the secrets $s_1, ..., s_n$ for any function $F_i()$ that $F_i(s_1, ..., s_n) = y$ for a public constant y. We borrow the model from [9] and assume the existence of the following building blocks:

1. **Commitment scheme** lets the prover $\mathcal{P}$ commit a number $0 \leq a < p_i$ in a commitment $A = \mathtt{commit}(a)$ in such a way that the verifier $\mathcal{V}$ cannot open the commitment to get the value a while the prover is unable to find two different values a corresponding to the same commitment A.
2. **Checking Protocol** allows $\mathcal{P}$ to prove in zero-knowledge to $\mathcal{V}$ the knowledge of the secret a for a given commitment A.
3. **Addition Gate Protocol** allows $\mathcal{P}$ to convince $\mathcal{V}$ in zero-knowledge that a secret is the sum of two other secrets given three commitments computed over Z_{p_i}.

4. **Multiplication Gate Protocol** allows $\mathcal{P}$ to convince $\mathcal{V}$ in zero-knowledge that a secret is the product of the two other secrets given three commitments computed over Z_{p_i}.
5. **Exponentiation Gate Protocol** allows $\mathcal{P}$ to convince $\mathcal{V}$ in zero-knowledge the exponentiation relation $c = a^b \bmod q_i (q_i = p_j)$ for the secrets a, b, c concealed in the commitments A, B, C respectively. A, C are computed in Z_{p_i} and B is computed in Z_{p_j}.
6. **Modulo Gate Protocol** allows $\mathcal{P}$ to convince $\mathcal{V}$ in zero-knowledge that the secrets a_i, a_j concealed in commitment A_i and A_j respectively, equal modulo q_j, i.e., $a_i - a_j = 0 \bmod q_j$. A_i is computed in Z_{p_i} while A_j is computed in Z_{p_j} ($q_i > q_j$, both are members of $\{p_i | i = 0, ..\}$).
7. **Equality Protocol** allows $\mathcal{P}$ to prove the equality of the two secrets concealed in two commitments, each computed in different finite fields and/or different multiplicative groups.

Then the protocol is executed as follows:

Protocol Description

STEP 0
First, $\mathcal{P}$ and $\mathcal{V}$ agree upon a circuit $F_i()$. This circuit is known and verified by both parties such that the circuit is correctly formed from the function $F_i()$. For convenience, let $L_1, ..., L_u$ be the leave and $G_1, ..., G_v$ be the gates of the circuit $F_i()$, G_v is the final gate and its output is y.

STEP 1
$\mathcal{P}$ makes u commitments $U_1, ..., U_u$ and v commitments $V_1, ..., V_v$ such that U_j contains the value of L_j and V_i contains the output value of G_i. All commitments are then sent to $\mathcal{V}$. $\mathcal{P}$ then uses the checking protocol for the commitments $U_1, ..., U_u$ to prove to $\mathcal{V}$ that $\mathcal{P}$ can open all of them[1].

STEP 2
For each gate, $\mathcal{P}$ proves the relation between the secrets concealed in the commitments representing the inputs and output of the gate using one of Addition, Multiplication, Exponentiation and Modulo Gate protocols.

STEP 3
If the two commitments U_i, U_j concealed the same secret $s_i = s_j$ computed in two different finite fields Z_{p_i} and Z_{p_j}, $\mathcal{P}$ uses the equality protocol to prove the equality of the secrets.

STEP 4
$\mathcal{P}$ opens the commitment V_v representing the final gate G_v to show the concealed secret y. $\mathcal{V}$ accepts the proof if and only if the circuit is correctly formed in Step 0, all the proofs in Steps 1-3 are valid and $\mathcal{P}$ opens the commitment V_v to reveal y.

---

[1] [9] requires $\mathcal{P}$ to prove for commitments $V_1, ..., V_v$ as well. This is unnecessary.

PROOF:
The completeness of the protocol is straightforward from the inspection of the protocol. The proof of soundness appears in Appendix A.

3 A Realization for Discrete Logarithm Setting

To give an implementation of our protocol, we show the realization of the building blocks used in the general construction. This section presents those building blocks in computational zero-knowledge mode for discrete logarithm settings. Here p_i are all primes larger than 2^l for some security parameter l ($l > 80$).

We now give the constructions of the commitment scheme, checking, addition-gate, multiplication-gate, exponentiation-gate, modulo-gate and equality protocols. Two protocols for special cases: *multiplication with a constant factor* and *exponentiation with a constant base* of multiplication-gate and exponentiation-gate, are also given for efficiency. For presentation's sake, we give the protocols that demonstrate the relation between secrets of every gate in three different blocks according to the relations demonstrated, namely modulo, polynomial and exponentiation .

In this section, let p, q denote some generic prime numbers chosen from $\{p_i | i = 0, 1, ...\}$ such that there is a multiplicative group G_q of order q over Z_p. The constants g, h are chosen randomly in G_q such that $\log_g(h)$ in Z_p is not known to the prover. The symbol $\in_R$ means a uniformly random choice.

3.1 Commitment Scheme

A commitment of a value x in Z_p is constructed as $\mathtt{commit}(x, r) = g^x h^r \bmod p$ to $\mathcal{V}$, here $r \in_R Z_p$. There are two phases in this scheme:

Commit to a number x in Z_p, $\mathcal{P}$ simply sends $\mathtt{commit}(x, r) = g^x h^r \bmod p$ to $\mathcal{V}$.

Open a commitment $\mathtt{commit}(x, r)$ is done by sending x, r to $\mathcal{V}$. $\mathcal{V}$ can easily check the correctness of the committed values x, r with the commitment.

This scheme is unconditionally hiding. Its security is proven to be equivalent to the intractable discrete logarithm problem[22]. Unless $\mathcal{P}$ knows $\log_g(h)$ in Z_p, it is infeasible to find two different sets of input ($x, r \in Z_q$) that produce the same output $\mathtt{commit}(x, r)$.

3.2 Checking Protocol

The checking protocol is zero-knowledge and convinces $\mathcal{V}$ that $\mathcal{P}$ knows the secret x for the given commitment $w = \mathtt{commit}(x, r)$.

Protocol

1. $\mathcal{P}$ chooses $\alpha, \beta \in_R Z_q$ and sends $W = g^\alpha h^\beta \bmod p$ to $\mathcal{V}$.
2. $\mathcal{V}$ chooses $c \in_R Z_q$ and sends c to $\mathcal{P}$.
3. $\mathcal{P}$ sends $\mathcal{V}$ two responses $u = \alpha - cx$ and $v = \beta - cr \bmod q$.
4. $\mathcal{V}$ accepts the protocol if and only if $W = w^c g^u h^v \bmod p$.

This protocol is well-known and its security is no weaker than that of Schnorr's signature[24].

3.3 Modulo Related Protocols

Modulo related protocols demonstrate modulo and equality relations between secrets. Here commitments are computed in different group orders/finite fields. We now present the range-check protocol given in [21] and the equality proof protocol with an unknown group order. They are used in the construction of the modulo gate and equality protocols.

Range Check Protocol

Given: commitment $w = \text{commit}(x, r)$, a value $t < q$.
Prove: $x \in [0, t]$.

Protocol (run in parallel l times)

1. First both parties agree upon a suitable positive integer e, roughly equals $t/3 - 1$.
2. $\mathcal{P}$ chooses $\alpha_1 \in [0, e]$, and sets $\alpha_2 = \alpha_1 - e$. $\mathcal{P}$ sends to $\mathcal{V}$ the unordered pair of commitments $W_1 = \text{commit}(\alpha_1, \rho_1)$ and $W_2 = \text{commit}(\alpha_2, \rho_2)$.
3. $\mathcal{V}$ challenges $\mathcal{P}$ by $c \in_R \{0, 1\}$.
4. If $c = 0$, $\mathcal{P}$ sends to $\mathcal{V}$ the values of $\alpha_1, \alpha_2, \rho_1, \rho_2$.
   If $c = 1$, $\mathcal{P}$ sends to $\mathcal{V}$ the value of $x + \alpha_j, r + \rho_j$ such that $x + \alpha_j \in [e, t - e]$.
5. $\mathcal{V}$ verifies $W_j = \text{commit}(\alpha_j, \rho_j)$ $(j = 1, 2)$ in the former and $wW_j = \text{commit}(x + \alpha_j, r + \rho_j)$, $x + \alpha_j \in [e, t - e]$ in the latter case.

This protocol allows $\mathcal{P}$ to convince $\mathcal{V}$ that the commitment $w = \text{commit}(x, r)$ indeed satisfies $0 \leq x \leq t$ for some value t. It works regardless whether the order q of the multiplicative group of $[g, h]$ is known. If the order q of the group is unknown, virtually the same protocol can be used[11,21][2]. We refer to this particular version of protocol as *Range Check protocol with unknown group order*. Here the corresponding commitment $w = \text{commit}_N(x, r)$ is also computed as $w = g^x h^r \bmod N$ but x, r are chosen in the integer field. Of course, the factoring of N is not known to $\mathcal{P}$.

---

[2] Another relevant protocol is given in [6]. This protocol is much more efficient. However, it is not suitable for our task as the range of valid secrets is not identical to the range that prover can prove.

Equality Protocol with an unknown group order

Given: two commitment $w_p = \text{commit}_p(x_p, r_p)$ and $w_N = \text{commit}_N(x_N, r_N)$. Here N is a RSA modulo, much larger than p and x_N, r_N are chosen over the positive integer field.

Prove: $x_p = x_N = x$ $(x \in Z_q)$.

Protocol (run in parallel l rounds)

1. First both parties agree upon a suitable positive integer e which roughly equals $q/3 - 1$.
2. $\mathcal{P}$ chooses $\alpha_1 \in [0, e]$, and sets $\alpha_2 = \alpha_1 - e$. $\mathcal{P}$ sends to $\mathcal{V}$ the unordered set of commitments (W_1, S_1), (W_2, S_2) computed as $W_i = \text{commit}_p(\alpha_i, \rho_i)$ and $S_i = \text{commit}_N(\alpha_i, \rho_i)$ $(i = 1, 2)$.
3. $\mathcal{V}$ challenges $\mathcal{P}$ a bit $c \in_R [0, 1]$.
4. If $c = 0$, $\mathcal{P}$ sends to $\mathcal{V}$ the values of $\alpha_1, \alpha_2, \rho_1, \rho_2$.
   Otherwise, $\mathcal{P}$ sends to $\mathcal{V}$ the values of $x + \alpha_j, r + \rho_j$ such that $x + \alpha_j \in [e, q-e)$.
5. $\mathcal{V}$ verifies
   If $c = 0$, $W_j = \text{commit}_p(\alpha_j, \rho_j)$, $S_j = \text{commit}_N(\alpha_j, \rho_j)$, $\alpha_j \in [0, e]$ and $\rho_j > 0$ $(j = 1, 2)$.
   If $c = 1$, $w_p W_j = \text{commit}_p(x + \alpha_j, r_p + \rho_j)$, $w_N S_j = \text{commit}_N(x + \alpha_j, r_N + \rho_j)$ and $x + \alpha_j \in [e, q - e)$.

Security Proof

The **completeness** is straightforward by observing that if $\alpha_2 = \alpha_1 - e$, and $\alpha_1 \in [0, e]$, for at least one $i : x + \alpha_i \in [e, q)$. For the **soundness**, note that in each round $c = 0$, $\mathcal{V}$ is convinced that the same α_k is concealed in both W_i and S_i $(i = 1, 2)$. Because the choice of c is independent of $\mathcal{P}$, in each round $c = 1$, with the probability of $1/2$ the verifier is convinced that the commitments W_i, S_i conceal the same secret α_i. Also in the same round $c = 1$, $\mathcal{V}$ is convinced that $w_p W_i$, $w_N S_i$ conceal the same value $x + \alpha_i \in [e, q - e)$. Commitments $w_N S_i$ is computed in Z_N, thus $x + \alpha_i = \alpha_i + x_N$ or $x = x_N$. In the other hand, $w_p W_i$ is computed in Z_p, thus $x + \alpha_i = x_p + \alpha_i \mod q$. Moreover $x_p < q$, $\alpha_i \le e$, thus $x_p + \alpha_i < p + e$. Therefore, $x + \alpha_i \in [e, p - e)$ if and only if $x + \alpha_i = x_p + \alpha_i$ or $x = x_p$. This shows that $\mathcal{V}$ is convinced $x_p = x_N$ with the probability of $1/2$ in each round $c = 1$. With l rounds run in parallel, the verifier is convinced that w_p, w_N conceal the same secret with the probability of roughly $1 - 2^{l/2}$, assuming c is chosen over the toss of a coin.

In the protocol, the view of the verifier is the same as in the range-check protocol. Thus it is zero-knowledge. Formal proof can easily be constructed with a standard simulator.

Using the range check protocol and equality protocol with an unknown group order, we can construct the modulo gate and the equality protocol respectively as follows:

Equality Protocol

Given: $w_i = \text{commit}_i(x_i, r_i)$ and $w_j = \text{commit}_j(x_j, r_j)$. Here commit_i and commit_j are commitments computed in Z_{p_i}, Z_{p_j} respectively.
Prove: $x_i = x_j$

Protocol

1. $\mathcal{P}$ and $\mathcal{V}$ agree upon a RSA modulo N that is larger than both p_i, p_j. In computational zero-knowledge model, N is set by a trusted third party. The factoring of N should not be known to $\mathcal{P}$.
2. $\mathcal{P}$ computes $w_N = \text{commit}_N(x, r_N)$ $(x = x_i = x_j)$ and sends w_N to $\mathcal{V}$.
3. $\mathcal{P}$ runs two instances of equality protocol with an unknown group order to prove that the secrets concealed in w_i, w_N and w_j, w_N are respectively the same.
4. $\mathcal{V}$ accepts the proof if and only if $\mathcal{V}$ is convinced in step (3).

Modulo Gate Protocol

Given: $w_i = \text{commit}_i(x_i, r_i)$ and $w_j = \text{commit}_j(x_j, r_j)$ and a number t. commit_i and commit_j are defined as for the equality protocol.
Prove: $x_i = x_j \bmod t$

Protocol

(without the loss of generality, let us assume that $x_i \geq x_j$)

1. $\mathcal{P}$ and $\mathcal{V}$ agree upon a RSA modulo N as for equality protocol.
2. $\mathcal{P}$ then computes $s_k = \text{commit}_N(x_k, \rho_k)$ $(k = i, j)$, $s = \text{commit}_N(d, \rho)$ $(d = (x_i - x_j)/t)$ and sends s_i, s_j, s to $\mathcal{V}$.
3. $\mathcal{P}$ runs two instances of the equality protocol with unknown group order to prove the equality of the secrets concealed in w_k, s_k $(k = i, j)$.
4. $\mathcal{P}$ then runs an instance of the range-check protocol to prove that the secret concealed in s is in $[0, q_i/t)$.
5. $\mathcal{V}$ accepts the proof if and only if $s_i = s_j s^t \bmod N$.

3.4 Polynomial Related Protocols

Polynomial related protocols demonstrate addition and multiplication relations. Addition and multiplication with a constant factor relations are proven by simply choosing the commitments corresponding to the input(s) and the output of the gate according to the relation. In an addition gate, let $w_1 = \text{commit}(x_1, r_1), w_2 = \text{commit}(x_2, r_2)$ and $w_3 = \text{commit}(x_3, r_3)$ be the commitments respectively representing the two inputs and the output of the gate. As the commitment scheme is homomorphic, thus by choosing $r_3 = r_1 + r_2$ the prover has already proved the addition relation for the gate. The verifier can verify the correctness of the proof by checking $w_3 = w_1 w_2 \bmod p$. Similarly, in the case of the multiplication gate with a constant factor c, the prover chooses $r_3 = c r_1 \bmod q$. The verifier can verify the relation with the check $w_3 = w_1^c \bmod p$. It is straightforward to see that both demonstrations are sound, complete and zero-knowledge.

The remaining is a protocol to demonstrate the relation for multiplication gate with both inputs are secret inputs. It is given as follows:

Multiplication Gate Protocol

Given: $w_1 = \mathtt{commit}(x_1, r_1)$, $w_2 = \mathtt{commit}(x_2, r_2)$ and $w_3 = \mathtt{commit}(x_3, r_3)$ that respectively correspond to the two inputs and the output of the gate.
Prove: $x_3 = x_1 x_2 \bmod q$

Protocol:

1. $\mathcal{P}$ chooses $\alpha, \beta \in_R Z_{i+1}$, sends $W_3 = w_2^\alpha h^\beta$ and $W_2 = g^\alpha h^\beta \bmod p$ to $\mathcal{V}$.
2. Upon receiving W_3, W_2, $\mathcal{V}$ issues a challenge $c \in_R Z_q$ to $\mathcal{P}$.
3. $\mathcal{P}$ computes the responses $u = \alpha - cx_1$, $v_3 = \beta - c(r_3 - r_2 x_1)$ and $v_2 = \beta - cr_1 \bmod q$, then sends u, v_3, v_2 to $\mathcal{V}$.
4. $\mathcal{V}$ accepts the proof if and only if $W_3 = w_3^c w_2^u h^{v_3}$ and $W_2 = w_1^c g^u h^{v_2} \bmod p$.

This is an instance of Chaum-Pedersen[8] proof of equality of discrete logarithms. Here the protocol proves the equality of the discrete logarithm of w_3 to the base w_2, and of w_1 to g in respective representation of w_3 to $[w_2, h]$ and of w_1 to $[g, h]$. This shows that the discrete logarithm of w_3 to g in the representation of w_3 to $[g, h]$ is the product of the discrete logarithms of w_1, w_2 to g in the respective representations of w_1, w_2 both to $[g, h]$, i.e., $x_3 = x_1 x_2$.

3.5 Exponentiation Related Protocols

Exponentiation related protocols demonstrate the relations between commitments corresponding to exponentiation gates. In an exponentiation gate, the exponent is not computed in the same finite field as the base. However, in order to maintain the algebraic relation the value of the base has to be in the multiplicative group of order k over Z_q if the exponent is computed in Z_k ($k = p_t$ for some t in the general setting). Here the exponentiation is calculated in Z_q.

For clarity, we denote the commitments computed in Z_q and Z_p respectively as $\mathtt{commit}_p$ and $\mathtt{commit}_q$. Secrets committed in $\mathtt{commit}_p$ and $\mathtt{commit}_q$ are computed in Z_q and Z_k respectively. Similar indices are also used for g, h. Note that for the two different exponentiation gates computed with the same finite field Z_q, it is perfectly legal to have two different values of k in our model.

The base in an exponentiation gate can either be a (public) constant or a secret concealed in a commitment. There is a different protocol for each case.

Exponentiation Gate with Constant Base Protocol

Given: a constant b, commitments $w_k = \mathtt{commit}_q(x_k, r_k)$ and $w_q = \mathtt{commit}_p(x_q, r_q)$
Prove: $x_q = b^{x_k} \bmod q$

Protocol (run in parallel, l rounds)

1. $\mathcal{P}$ chooses $\alpha_k, \beta_k \in_R Z_k, \beta_q \in_R Z_q$. $\alpha_q = b^{\alpha_k} \bmod q$, sends $W_k = \text{commit}_q(\alpha_k, \beta_k)$ and $W_q = \text{commit}_p(\alpha_q, \beta_q)$ to $\mathcal{V}$.
2. $\mathcal{V}$ then challenges $\mathcal{P}$ with $c \in_R [0, 1]$
3. $\mathcal{P}$ returns to $\mathcal{V}$ the responses u_k, v_k, v_q computed as $u_k = \alpha_k - c x_k$, $v_k = \beta_k - cr_k \bmod k$ and $v_q = \beta_q - cr_q \bmod q$
4. $\mathcal{V}$ is convinced if and only if $W_k = w_k^c g_q^{u_k} h_q^{v_k}$ and $W_q = w_q^c g_p^{b_k^{u}} h_p^{v_q}$.

This is a variant of Stadler's protocol[25] to prove the knowledge of double discrete logarithm. Here the protocol is given for the commitment in the form of $w = g^x h^r$ instead of $w = g^x$. Both of these forms are instances of the representation problem of w to a set of bases (e.g. g, h in the former and g in the latter). The security of the representation problem is equivalent to that of the discrete logarithm problem[2].

Exponentiation Gate Protocol

Given: $w_1 = \text{commit}_p(x_1, r_1)$, $w_2 = \text{commit}_q(x_2, r_2)$ and $w_3 = \text{commit}_p(x_3, r_3)$ that correspond to the base, exponent and the output of the gate.
Prove: $x_3 = x_1^{x_2} \bmod q$

Protocol (run in parallel, l rounds)

1. $\mathcal{P}$ chooses $\alpha, \beta \in_R Z_k$, forms $u_1 = g_p^{h_q^{\alpha}}$ and $u_2 = g_p^{h_q^{\beta}}$ and sends u_1, u_2 to $\mathcal{V}$.
2. $\mathcal{V}$ challenges $\mathcal{P}$ a bit $c \in_R [0, 1]$.
3. If $c = 0$, $\mathcal{P}$ sends to $\mathcal{V}$: α, β. Otherwise $\mathcal{P}$ sends to $\mathcal{V}$: $v_1 = x_1 h_q^{\alpha}$ and $v_2 = x_1^{x_2} h_q^{\beta} \bmod q$.
4. $\mathcal{V}$ (with the help of $\mathcal{P}$) processes as follows:
   (a) If $c = 0$, $\mathcal{V}$ checks the correctness of u_1, u_2 with α and β
   (b) If $c = 1$, $\mathcal{P}$ runs two instances of multiplication protocol to prove v_1 (resp. v_2) is the product of secrets concealed in u_1 and w_1 (u_2 and w_3). $\mathcal{P}$ also runs the **sub-protocol** to prove that the same secret x_2 is concealed in both w_2 and $v_2 = \text{commit}(x_2, \beta - x_2\alpha)$. Here $\text{commit}(x, r) = v_1^x h_q^r \bmod q$ is a commitment scheme computed in Z_q. This scheme is legitimate since the discrete logarithm of v_1 to h is unknown and they are both in the multiplicative group G_k of order k in Z_q.
5. $\mathcal{V}$ accepts the proof if $\mathcal{V}$ is convinced at every step.

The sub-protocol

Given: $w_2 = g_q^{x_2} h_q^{r_2}$ and $v_2 = v_1^{x_2} h_q^r \bmod q$
Prove: w_2, v_2 are commitments of the same value x_2

Process:

1. $\mathcal{P}$ chooses $\alpha, \rho_w, \rho_v \in Z_k$, sends $w = g_q^{\alpha} h_q^{\rho_w}$ and $v = v_1^{\alpha} h_q^{\rho_v} \bmod q$ to $\mathcal{V}$.

2. $\mathcal{V}$ challenges $\mathcal{P}$ a number $c \in_R Z_k$.

3. $\mathcal{P}$ responses with $u = \alpha - cx_2$, $r_w = \rho_w - cr_2$ and $r_v = \rho_v - cr$ mod q.

4. $\mathcal{V}$ accepts the proof if and only if $w = w_2^c g_q^u h_q^{r_w}$ and $v = v_2^c v_1^u h_q^{r_v}$ hold.

This sub-protocol is an instance of Chaum-Pedersen's proof of equality of discrete logarithms. It proves that the discrete logarithm part of w to g_q in the representation of w to the base $[g_q, h_q]$ equals the discrete logarithm part of v to v_1 in the representation of v to $[v_1, h_q]$. Note h_q is chosen such that $\log_{h_q}(v_1)$ is not known.

Security Proof

The completeness comes straight from the inspection of the protocol. For the soundness, note that in each round $c = 0$, the verifier is convinced with the probability of 1 that $\mathcal{P}$ knows the double discrete logarithm of both u_1, u_2. Because c is chosen by $\mathcal{V}$ over the toss of a coin, in each round $c = 1$ $\mathcal{V}$ is convinced with probability of $1/2$ that $\mathcal{P}$ knows the discrete logarithm α, β of both $u_1 = g_p^{h^\alpha}$ and $u_2 = g_p^{h^\beta}$.

The two instances of the multiplication protocol convince $\mathcal{V}$ that v_1 (resp. v_2) is the product of the secrets committed in w_1 and u_1 (w_3 and u_2), i.e., $v_1 = x_1 \log_{g_p} u_1 = x_1 h^\alpha$ and $v_2 = x_3 \log_{g_p} u_2 = x_3 h^\beta$. Since the discrete logarithm of x_1 to h is not known, v_2 is a commitment of $\log_{x_1} x_3$.

The *sub-protocol* then proves that the secrets concealed in both v_2 and w_2 are the same. This means $\mathcal{V}$ is convinced that x_2 equals the discrete logarithm of x_3 to x_1 or $x_3 = x_1^{x_2}$ mod q with the probability of $1/2(1 - 2^l)$ for each round ($c = 1$). After l rounds, $\mathcal{V}$ is convinced with the probability of roughly $1 - 2^{l/2}$ that $x_3 = x_1^{x_2}$, assuming c is chosen uniformly as either 1 or 0.

In each round for $c = 0$, the verifier learns nothing about the secrets because all information shown by the prover, is independent of w_1, w_2, w_3. In each round for $c = 1$, $\mathcal{P}$ executes an instance of equality protocol and two instances of multiplication protocol. The multiplication protocol and the sub-protocol are zero-knowledge and reveals no useful information to the verifier. All instances are independent of others. The rest of verifier's view is v_2, v_1. As α, β are chosen at random, $v_1 = x_1 h^\alpha$ and $v_2 = x_1^{x_2} h^\beta$ are witness-indistinguishable. Thus at no stage, the verifier learns any useful information about the secrets.

3.6 Discussion

This construction is unconditionally hiding and computational interactive honest-verifier zero-knowledge. Non-interactive zero-knowledge is achieved by forming the challenge c as the hash value of information sent to $\mathcal{V}$ by $\mathcal{P}$. Unconditionally binding, in which the computational power of the prover is not limited, is achieved by applying the technique of [9]. Honest-verifier can also be loosen by standard techniques[12].

With some slight modification, similar building blocks can be built in RSA setting[3] for any mode of proofs. The technique of [15] can be used to construct our building blocks for multiplication, addition, modulo gate for the commitments computed in multiplicative groups of finite fields with unknown group order(s). Moreover, if the secret is a single value, i.e., there is no descendant gate(s) of the corresponding gate, exponentiation gates with constant base can also be demonstrated for such commitments. It is possible to add boolean relations to our general circuit using the construction of [3]. Especially AND relation can be shown at no extra cost. Other arithmetic relations, such that div, mod with secret modulus, can be achieved with similar techniques.

4 Zero-knowledge Proof of Possession of Digital Signatures

Zero-knowledge proof of possession of digital signatures can be constructed if the underlying signature scheme is in $\mathcal{L}$. It is done by first forming an arithmetic circuit corresponding to the underlying signature scheme and then using the general technique to realize zero-knowledge proofs.

Signature schemes that do not require hash functions can easily be shown to be members of $\mathcal{L}$. They include RSA, ElGamal[13], DSS[18] and Nyberg-Rueppel[20] signature schemes. Let $f()$ be the signature function, m be the message and S be the signature. The function $f()$ takes S as its input and outputs m or some isomorphic value of m under some one-way function. In RSA, $\mathcal{P}$ proves the knowledge of $S = s$: $f(s) = m$ for $f(x) = x^d$ and d is the public key. In ElGamal signature, $\mathcal{P}$ proves the knowledge of $S = (r, s)$: $f(r, s) = g^m$ for $f(\alpha, \beta) = y^\alpha \alpha^\beta \bmod p$ and y is the public key. In Nyberg-Rueppel signature, $\mathcal{P}$ proves the knowledge of $S = (r, s)$: $f(r, s) = m$ for $f(\alpha, \beta) = g^\alpha \bmod q \, y^\beta \beta$. See the appendix for detailed circuit constructions of these signature schemes

4.1 Comparison to previous works

RSA-based signatures can be proven in zero-knowledge without modulo related nor exponentiation related protocols. Our construction achieves virtually the same efficiency as those presented in [5,15].

Kilian and Petrank gave the only zero-knowledge proof of possession of discrete-logarithm based digital signatures in [17]. Their protocol is specifically designed for ElGamal signature scheme, while ours are applicable for many signature schemes. Nevertheless, with the security parameter l, Kilian and Petrank protocol requires $O(l^2)$ rounds. No protocol in our overall construction requires more than $O(l)$ rounds. All (sub) protocols are independent and thus can be done in parallel. Therefore the overall complexity of our protocol to prove the possession of digital signatures is $O(l)$. This is an improvement of the order $O(l)$ over Kilian-Petrank protocol. Recently Camenisch and Michels[4] proposed a new general construction that can prove any arithmetic relation. But in proving possession of digital signatures, their proposal requires a minimum of $O(l)$ rounds and is no better efficient than Kilian-Petrank protocol. It is still highly desirable to construct protocols with $O(1)$ rounds. This is achieved if cut-and-choose is eliminated in modulo and exponentiation related protocols. Such solutions are unfortunately not found yet.

4.2 Applications

Our zero-knowledge proofs of possession of digital signatures have many applications. An obvious application is identity-escrow. Here identity certificates can be of any mentioned signature schemes. The identity escrow then consists of two parts. One is the zero-knowledge proof of possession of a valid identity certificate, i.e., a signature. The other is the ciphertext of some information that uniquely identify the certificate under some trusted authority public key. Similarly, our proof can realize fix-size group signature schemes. So far, fix-size groups signatures exist only for RSA-based group certificates. Those group signatures are non-interactive proofs of possession of RSA signatures. Our construction allows group signatures to be constructed with group certificates being any mentioned signature.

Another significant application is fair-exchange of digital signatures. The standard method requires the encryption of the digital signatures under some trusted authority public keys. The sender then proves the validity of the ciphertext to the receiver. This is proposed by Asokan, Shoup and Waidner[1]. In their protocols, only "half" of the signature is encrypted and proven in zero-knowledge, the other half is shown directly to the receiver. It could be undesirable if the sender is required to be anonymous/unlinkable when the exchange fails. Our technique facilitates complete hiding. It is done by encrypting every part of the signature under the trusted authority public key. Each ciphertext is then considered as a unconditionally binding commitment. Then the sender proves in zero-knowledge that the corresponding plaintext is a valid signature of a specific message. The whole signature is not revealed unless the exchange succeeds.

References

1. N. Asokan, V. Shoup and M. Waidner, *Optimistic fair exchange of digital signatures*, Proceedings of EUROCRYPT'98, LNCS 1403, pp.591-606.
2. S. Brands, *Untraceable off-line cash based on the representation problem*, Technical Report CS-R9323, Centrum voor Wiskunde en Informatica, April 1993.
3. S. Brands, *Rapid Demonstration of Linear Relations Connected by Boolean Operators*, Proceedings of EUROCRYPT'97, LNCS 1223, pp.318-333.
4. J. Camenisch and M. Michels, *Proving in Zero-Knowledge that a Number is the Product of Two Safe Primes*. BRICS Technical Report RS-98-29, To appear in Eurocrypt'99.
5. J. Camenisch and M. Stadler, *Efficient Group Signatures for Large Groups*, Proceedings of Crypto'97, LNCS 1294, pp. 465-479.
6. A. Chan, Y. Frankel and T. Tsiounis, *Easy come-easy go divisible cash*, Proceedings of EUROCRYPT'98, LNCS, Finland, pp. 561-575.
7. D. Chaum, E. van Heijst and B. Pfitzmann, *Cryptographically Strong Undeniable Signature, Unconditionally Secure for the Signer*, Proceedings of Crypto'91, LNCS 576, pp.204-212.
8. D. Chaum and T.P. Pedersen, *Wallet databases with observers*, Proceedings of Crypto'92, LNCS 740, pp. 89-105.
9. R. Cramer and I. Damgard, *Zero-Knowledge Proofs for Finite Field Arithmetic or: Can Zero-Knowledge be for Free?*, Proceedings of Crypto'98, to appear.

10. R. Cramer and V. Shoup, *A Practical Public Key Cryptosystem Provably Secure against Adaptive Chosen Ciphertext Attack*, Proceedings of Crypto'98, to appear.

11. I. Damgard, *Practical and Provably Secure Release of a Secret and Exchanges of Signatures*, Journal of Cryptology, vol. 8, n. 4, 1995, pp.201-222.

12. G. Di Crescenzo, T. Okamoto and M. Yung *Keeping the SZK-Verifier Honest Unconditionally*, Proceedings of Crypto'97, LNCS 1294, pp.31-45.

13. T. ElGamal, *A Public-Key Cryptosystem and a Signature Scheme Based on Discrete Logarithms*, IEEE Transactions on Information Theory, v. IT-31, n. 4, 1985, pp. 469-472.

14. U. Feige, A. Fiat and A. Shamir, *Zero-knowledge proofs of Identity*, Journal of Cryptology 1988, vol. 1, pp.77-94.

15. E. Fujisaki and T.Okamoto, *Statistical Zero-Knowledge Protocols to Prove Modular Polynomial Relations*, Proceedings of Crypto'97, LNCS 1294, pp. 16-30.

16. O. Goldreich, S. Micali and A. Wigderson, *Proofs that Yield Nothing But their Validity and a Methodology of Cryptographic Protocol Design*, Proceedings of Foundation of Computer Science 1986, pp. 174-187.

17. J. Kilian and E. Petrank, *Identity Escrow*, Proceedings of Crypto'98, to appear.

18. National Institute of Standards and Technology, NIST FIPS PUB 186, *Digital Signature Standard*, US Department of Commerce, May 1994.

19. A. Menezes, P. van Oorschot and S. Vanstone *Handbook of Applied Cryptography*, CRC Press, 1997.

20. K. Nyberg and R.A Rueppel, *Message Recovery for Signature Schemes Based on the Discrete Logarithm Problem*, Proceedings of EUROCRYPT'94, LNCS 950, pp.182-193.

21. T. Okamoto, *An efficient divisible electronic cash scheme*, Proceedings of Crypto'95, LNCS 963, pp.438-451.

22. T. Pedersen, *Non-Interactive and Information Theoretic Secure Verifiable Secret Sharing*, Proceedings of Crypto'91, LNCS 576, pp. 129-140.

23. B. Schneider, *Applied Cryptography*, 2nd edition, John Wiley & Sons, Inc, 1996.

24. C. Schnorr, *Efficient Signature Generation for Smart Cards*, Proceedings of Crypto 89, LNCS 435, pp. 239-252.

25. M. Stadler, *Publicly Verifiable Secret Sharing*, Proceedings of EUROCRYPT'96, LNCS 1070, pp.190-199.

A Soundness Proof for Protocol in section 2.2

To prove the soundness of the protocol proposed in section 2.2, we make three assumptions:

1. The circuit is constructed correctly from the function $F_i()$ with the cheating probability of 0. This is plausible as such construction is public and $F_i()$ is public.

2. The circuit is of polynomial size, i.e., both number of gates and leaves are of polynomial order.

3. All the building blocks are secure and have a negligible cheating probability.

We now prove the cheating probability of the protocol ϵ is negligible.

Lemma 1. *For each building block, let the combined cheating probability of the input(s) be ϵ_I, the cheating probability of the output be ϵ_O and of the protocol be ϵ_G, The following inequality holds:*

$$\epsilon_O \leq \epsilon_I + \epsilon_G$$

This is because for the inputs are correct for with the probability of $(1 - \epsilon_I)$. Accordingly the cheating probability of the output when the inputs are correct, is $\epsilon_O = (1 - \epsilon_I)\epsilon_G$. Combined this with the cheating probability ϵ_I of inputs, we have:

$$\epsilon_O = \epsilon_I + (1 - \epsilon_I)\epsilon_G \leq \epsilon_I + \epsilon_G$$

Here for any probability ϵ, it satisfies $0 \leq \epsilon \leq 1$. Similarly, we can also obtain the following lemma:

Lemma 2. *Let the cheating probability of the inputs A, B be ϵ_A, ϵ_B, the combined cheating probability of the inputs be ϵ_I, the inequality holds:*

$$\epsilon_I \leq \epsilon_A + \epsilon_B$$

As in each of our building block there are at most two inputs, the cheating probability of the output is always no greater than the sum of all the cheating probabilities of the inputs and the protocol itself for any stance of the building blocks. This comes straightforward from two lemmas above.

In our circuit $F_i()$, the inputs of each gate are the outputs of other gates unless they are the leaves. The final output of all the gates is G_v and all the non-reducible inputs are $U_1, ..., U_u$. Each gate G_i is associated with a protocol $\mathcal{G}_i$ to prove the relation between the secrets concealed in the input and the output commitments of the gate. Each leaf L_j is associated with a protocol $\mathcal{L}_i$ proving the knowledge of the secret concealed in the associated commitment U_j. Hence, we have the following result:

Lemma 3. *Let θ_i, λ_j, ε and ϵ be the cheating probability of $\mathcal{G}_i$, $\mathcal{L}_j$, the opening commitment required in step 4 and the whole protocol respectively. The following inequality holds*

$$\epsilon \leq \varepsilon + \sum_{i=1}^{v} \theta_i + \sum_{j=1}^{u} \lambda_j$$

From assumption (2) and (3), the cheating probability of the building blocks are negligible, $u + v$ are of polynomial size, thus the cheating probability of the protocol (ϵ) is no greater than a polynomial order of a negligible value. That proves the soundness of the protocol.

B Arithmetic circuits of digital signature schemes

Let $f()$ be the signature function, m be the message. The function $f()$ takes the signature as it input and outputs m or some isomorphic value of m under some one-way function. The construction of the arithmetic circuit corresponding to $f()$ is as follows:

The case of RSA family

Function $f()$: $f(x) = x^d \bmod n$
Verification: $f(s) = m$, s is the signature
Circuit construction: $d = d_1..d_k$ $(d_i \in [0,1])$.

$$f(x) := \sum_{i=1}^{k} (f_i(x))^{d_i} \bmod n$$
$$f_i(x) := (f_{i-1}(x))^2 \bmod n$$
$$f_1(x) := x$$

The case of ElGamal digital signature scheme[13]

Function $f()$: $f(\alpha, \beta) = y^\alpha \alpha^\beta \bmod p$ (y is public key)
Verification: $f(r, s) = g^m$, (r, s) is the signature
Circuit construction:

$$f(\alpha, \beta) := f_1(\alpha, \beta) f_2(\alpha, \beta) \bmod p$$
$$f_1(\alpha, \beta) := y^{f_3(\alpha,\beta)} \bmod p$$
$$f_2(\alpha, \beta) := f_4(\alpha, \beta)^{f_5(\alpha,\beta)} \bmod p$$
$$f_3(\alpha, \beta) := \alpha \bmod (p-1)$$
$$f_4(\alpha, \beta) := \alpha$$
$$f_5(\alpha, \beta) := \beta$$

The circuit construction for Digital Signature Algorithm[18] is identical to that of ElGamal. The only difference in the circuit is that $f_3(\alpha, \beta)$ is computed as $\alpha \bmod q$, not $\alpha \bmod (p-1)$. Other variants of ElGamal[19] could be proven using similar circuit.

The case of Nyberg-Rueppel digital signature scheme[20]

Function $f()$: $f(\alpha, \beta) = g^{\alpha \bmod q} y^\beta \beta$
Verification: $f(r, s) = m$, (r, s) is the signature
Circuit construction:

$$f(\alpha, \beta) := f_1(\alpha, \beta) f_2(\alpha, \beta) f_3(\alpha, \beta) \bmod p$$
$$f_1(\alpha, \beta) := g^{f_4(\alpha,\beta)} \bmod p$$
$$f_2(\alpha, \beta) := y^{f_3(\alpha,\beta)} \bmod p$$
$$f_3(\alpha, \beta) := \beta \bmod q$$
$$f_4(\alpha, \beta) := \alpha$$

Note that the checks $0 \leq \alpha < q$ and $0 < \beta < 0$ are also satisfied. This is due to their commitments are computed in respective finite fields. To prove $\beta > 0$, the modulo protocol should use the range $[1, q-1]$ instead of $[0, q-1]$ as presented in section 3.3.

Signature Scheme for Controlled Environments*

Kapali Viswanathan, Colin Boyd and Ed Dawson

Information Security Research Centre
Queensland University of Technology
Brisbane, Australia.
{viswana,boyd,dawson}@fit.qut.edu.au

Abstract. We propose a signature system that will be very useful for controlled environments, like corporate networks, where regulation of signatures generated using the *certified* public/private key pairs is essential. We then demonstrate the usefulness of the system by extending it to realise a proxy signature system *with revocable delegation*.
Keywords: Controlled environment, joint signature, proxy signature, revocation, traceability, enforceability.

1 Introduction

Traditionally cryptography was used to provide *unconditional* services to various users, but evolving trends in research suggests that in the real (or commercial) world scenario a *conditional* service is more useful than an unconditional service. This is a result of the growing need for compliant systems [12, 2], as they closely emulate real time requirements. Several systems for conditional confidentiality (key escrow) and anonymity (electronic cash, voting, etc.) services have been proposed. In this paper we propose a system that can be used for a *controlled* non-repudiation service. This system, for example, can be used when an organisation that certifies a public key for an individual as its manager will require the individual not to use the power (as the manager of the organisation) for personal purposes or to abuse the public key.

The desirable properties of such a controlled signature system would be;

CR 1 The organisation must be able to regulate and control the signature generation and verification functions.

CR 2 The owner of the public key must be assured that the organisation or any other outside attacker cannot form his/her signature.

Most currently available signature systems [24, 20, 8] provide **CR 2**. In this paper we propose a controlled signature system called *joint signature* system that additionally accomplishes **CR 1**. It is based on the Schnorr signature scheme [8] and a variation of the ElGamal signature scheme [24, 13]. For a more elaborate discussion on **CR 1** we refer to the survey on threshold cryptosystems by Desmedt [10].

* Research supported by the Australian Research Council grant A49804059

Certification of public keys is essential for public key protocols to work as intended in open networks. Public key infrastructures were propounded to accomplish the certification task. It is being realised that this may not be an easy task when certificate revocation is essential. The revocation task becomes more difficult when certification models consider *off-line* certificate authorities. If a signature scheme has property **CR1**, it could accomodate *on-line* certificate authorities and this will make revocation of certificates much simpler. This will be a major advantage of joint signature systems.

There are numerous application areas for controlled signature systems. One area where such a system would be of great use would be in proxy signature systems *with revocable delegation*. One of the major drawbacks of public key cryptography is the difficulty in revocation of privileges (certified public keys). This problem becomes even greater in proxy signature schemes where a proxy can form valid signatures for which the delegator can be given credit or held accountable. We present a proxy signature system using the joint signature system in which revocation can be centrally effected by a trusted central authority.

The paper is organised so that Section 2 provides the necessary background for the discussion, Section 3 contains a proposal for the joint signature system, Section 4 presents an application of the joint signature to proxy joint signature system with revocable delegation and Section 5 concludes the paper.

2 Background

Some types of signature systems proposed in the literature are:

General purpose system [24, 8, 20]: An entity can sign a message in a manner that would allow *any* verifier to check the signature using the entity's public key.

Multi-signature system [3, 14]: A group of entities can sign a message in a manner that would enable a verifier to check the signature using the group's public key and give credit to, or hold accountable, *all* the entities in the group. The joint signature scheme, to be proposed, is a sub-type of multi-signature scheme.

Designated verifier system [4, 15]: An entity can sign a message in a manner that would allow only a *designated* verifier (confirmer) to check (assist in checking) the signature using the entity's public key and the verifier's (confirmer's) private key.

The two general purpose signature systems used in the proposed joint signature are the Schnorr signature scheme [8] and a variation of the ElGamal signature scheme (see scheme EG I.4 in [13]). The ElGamal signature variant will henceforth be termed as the HoP signature scheme. The Schnorr signature is used by the initiator of the protocol and the HoP signature scheme by the other signer.

The joint signature system makes use of concepts from multi-signature systems and designated verifier systems to accomplish its goal. We would like to bring out certain differences between some of the proposed multi-signature systems:

1. The primary objective of the joint signature scheme is to provide a controlled signature environment, as opposed to collective identification or message authentication (which is an additional property). For this reason, the partial signatures generated by individual signers must not be globally verifiable. In the multi-signature scheme proposed by Horster *et al.* [14], the central clerk was required to verify the validity of the partial signatures generated by participating entities. In the joint signature only the next signer in the line can check the validity of the partial signature generated by the previous signer(s), and the signature will be globally verifiable only after every signer signs the message.

2. Horster *et al.* [14] proposed multi-signature systems based on the discrete logarithm problem using the ElGamal signature scheme. Their scheme needs an additional clerk to *combine* partial signatures to obtain the multi-signature – the signers perform signature operations in a pseudo-parallel fashion that needs interaction between the clerk and the signers. In the joint signature system such clerks are not needed as the signers perform the signature operations in a serial fashion that eliminates *multiple interactions* between entities (signer i must wait for the outputs from signer $i - 1$ before proceeding with its signature operations).

3. The digital multi-signature scheme in [3] has a serial operation similar to the joint signature scheme and uses a RSA based cryptosystem. The original scheme required an organisation (trusted party) to generate the composite modulus, which essentially meant that the organisation possessed the potential to forge any signature it wanted. But there are schemes, that are now known [9, 6, 7, 22], for the joint generation of RSA modulus and private keys. Also, some form of threshold RSA scheme [11, 10] can be employed in their scheme.

In the joint signature system, the signer who initiates the signature uses the Schnorr signature scheme in the designated verifier mode. The technique used is similar to the techniques used in [4] and [15].

The properties of the joint signature scheme may be interesting for application to different problem areas. A listing of the properties are as follows.

1. It is an extension of the Schnorr signature scheme, thereby based on the the Discrete Logarithm problem.

2. The signature operations are serial in nature, as opposed to parellel operations. This minimises the communication overhead that is common among known discrete logarithm based multi-signature schemes.

3. It achieves the protocol goals by relying only on protocol constructs and does not rely on the existance of tamperproof mechanisms.

4. It can support on-line certificate authorities, thus providing an effective mechanism for applications needing revocation such as certificate revocation, proxy revocation, membership revocation.

5. It is extensible, in that it will be possible to extend the joint signature scheme to achieve other applications of digital signatures, such as proxy signature or group signature with minimum computational overhead.

The proxy signature scheme, that will be introduced to demonstrate the use of joint signature, employs the delegation scheme introduced by Mambo, Usuda and Okamoto [18]. Extensions to the other versions of their signature is trivial.

3 Joint signature

In this section we propose a joint signature scheme that uses the traceability architecture proposed in [2], the Schnorr signature scheme in designate verifier mode [8] and the HoP [13] signature scheme to provide a *controlled environment* for the generating digital signatures.

3.1 System entities

The joint signature system consists of a group of signers, labelled as $1, \cdots, l$ for an integer l. We focus on the case when $l = 2$. The group of l signers will labelled as $S = \{s_1, \cdots, s_l\}$. Any global verifier can know the identity of S from the public key certificate and cannot obtain any information about the individual signers, if the information is not certified. A certificate authority, CA, maintains the public key certificate repository.

3.2 System description

The system consists of signers who share the private key corresponding to the certified public key. For clarity we assume that two entities share a public key - extensions to accommodate more than two signers can be made by incorporating minor modifications. The signers, or a certificate authority, choose a large prime p as the modulus, such that $p = 2q + 1$ (*C.f.* [2]) and a random generator g. The registration phase uses the protocol explained in [2], which is a variant of the Diffie-Hellman protocol with constraints on the participants.

- $y \equiv g^{x_1 x_2} \bmod p$ is the certified public key of entity s. x_1 is the secret key of signer 1 and x_2 is the secret key signer 2, corresponding to the public key y.
- Entity 1 knows the value $w \equiv g^{x_2} \equiv y^{x_1^{-1}} \bmod p$. Entity 2 possess a similar knowledge.

Henceforth, all operations will be performed in the congruence class mod p unless stated otherwise.

3.3 Signature process

The signature process is distributed and consists of the following algorithms:

1. partSign with inputs (m, x_i, w, q) and outputs (c, d): generates a partial signature tuple (c, d) on the message m using the secret key share x_i. It employs the Schnorr signature generation algorithm in the designated verifier mode.
2. partCheck with inputs (m, c, d, x_j, y, g, p) and outputs $(1 \vee 0)$: checks the correct formation of the tuples (c, d) on the message m using the other secret share $(x_i \mid i \neq j)$ and, outputs 1 for success or 0 for failure. Note that the 3-tuple (m, c, d) can be checked only with the knowledge of $(x_j \mid i \neq j)$. It employs the Schnorr signature verification algorithm in the designated verifier mode.
3. completeSign with inputs (d, x_j, g, q) and outputs (d', r): completes the partial signature so that the 4-tuple (m, c, d', r) can be successfully checked by a universal verifier, if the 3-tuple (m, c, d) was successfully checked using the function desigCheck. It uses the HoP signature generation algorithm.
4. completeCheck with inputs (m, c, d', r, y, g, p) and outputs $(1 \vee 0)$: checks if the 3-tuple (c, d', r) is a valid signature on m by the certificate holders of the public key y. It uses a combination of the Schnorr (in designated verifier mode) and HoP signature verification algorithms.

Figure 1 shows a graphical representation of the message dynamics in the joint signature system. The protocol and algorithms are shown in Table 1. The signature scheme is valid because:

Signer 1	Signer 2	Verifier
partSign:		
$k_1 \in_R \mathcal{Z}_q^*$		
$c = \mathcal{H}(m \| w^{k_1})$		
$d = k_1 - c x_1 \bmod q$		
$\xrightarrow{\quad c,d,m \quad}$		
	partCheck:	
	$c \stackrel{?}{=} \mathcal{H}(m \| y^c g^{x_2 d})$	
	completeSign:	
	$k_2 \in_R \mathcal{Z}_q^*$	
	$r = g^{k_2}$	
	$d' = d x_2 - k_2 r \bmod q$	
	$\xrightarrow{\quad c,d',m,r \quad}$	
		completeCheck:
		$c \stackrel{?}{=} \mathcal{H}(m \| y^c g^{d'} r^r)$

Table 1. Joint signature scheme

$$w^{k_1} = g^{x_2 k_1} \tag{1}$$

$$g^{x_2 k_1} = g^{x_2(c x_1 + d)} \tag{2}$$

$$g^{x_2 x_1 c + x_2 d} = y^c g^{x_2 d}$$

Thus, $w^{k_1} = y^c g^{x_2 d}$ is used in partCheck. Now,

$$y^c g^{x_2 d} = y^c g^{d' + k_2 r} \tag{3}$$
$$y^c g^{d' + k_2 r} = y^c g^{d'} r^r$$

Thus, $w^{k_1} = y^c g^{d'} r^r$ is used in the function completeCheck.

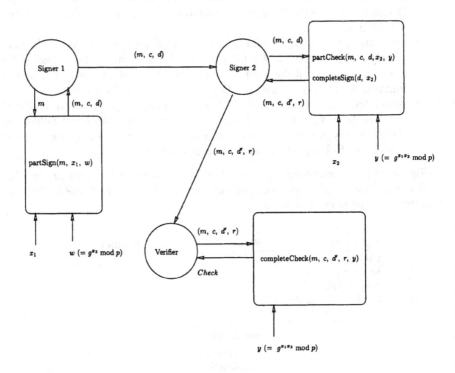

Fig. 1. Message dynamics in the joint signature system

3.4 Security analysis

Formal security analysis for signature schemes has been very difficult, which is the reason that many signature systems believed to be secure do not have a concrete proof of security but only in specialised models of security [21] working under strict assumptions [19]. We would ideally like the proof of security to prove the equivalence of "breaking" a signature scheme to solving a known hard problem (There seem to be many notions of breaking and the number of these notions make proving even harder; for a notion on breaking signature schemes see [23]). We present security arguments of our scheme that would facilitate a better understanding of the system, so that attempts to prove its security could be efficient and successful. This section will discuss the desirable properties of joint signature schemes in general followed by an analysis on a model of the joint signature scheme. In order to improve clarity, the analysis will assume only two signers sharing a public key.

Security properties The desirable security properties of the joint signature scheme (see figure 1) are stated in this section.

Security Property 1 *Only signer 1 (signer 2) can compute the function part-Sign to obtain valid signature tuples that can be verified by signer 2 (signer 1).*

Security Property 2 *Only signer 2 (signer 1) can compute the function partCheck to check the validity of partial signatures when signer 1 (signer 2) computes the function partSign.*

Security Property 3 *Function completeCheck can be universally checked and outputs 1 when valid outputs of the completeSign are given.*

Security Property 4 *Signer 2 (signer 1) cannot form valid universally verifiable signatures using the function completeSign without the outputs of the function partSign as computed by signer 1 (signer 2).*

Security Property 5 *Only signer 2 (signer 1) can compute the function completeSign to obtain valid joint signature tuples when the outputs of the function partSign as calculated by signer 1 (signer 2) are given.*

Theorem 1 *The proposal for the joint signature system satisfies:*

1. *security property 1, if the Schnorr signature is not forgeable,*
2. *security properties 2, and,*
3. *security property 3.*

Proof:

1. The algorithm for the function partSign is a direct application of the Schnorr signature scheme with the base for generating the challenge as $w = g^{x_2}$ (for signer 1) instead of g, as is the case in the Schnorr signature scheme.
2. Following the previous arguments, to check the validity of the Schnorr signature triplets the verifier should possess the knowledge of the discrete logarithm $\log_g w \bmod p$. This is because the values x_1 and x_2 are chosen at random and no universal verifier can correlate the relation between w and y due to the decisional Diffie-Hellman problem.
3. The algorithm for the function completeCheck does not require any private data as its input. Correctness follows from equation 3. $\square$

It is interesting to note that in the case of standard Schnorr signature $w = g$ mod p, so that everyone knows a solution for the discrete logarithm $\log_g w$ mod p.

Security property 4 states that universally verifiable signatures can be formed only when both the signers co-operate. Informal analysis suggests that the proposal satisfies this property. We conjecture that the proposal relies on the security of the Schnorr signature scheme against existential forgery to achieve this property.

Security property 4 aims at providing security for a signer when the other signer is the attacker. Security property 5 provides a stronger notion of security, in that it provides security against an external attacker. We conjecture that the proposal satisfies this property when the Schnorr signature scheme is secure against existential forgery.

The next section will discuss an analysis on a model of our signature system.

Analysis on a model The modes of attack on any signature scheme [23] are:

- **Total Break:** Given a set of well formed signatures, it will be possible to compute the secret key.
- **Universal Forgery:** Finding an algorithm that will form a valid signature on a message without the knowledge of the secret key corresponding to the public key.
- **Selective Forgery:** Forge a signature for a particular chosen message, with or without interacting with the original signer.
- **Existential Forgery:** Forge at least one message without the knowledge of the private key. Note that the message value could be random without any meaning [23].

One approach for proving the security of a signature scheme could be to show that existential forgery is not possible, which being the weakest form of attack will provide the strongest notion of security. Unfortunately it seems difficult to achieve this proof. This section provides a security analysis of the proposal against a special form of existential forgery on the Schnorr signature system and its derivatives called (m, c)-forgery.

In the rest of this security analysis, all arithmetic operations are assumed to be performed in a pre-defined group of integers, Z_p, unless stated otherwise. The Schnorr signature verification process can be defined as:

Definition 1 *The tuple (c, d) is said to be a valid Schnorr signature on a message m by the entity owning the public key y, if the following equation is verified:*

$$c \stackrel{?}{=} \mathcal{H}(m \| y^c g^d) \tag{4}$$

Let S denote the set of algorithms that can be used to implement (m, c)-forgery attacks on the Schnorr signature scheme. Then the attack algorithms in set S can be defined as:

Definition 2 *The algorithms in set S can be modeled using a probabilistic polynomial time Turing machine that on inputting d will return, with a high probability, m and c such that the tuple (c, d) is a valid Schnorr signature on m, verifiable using the public key y.*

The verification process for the joint signature scheme can be defined as:

Definition 3 *The tuple (c, d', r) is said to be a valid joint Schnorr signature on a message m by the entity owning the public key y, if the following equation is verified:*

$$c \stackrel{?}{=} \mathcal{H}(m \| y^c g^{d'} r^r) \tag{5}$$

Let J denote the set of algorithms that can be used to implement (m, c)-forgery attacks on the joint signature scheme. Then the attack algorithms in set J can be defined as:

Definition 4 *The algorithms in set J can be modeled using a probabilistic polynomial time Turing machine that on inputting (d', r) will return, with a high probability, m and c such that the tuple (c, d', r) is a valid Schnorr signature on m, verifiable using the public key y.*

Note that algorithms in the sets S and J appear to be related. This is because a common strategy to implement the algorithms in sets S and J could be to attack the algorithm used for the hash function. That is given the partial sequence of the input, d or (d', r), find the remaining input, m, and the corresponding output, c.

The (m, c) attack concentrates on the structure of the universal verification equation, which is an essential part of any signature scheme, and tries to form valid signature n-tuples on (random) messages that could be verified using the public key *without* the knowledge of the corresponding private key. Note that this attack is stronger than existential forgery, thereby proof of security against (m, c)-attack provides a weaker notion of security than a proof of security against existential forgery. So, if there exists an algorithm to perform existential forgery on a signature scheme, an algorithm to implement the (m, c)-attack may also exist, but not necessarily the other way around.

Now, let us analyse the security of the HoP signature scheme. There exists a polynomial time algorithm to perform existential forgery on the ElGamal signature system [5] and its variants.

Lemma 1 *The HoP signature system is existentially forgeable.*

The HoP signature tuple (r, s) on a message m and public key y must satisfy the equation $g^s = y^m r^r \mod p$, where the private key $x = \log_g y$. To do this an attacker:

1. Chooses arbitrary integers b and c.
2. Computes $r = g^b y^c \mod p$, $s = br \pmod{p-1}$ and $m = -cr \pmod{p-1}$.

The tuple (r, s) will then be a valid signature on the message m verifiable using the public key y.

The joint signature system can be viewed as a cascade of the Schnorr signature and HoP signature schemes as shown below:

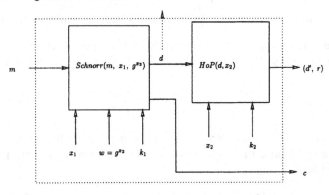

Theorem 2 *If the Schnorr signature scheme can be (m, c)-forged employing algorithms from the set S, then the joint signature scheme can be (m, c)-forged employing algorithms from the set J.*

Proof: The figure presented below will illustrate the approach of this proof:

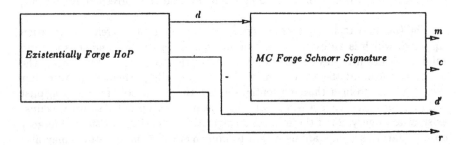

Suppose that signer 2 is mounting an attack on signer 1. Let $w = g^{x_1}$, so that $\log_{g^{x_2}} y = \log_g w$. Assume that the Schnorr signature scheme is (m, c)-forgeable. Since the HoP signature scheme is existentially forgeable, the attacker can obtain a valid signature tuple (d', r) on some message value d that can be verified using w as the public key. Since the Schnorr signature is assumed to be (m, c)-forgeable, the tuple (m, c) can be formed when given d, so that the 3-tuple (m, c, d) satisfies the equation $c \stackrel{?}{=} \mathcal{H}(m\|y^c w^d)$. But the HoP signature on d is (d', r) so that $w^d = g^{d'} r^r$. Thus $c \stackrel{?}{=} \mathcal{H}(m\|y^c g^{d'} r^r)$, which is the signature verification scheme for the joint signature scheme. Thus the 4-tuple (m, c, d', r) is a (m, c)-forged signature on the joint signature scheme. $\square$

Theorem 3 *If the joint signature scheme can be (m, c)-forged employing algorithms from the set J, then the Schnorr signature scheme can be (m, c)-forged employing algorithms from the set S.*

*Proof:*The figure presented below will illustrate the approach of this proof:

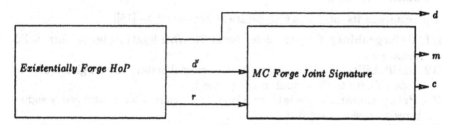

Suppose that signer 2 is mounting an attack on signer 1. Let $w = g^{x_1}$, so that $\log_{g^{x_2}} y = \log_g w$. Assume that joint signature scheme is (m, c)-forgeable. There exists an existential forgery on the HoP signature scheme. Find existentially forged signatures (d', r) on d such that $w^d = g^{d'} r^r$. Since joint signature scheme is assumed to be (m, c)-forgeable, find (m, c) so that (m, c, d', r) is a valid joint signature satisfying the equation $c \stackrel{?}{=} \mathcal{H}(m\|y^c g^{d'} r^r)$. Thus $c \stackrel{?}{=} \mathcal{H}(m\|y^c w^d)$, which is a (m, c)-forged Schnorr signature tuple (d, c) on m. $\qquad\square$

Lemma 2 *If signer 2 can form signature tuples without the help from signer 1, then so can a universal attacker without the help of signer 1 and signer2.*

Note that the public key is of the form $y = g^{x_1 x_2} = g^x$. Thus, $x_2' = 1$ is a valid share because $y = g^x = g^{x x_2'}$.

Corollary 1 *Joint signature scheme is (m, c)-forgeable if and only if Schnorr signature scheme is (m, c)-forgeable.*

Maurer and Massey [16] presented a folk theorem that suggested that cascaded ciphers will be at least as difficult to break as any of its component ciphers. Similarly, the following observation on the joint signature scheme, which is a cascaded signature system, can be made as:

Observation 1 *A joint signature scheme's security against (m, c)-forgery will not be any more secure than the most secure (against (m, c)-forgery) signature scheme in the cascade.*

4 Proxy joint signature

In this section we present the design for proxy signature system with revocable delegation employing the joint signature scheme. Proxy signature systems [18, 17] allow a designated proxy, *delegate*, to sign documents on behalf of an actual signer, *delegator*. Mambo et al. [18] identify three types of delegation as:

1. **Full Delegation:** The *delegator* gives his/her secret to the *delegate*.
2. **Partial Delegation:** The *delegator* creates a new secret x_{proxy} using a suitable function of the actual secret x, and gives x_{proxy} to the *delegate*.
3. **Delegation by warrant:** Here the *delegator* certifies the *delegate* as the authorised proxy signer by certifying appropriate public values, the secret of which is known to the *delegate*.

The requirements for a proxy signature system could be [18]:

R1 Unforgeability: Only the *delegator* and his/her legal proxies can form valid signatures.

R2 Verifiability: A valid signature or a proxy signature should give credit or responsibility to the actual owner (signer).

R3 Proxy signature deviation: The *delegate* cannot form valid proxy signatures and be unidentified.

R4 Distinguish-ability: Anyone can distinguish between valid signature formed by the *delegator* and the valid proxy signature formed by the *delegate*.

R5 Identifiability: The *delegator* can determine the identity of the *delegate*, given a proxy signature, in polynomial time.

R6 Private-key dependence: A new secret x_{proxy} can be computed from the original secret x.

R7 Non-repudiation: The *delegate* cannot repudiate valid proxy signatures that he/she formed.

Our design will use partial delegation to achieve properties R1, R2, R5, R6 and effective revocation.

4.1 System settings

The system parameters remain essentially similar to the joint signature scheme. In the proxy signature scheme the role of Signer 2 will be played by an on-line Trustee T. This scenario is appropriate for a corporate environment in which authorised personell may need to delegate and revoke signing capabilities at unpredictable times. This constrains all the users in the domain of T to share their certified public keys with T, in such a way that neither users nor T can form signatures alone. Thus, for the public key of the form $y \equiv g^{x_1 x_2} \bmod p$, the user will know x_1 and the trustee x_2. An elegant technique that could be used for this set up is described in [2]. In order to sign a message, user U (Signer 1) must employ the protocol prescribed in table 1 with T (Signer 2).

4.2 Delegation and proxy signature generation

In this section we present a scheme that will allow an entity s to delegate to a proxy s' the power to sign on its behalf, using the proxy delegation technique suggested by Mambo *et al.* [18]. Entity s calculates the proxy key as follows:

$$l_{s'} \in_R \mathcal{Z}_p \tag{6}$$

$$L_{s'} = g^{l_{s'}} \tag{7}$$

$$x_{1'} = x_1 + l_{s'} L_{s'} \bmod q$$

s sends $x_{1'}$ to s' securely and sends $(L_{s'}, ID_{s'}, ID_s)$ to T. It also securely stores $(l_{s'}, L_{s'}, ID_{s'})$.

To sign on behalf of s, s' engages in a protocol described in table 2 with T. This signature scheme can be derived from the equation sets 1 and 3 by replacing

Proxy	Trustee	Verifier
formProxyPartialSignature:		
$k_{s'} \in_R Z_q^*$		
$c = \mathcal{H}(m \| w^{k_{s'}})$		
$d = k_{s'} - cx_{1'} \bmod q$		
$\xrightarrow{c, d, m, ID_s, ID_{s'}}$		
	checkProxyPartialSignature:	
	$c \stackrel{?}{=} \mathcal{H}(m \| y^c g^{x_2 d} L_{s'}^{cL_{s'}})$	
	formCompleteSignature:	
	$k \in_R Z_q^*$	
	$r = g^k$	
	$d' = dx_2 - kr \bmod q$	
	$\xrightarrow{c, d', m, r, ID_s, L_{s'}}$	
		checkProxyCompleteSignature:
		$c \stackrel{?}{=} \mathcal{H}(m \| y_s^c g^{d'} r^r L_{s'}^{cL_{s'}})$

Table 2. Proxy joint signature scheme

x_1 with $x_{1'} = x_1 + l_{s'} L_{s'}$.

4.3 Revocation

To revoke the delegation, s informs T about this intent using a suitable protocol or procedure. T then *deletes* the 3-tuple $(L_{s'}, ID_{s'}, ID_s)$ from its database. After this event, s' cannot use the proxy key to generate valid signatures on behalf of s. It is noted that T must be trusted to delete the 3-tuple and not to engage in proxy joint signature protocol with s' to generate valid signatures on behalf of s. In order to reduce the level of trust robust time-stamping techniques can be used as an essential requisite of the signature generation and verification procedures, so that such collusions can be detected if they occur.

4.4 Observations

The following observations can be made on the proxy signature scheme:

1. The signer s can identify the proxy s' given the 5-tuple $(c, d', m, r, L_{s'})$.
2. The trustee cannot identify s' given $(c, d, m, ID_{s'})$, because it could have been formed by s or $ID_{s'}$ could be an arbitrary identification number that

was generated by the delegator (ID_s). Note that this property can be used to realise a group signature scheme with *on-line* membership revocation, when s is the revocation manager, the trustee is the group manager and many users assuming the role of s' with different values of $x_{1'}$. Also, suitable techniques can be employed so that the trustee can uniquely identify s' given a partial signature from the proxy s', this would mean that s should not know $x_{1'}$.

3. The universal signature verification function is modified. This allows the verifier to know that some authorised proxy has signed for the entity s and s has to be given credit or responsibility due for the message. This is a usual practice in the commercial world were the verifier can see words like "for" or "pp" in case of handwritten proxy signatures with an affixed seal.

4.5 Security analysis

The proxy-joint signature scheme can be considered to be a cascaded signature scheme as shown in figure 2. Internal details of the joint proxy signature are not show in the figure to improve clarity. For example, the trustee (not shown in the figure) is assumed to carry out the prescribed checks on the partial signature before completing the signature.

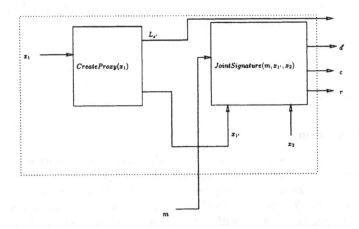

Fig. 2. Conceptual representation of proxy joint signature system

In the subsequent analysis, the proxy is assumed to be reliable and will not collude with the attacker. This is a valid assumption because the proxy signer should be *unconditionally* trusted [1] by the signer before delegating the power.

The *CreateProxy* function (adopted from the proposal by Mambo *et al.* [18]) can be analysed as a HoP signature on a *unity* message , $m = 1 \bmod q$ (see scheme EG I.4 in [13]). The HoP signature generation equation is $s = mx + kr \bmod q$, where s is the signature, x is the private key, k is a random number and $r = g^k$. In the case of unity message the equation becomes $s = x + kr \bmod q$, which is the same as the proxy equation suggested in [18]. Existential forgery on this scheme seems to be difficult because the forgery has to generate a signature tuple for a *fixed* message.

When the CreateProxy function is treated as a signature scheme (on unity message) the proxy joint signature scheme can be analysed as a cascade of two signature systems, namely, the CreateProxy function and the joint signature scheme. Thus the security of the proxy joint signature will depend on the CreateProxy function and the joint signature scheme.

5 Conclusion

A signature scheme that will be very useful for controlled environments was presented. We believe that the security of the new scheme against existential forgery should be similar to the Schnorr signature scheme.

Although not proposed in this paper, we believe that an effective group signature scheme *with membership revocation* can be realised using the proxy joint signature system.

6 Acknowledgement

We thank the anonymous refree for constructive reviews on the previous version of security analysis for the joint signature scheme.

References

1. Matt Blaze, Gerrit Bleumer, and Martin Strauss. Divertible protocols and atomic proxy cryptography. In Kaisa Nyberg, editor, *Advances in Cryptology – EURO-CRYPT'98*, number 1403 in Lecture Notes in Computer Science, pages 127–144. Springer, 1998.
2. Colin Boyd. Enforcing traceability in software. In *Information and Communication Security - First International Conference, ICICS'97*, pages 398–408. Springer, 1997.
3. C.Boyd. Digital multisignatures. In Henry J. Beker and F. C. Piper, editors, *Cryptography and Coding - 1986*, Oxford Science Publications, pages 241–246. Clarendon Press, Oxford, 1989.
4. David Chaum. Designated confirmer signatures. In *Advances in Cryptology – EUROCRYPT'94*, pages 86–91. Springer, 1994.
5. C.J.Mitchell, F.Piper, and P.Wild. *Contemporary Cryptology: The Science of Information Integrity*, chapter Digital Signatures, pages 325–378. IEEE Press, 1992.
6. Clifford Cocks. Split Knowledge Generation of RSA Parameters. Available at http://www.cesg.gov.uk/pubs/math, August 1997.
7. Clifford Cocks. Split Knowledge Gemeration of RSA Parameters with Multiple Participants. Available at http://www.cesg.gov.uk/pubs/math, February 1998.
8. C.P.Schnorr. Efficient signature generation for smart cards. *Journal of Cryptology*, 4:161–174, 1991.

9. D.Boneh and M.Franklin. Efficient Generation of Shared RSA Keys. In *Advances in Cryptology - CRYPTO'97*, number 1294 in Lecture Notes in Computer Science, pages 425–439. Springer-Verlag, 1997.

10. Yvo Desmedt. Threshold cryptosystems. In Jennifer Seberry and Yuliang Zheng, editors, *Advances in Cryptology - AUSCRYPT'92*, number 718 in Lecture Notes in Computer Science, pages 3–14. Springer-Verlag, December 1992.

11. Yvo Desmedt and Yair Frankel. Threshold cryptosystems. In Brassard .G, editor, *Advances in Cryptology - CRYPTO'89*, Lecture Notes in Computer Science, pages 307 – 315. Springer-Verlag, 1990.

12. Yair Frankel and Moti Yung. Escrow encryption systems visited: Attacks, analysis and designs. In *Advances in Cryptology - CRYPTO'95*, Lecture Notes in Computer Science, 1995.

13. Patrick Horster, Markus Michels, and Holger Petersen. Generalized ElGamal signatures for one message block. Technical Report TR-94-3, Department of Computer Science, University of Technology Chemnitz-Zwickau, May 1994.

14. Patrick Horster, Markus Michels, and Holger Petersen. Meta-multisignature schemes based on the discrete logarithm problem. Technical Report TR-94-12-F, Department of Computer Science, University of Technology Chemnitz-Zwickau, September 1994.

15. Markus Jakobsson, Kazue Sako, and Russell Impagliazzo. Designated verifier proofs and their applications. In *Advances in Cryptology - EUROCRYPT'96*, 1996.

16. Ueli M. Maurer and James L. Massey. Cascade ciphers: The importance of being first. *Journal of Cryptology*, 6(4):55–61, 1993.

17. Masahiro Mambo and Eiji Okamoto. Proxy cryptosystems: Delegation of the power to decrypt ciphertexts. In *IEICE Trans. Fundamentals*, volume E80-A, January 1997.

18. Masahiro Mambo, Keisuke Usuda, and Eiji Okamoto. Proxy signatures: Delegation of the power to sign messages. In *IEICE Trans. Fundamentals*, volume E79-A, September 1996.

19. M.Bellare and P.Rogaway. Random oracles are practical: a paradigm for designing efficient protocols. In *Proceedings of the 1st ACM Conference on Computer and Communications Security*, pages 62–73, 1993.

20. National Institute of Standards and Technology, Federal Information Process. *Standard FIPS Pub 186: Digital Signature Standard (DSS)*, 1991.

21. David Pointcheval and Jacques Stern. Security proofs for signature schemes. In Ueli Muarer, editor, *Advances in Cryptology - EUROCRYPT'96*, number 1070 in Lecture Notes in Computer Science, pages 387–398. Springer-Verlag, 1996.

22. Guillaume Poupard and Jacques Stern. Generation of Shared RSA Keys by Two Parties. In Kazuo Ohta and Dingyi Pei, editors, *Advances in Cryptology - ASIACRYPT'98*, number 1514 in Lecture Notes in Computer Science. Springer, 1998.

23. S.Goldwasser, S.Micali, and R.Rivest. A digital signature scheme secure against adaptive chosen-message attacks. *SIAM journal of computing*, 17(2):281–308, April 1998.

24. T.ElGamal. A public key cryptosystem and a signature scheme based on discrete logarithms. *IEEE Transactions on Information Theory*, IT-30(4):469–472, July 1985.

On the Cryptographic Value of the qth Root Problem

Cheryl L. Beaver[1], Peter S. Gemmell[2], Anna M. Johnston[1], and William Neumann[2]

[1] Sandia National Laboratories, Albuquerque, New Mexico 87185-0449, USA
ajohnst@sandia.gov or cldelor@sandia.gov
[2] The University of New Mexico, Albuquerque, New Mexico 87108, USA
gemmell@cs.unm.edu or wdnx@unm.edu

Abstract. We show that, for a prime q and a group G, if $ord(G) = q^k r, k > 1$, and r is smooth, then finding a qth root in G, is equivalent to the discrete logarithm problem over G (note that the discrete logarithm problem over the group G reduces to the discrete logarithm problem over a subgroup of order q – see reference [5]). Several publications describe techniques for computing q^{th} roots (see [3] and [1]). All have the stated or implied requirement of computing discrete logarithm in a subgroup of order q.

The emphasis here will be on demonstrating that with a fairly general q^{th} root oracle, discrete logarithms in a subgroup of order q may be found, describing the cryptographic significance of this problem, and in introducing two new public key signature schemes based on it.

1 Introduction

The discovery of new apparently hard problems can be an important cryptographic event in at least the following two senses. First, the new hard problem may lend itself to more pleasing public key algorithms. A public key algorithm may be more pleasing if it is more efficient for the intended application, if it reduces trapdoor fears, or if it satisfies numerous other technical or political concerns or needs. Secondly, the new problem may yield insight into the complexity of other apparently intractable – though apparently not NP-complete – problems widely used in cryptography.

The qth root problem appears to be one of these new hard problems. We believe that finding q^{th} roots in a cyclic group whose order is divisible by q^2 is as difficult as finding discrete logarithms in a subgroup of order q. If discrete logarithms in this subgroup are difficult, then so is the problem of finding q^{th} roots in this group. Notice that though we speak of discrete logarithms on a subgroup of order q, the reference [5] demonstrates that the discrete logarithm problem over any cyclic group G of order $q^k r$, where k is greater than one and r is smooth, can be reduced to the discrete logarithm problem on the subgroup of order q. Because of this reduction and for simplicity, the remainder of the paper will focus on groups of order q^2.

To show this equivalence relation we must demonstrate two things:

1. Given the ability to take discrete logarithms we can compute q^{th} roots;
2. Given the ability to take q^{th} roots we can compute discrete logarithms.

The first part of this proof is trivial. The algorithm described in [3] describes how to find q^{th} roots, and requires the ability to find discrete logarithms in the subgroup of order q (assuming q^2 divides the order of the group). The algorithm in [1] is very similar and in the end, also requires the ability to find discrete logarithms in the subgroup of order q.

The second part of the proof is more difficult. It is this part, demonstrating that with the ability to find q^{th} roots we can take discrete logarithms, along with the cryptographic impact of this problem, that the paper will focus on.

2 Background and Overview

Finding q^{th} roots (where q is prime) in a finite cyclic group G is not difficult if one of several conditions hold (see [3]):

- q is relatively prime to the order of G;
- q divides the order of G only once; or
- q is small.

If q is relatively prime to the order of G then the root can be found with the GCD algorithm and an exponentiation, similar to the processes in RSA (see reference [8]). If q divides the order of the group exactly once, then the roots can be found by computing an inverse, doing a few simple integer operations and doing one exponentiation in the group (see [3]). Finally, if q is small then the amount of work required to find the roots is $O((k-1)\sqrt{q})$, where k is the largest power of q which divides the order of G (any one of the order square root discrete logarithm attacks, such as the meet-in-the-middle or Pollard-Rho techniques – see reference [9] and [6]). Although this estimation holds for any q, it becomes computationally infeasible if q is too large.

What happens if none of these conditions hold? In particular, what if q is large and divides the order of G more than once (i.e., q^2 divides $|G|$)?

3 An Efficient Reduction from Discrete Logs to q^{th} Roots

In this reduction we will show how, with a fairly general q^{th} root oracle we can compute discrete logarithms. The reduction will operate over a general cyclic group G of order q^2. Note that both the qth root oracle and the reduction from discrete log problem described in this section can be extended to groups of order $q^k r$ where $k > 1$ and r is smooth (see reference [5] for a reduction of the discrete logarithm problem).

Let α be a generator of G and σ be a qth residue ($\sigma \in G$ such that $\sigma \equiv \alpha^{qx}$ for some $0 \leq x < q$). The reduction demonstrates an efficient discrete logarithm algorithm which requires a qth root oracle. Because q^{th} roots of a given residue are not unique in G we need to make some assumption on the form of the qth root that the oracle will return. We assume only that this oracle generates q^{th} roots of a plausible algebraic form, similar to the qth roots generated by the technique in [3]. This form is described in the subsection 3.1. Using this oracle and counting calls to the oracle as single operations (as is common practice in complexity), we are able to compute the discrete logarithm of any element in G in $O(2 \log_2(q))$ time.

3.1 The qth Root Oracle

A qth root oracle $M : G \rightarrow G$, determines a function mapping a qth residue to a particular q^{th} root. It acts on inputs of the form α^{qx} and will output $\alpha^{x+\delta_x q}$, where δ_x is some value modulo q:

$$M(\alpha^{qx}) = \alpha^{x+\delta_x q}.$$

The algebraic form assumed is: $\delta_x = xc$ where c is some constant. This assumption arises from the qth root technique in [3] using a group of order qr where q and r are relatively prime.

When the order of the group is qr, where r and q are relatively prime, this q^{th} root oracle can be implemented as:

$$M(\alpha^{qx}) = (\alpha^{qx})^{\frac{1+r(-r^{-1} \bmod q)}{q}}$$
$$= \alpha^{x+r(-xr^{-1} \bmod q)}$$
$$= \alpha^{x+\delta_x r}$$

where $\delta_x \equiv -xr^{-1} \bmod q$.

If the δ portion of the qth root oracle on G can be written as $\delta_x \equiv xc \bmod q$ where c is some (possibly unknown) constant, then this oracle can be used to compute a discrete log.

3.2 The Reduction: Finding Discrete Logarithms Using the qth Root Oracle

Using the previously described oracle, we now describe how to compute the discrete logarithm of an element in G. Recall that G is a cyclic group of order q^2, $\alpha \in G$ is a generator, and the oracle returning q^{th} roots has the following form:

$$M(\alpha^{qx}) = \alpha^{x+qxc}$$

where c is some constant.

Let $\sigma = \alpha^t$ be an element for which we want to find the discrete logarithm (i.e., find t). Let $t = t_1 q + t_0$ with $0 \leq t_0, t_1 < q$. The discrete logarithm (t) is found by first isolating t_0 ($\sigma^q = \alpha^{qt_0}$), determining t_0 with the algorithm described below, then removing t_0 from σ ($\sigma^{-t_0} = \alpha^{qt_1}$), then computing t_1

with the same technique. Initial candidate values for both t_0 and t_1 range from 0 to $q - 1$. Given a range for a variable, the oracle will enable us to determine which half of the range the variable is in. The reduction works in stages. Each stage of the reduction uses the oracle to reduces the possible range of the variable by half. So after $\lg(q - 1)$ steps, the variable has been discovered.

Interestingly, such a reduction works with a group of order q^2 but not with a group of order qr, where $GCD(q, r) = 1$. The oracle as defined computes:

$$M(\alpha^{qt}) = \alpha^{t + \frac{|G|}{q}\delta_t}.$$

In groups of order q^2 both the t and δ_t portions of the equation are in Z_q. However in groups of order qr, the t is in Z_r and the δ_t is in Z_q. Removing the δ_t portion is very easy in the q^2 case (as we will show), but seems to require a discrete logarithm in the qr case.

3.3   Reducing the Range

Let $\sigma = \alpha^{gx}$, with $0 \le x < \frac{q}{2^k}$. The oracle gives us the ability to determine whether x is greater than or less than $\frac{q}{2^{k+1}}$. This determination is made by first raising σ to the 2^{k+1}, then sending it to the oracle. The original value σ is also sent to the oracle, and the returned result is raised to the 2^{k+1}. This process is illustrated in the following equations:

$$M(\sigma^{2^{k+1}}) = M(\alpha^{q(2^{k+1}x \bmod q)} = \alpha^{(2^{k+1}x - bq) + q(2^{k+1}x \bmod q)}$$

$$M(\sigma)^{2^{k+1}} = \alpha^{2^{k+1}x + q(2^{k+1}x \bmod q)}$$

These two equations are identical except for the variable b, which represents the modular reduction by q of $2^{k+1}x$. The value b must be either zero (x is less than or equal to $\frac{q}{2^{k+1}}$) or one (x is greater than $\frac{q}{2^{k+1}}$). Multiplying one of these equations by the inverse of the other will produce $\alpha^{\pm bq}$. If this result is one (the identity in G), b is zero and $0 \le x < \frac{q}{2^{k+1}}$. If the result is not one, then b is one and $\frac{q}{2^k} \le x < \frac{q}{2^{k+1}}$.

3.4   Finding the Discrete Logarithm

Our discrete logarithm technique uses the range finding technique described in 3.3 to iteratively reduce the range until the discrete logarithm is found.

Let $\beta = \alpha^t$ with $t = t_1 q + t_0$ and $0 \le t_1, t_0 < q$. The following algorithm will compute the discrete logarithm of β base α:

Discrete Logs using the q^{th}root Oracle

1. Isolate t_0 by raising β to the q^{th}power.

$$\beta^q \equiv \alpha^{qt_0}$$

2. Compute t_0 using FINDx (below)
3. Isolate t_1:

$$\beta\alpha^{-t_0} \equiv \alpha^{qt_1}$$

4. Compute t_1 using FINDx (below)
5. $t = t_1 q + t_0$.

Subroutine FINDx

1. Input is $\sigma = \alpha^{qx}$ with $0 \le x < q$.
2. Let $i = 0$, and $\sigma_i = \sigma$, $x_i = x$: $0 \le x_0 < \frac{q}{2^i}$
3. The final value for x will be stored in T. Initially $T = 0$.
4. For $i = 1$ to $i = \lceil \log_2(q) \rceil$:

   (a) Raise σ_{i-1} to the 2^i power and send it to the oracle:

   $$M(\sigma_{i-1}^{2^i}) = \alpha^{(2^i x_{i-1} - bq) + q(2^i x_{i-1} c \bmod q)}$$

   (b) Send σ_{i-1} to the oracle then raise the results to the 2^i power:

   $$M(\sigma_{i-1})^{2^i} = \alpha^{2^i x_{i-1} + q(2^i x_{i-1} c \bmod q)}$$

   (c) If these two results are equivalent then $b = 0$ and let

   $$\sigma_i = \sigma_{i-1}$$
   $$T = T$$
   $$x_i = x_{i-1}$$

   (d) If these two results are different then $b = 1$ and let

   $$\sigma_i = \sigma_{i-1}\alpha^{-q\left\lceil \frac{q}{2^i} \right\rceil}$$
   $$T = T + \left\lceil \frac{q}{2^i} \right\rceil$$
   $$x_i = x_{i-1} - \left\lceil \frac{q}{2^i} \right\rceil$$

   (e) At this point, $0 \le x_i < \frac{q}{2^i}$
5. Return $x = T$.

4 Public Key Protocols that use the qth Root Problem

The qth root problem over a group of order q^2 is parallel, in many ways, to the square root problem modulo N, where N is a two prime composite. both are equivalent to their respective hard problems of discrete log or factoring. Raising a value to the qth power in G is a one-way function as is squaring modulo N – assuming the factors of N are secret.

There is one major difference between these two problems. If the factors of N are known, square roots modulo N are easy to find. This amounts to a trapdoor in square root problem. No such trapdoor exists in the qth root problem. This difference has both positive and negative effects on the schemes generated by these problems. If a trapdoor exists, doing public key encryption is very simple. If the trapdoor does not exist then public key encryption is more complicated. However the non-existence of a trapdoor may make the schemes based on the qth root problem more appealing to many users, especially in situations with mutual distrust.

Computationally, squaring a value modulo N will almost always be more efficient than raising a value to the qth power in G. Carefully choosing G and q will minimize this difference. As for signature length, the flexibility on the choice of the group can make the lengths of signatures based on the qth root problem much less than the square root problem.

This paper contains two signature protocols. The first is similar to the Fiat-Shamir signature protocol and makes use of the parallelism between the two (q^{th} root and square root) problems. The second is closer to ElGamal. While our protocols may prove useful in their own right, perhaps more important is the idea that others may follow that employ the hardness of the qth root problem and which offer time and space improvements for public key operations.

5   The Exponential Knapsack Signature

This scheme is very similar to the Fiat-Shamir signature protocol (see [2]). A one-way hash function and access to random or pseudorandom bits are also necessary for this scheme.

- Problem Components:
    G A group of order q^2;
    k A fixed integer determining the maximum number of elements in the knapsack and the number of bits returned by the hash function;
    H A one-way hash function returning k bits;
    $\{x_i\}$ A set of k secret random elements in G, $i = 0, 1, ..., k-1$;
    $\{y_i\}$ A set of k public elements with $y_i = x_i^{-q}$, $i = 0, 1, ..., k-1$;
    m A message to be signed;
    r A one-time randomly generated secret element of G;
    t $t = r^q$, a public element of G;
    U $U = H(m||t)$;
    $\{u_i\}$ For $\{i = 0, 1, ..., k-1\}$, the k bits of U ($U = \sum_{i=0}^{k-1} u_i 2^i$).
- To Sign:
    1. Compute $s = r \prod_{i=0}^{k-1} x_i^{u_i}$
    2. Transmit: $[m, U, s]$
- To Verify:

1. Raise s to the q-power and strip off the knapsack elements. If all works well t will be recovered:

$$t' = s^q \prod_{i=0}^{k-1} y_i^{u_i} \stackrel{?}{=} r^q \prod_{i=0}^{k-1} x_i^{qu_i} x_i^{-qu_i} = r^q$$

2. Compute what should be U: $U' = H(m||U')$
3. If $U' = U$ then the message was correctly signed. Otherwise the signature fails.

The proof of security for this protocol is nearly identical to that of Fiat and Shamir's original protocol upon which our protocol is based (see reference [2]).

6 ElGamal-Like Signature Scheme

The following signature scheme requires much less storage than the exponential knapsack scheme because it uses one long term key instead of k of them. In some ways it is similar to ElGamal style signatures, though the secret components are in the group instead of the exponent. Notice that this scheme requires the verifier to compute an inverse modulo the order of the group. This prevents a similar square root style scheme from working (the order of the group in the square root problem must remain unknown or problem is easily broken). The required component list looks almost identical to that of the exponential knapsack. A long term secret, a one-time random secret and a good one-way hash function are all necessary components to both schemes.

- Problem Components:
  G A group of order q^2;
  H A one-way hash function returning k bits;
  x A long term secret with $x \in G$;
  y A long term public key with $y = x^{-q}$;
  m A message to be signed;
  r A one-time randomly generated secret element of G;
  t $t = r^q$, a public element of G;
  U $U = H(m||t)$, with $0 < U < q$.
- To Sign:
  1. Compute $s = xr^U$
  2. Transmit: $[m, U, s]$
- To Verify:
  1. Raise s to the q-power and multiply by y to remove the long term key, x^q. Raise this result to the $U^{-1} \bmod q$ power. If all works well t will be recovered:

$$t' = (s^q y)^{U^{-1}} \stackrel{?}{=} (x^q r^{qU} x^{-q})^{U^{-1}} = r^q$$

2. Compute what should be U: $U' = H(m||t')$
3. If $U' = U$ then the message was correctly signed. Otherwise the signature fails.

7 Conclusion

We have shown, modulo one assumption about which particular q^{th} root the oracle outputs, that finding q^{th} roots in a group whose order is q^2 is equivalent to finding discrete logarithms in that same group (which can be extended to groups of order $q^k r, k > 1$ where r is smooth – see reference [5]). The immediate implications are that we have either uncovered a new class of cryptographically weak primes or that we have opened up a new class of cryptographic algorithms.

However, there are still several open problems and much new work to be done in this area.

- Can the reduction presented in section 3.2 be modified so that the assumption on which q^{th} root the oracle outputs can be relaxed?
- Are these groups (where q^2 divides the order) suseptible to some new type of discrete logarithm technique?
- Are there other designs based on the qth root problem which offer significant improvement over current designs?.

References

1. Bach E., Shallit J.: *Algorithmic Number Theory, Volume I: Efficient Algorithms.* MIT Press, Cambridge, Massachusetts (1996) 160–163
2. Feige U., Fiat A., Shamir A.: Zero-knowledge proofs of identity. Journal of Cryptology 1 (1988), 77–94
3. Johnston A.: A Generalized q^{th} Root Algorithm. Proc. of the Symp. on Discrete Algorithms, January 1999, Baltimore Maryland.
4. Menezes A., van Oorschot P., Vanstone S.: *Handbook of Applied Cryptography.* CRC Press, 1996
5. Pohlig S., Hellman M.: An improved algorithm for computing logarithms over GF(p) and its cryptographic significance. IEEE Trans. on Information Theory 24 (1978), 106–110
6. Pollard J.: Monte Carlo methods for index computation mod p. Math. of Computation 32 (1978), 918–924
7. Rabin M.: Digitalized signatures and public key functions as intractable as factorization. MIT/LCS/TR-212, MIT Laboratory for Computer Science, 1979.
8. Rivest R., Hellman M., Adleman L.: A Method for Obtaining Digital Signatures and Public Key Cryptosystems. Comm. of the ACM v. 21, n. 2 (1978), 120–126
9. Shanks D.: Solved and unsolved problems in Number Theory. Washington, D.C., Spartan (1962)
10. Williams H.: A refinement of H.C. Williams' qth root algorithm. Math. Comp. 61 (1993), no.203, 475–483

KEYNOTE SPEECH

Protecting
Critical Information Systems

Professor Sushil Jajodia

Center for Secure Information Systems
George Mason University, USA

Abstract

This talk will discuss issues and methods for survivability of systems under malicious attacks. To protect from information attacks, it is necessary to take steps to prevent attacks from succeeding. At the same time, it is important to recognize that not all attacks can be averted at the outset; attacks that are successful to some degree must be recognized as unavoidable and comprehensive support for identifying and responding to attacks is required. This places increased emphasis on the ability to live through and recover from successful attacks. We have adopted a fault-tolerance approach that addresses all phases of survivability: attack detection, damage confinement, damage assessment and repair, and attack avoidance. This talk will describe our research dealing with each of these phases.

Delegation Chains Secure up to Constant Length

Masayuki ABE and Tatsuaki OKAMOTO

NTT Laboratories
Nippon Telegraph and Telephone Corporation
1-1 Hikarinooka, Yokosuka-shi, Kanagawa-ken, 239-0847 Japan
E-mail: {abe,okamoto}@sucaba.isl.ntt.co.jp

Abstract. In this paper we discuss how one can delegate his power to authenticate or sign documents to others who, again, can delegate the power to someone else. A practical cryptographic solution would be to issue a certificate that consists of one's signature. The final verifier checks verifies the chain of these certificates. This paper provides an efficient and provably secure scheme that is suitable for such a *delegation chain*. We prove the security of the scheme against an adaptive chosen message attack in the random oracle model.

Though our primary application would be agent systems where some agents work on behalf of a user, some other applications and variants will be discussed as well. One of the variants enjoys a threshold feature whereby one can delegate his power to a group so that they have less chance to abuse their power. Another application is an identity-based signature scheme that provides faster verification capability and less communication complexity compared to those provided by existing certificate-based public key infrastructure.

Keywords: Proxy Signatures, ID-based Signatures, Random Oracle

1 Introduction

Background: The ability to delegate the power of authentication or signing is needed in several kinds of network applications. Consider, for instance, a server-client system where a server broadcasts messages with his signature to a huge number of client terminals. The clients are supposed to accept messages that have a valid signature of the server. The server may, only for limited periods, designate remote sub-servers to broadcast messages on his behalf due to maintenance or the need to distribute traffic loads. If the server delegates his power of signing to the sub-servers by giving them a kind of certificate that contains his signature, the clients do not need to change the foundation of their trust, i.e., the public key of the main server.

Another, and more concrete example is an agent system wherein several agents work across electronic networks on behalf of users to get information from information providers. A user specifies an ACL (access control list), which contains "Issuer", "Subject", "Delegate", "Permit" and "Validity", and other application dependent items in a certificate with his signature [6]. The delegate (agent) sets off to perform the duty specified. Since the agent can work only within the networks that support his language (protocol), he hires other agents for other networks if "Permit" clears such an activity. The user agent gives the hired agent another certificate that contains possibly more restricted permissions. The last agent, who has the list of certificates, signs a document on behalf of the user. The information provider verifies the signature and all the certificates one by one.

To provide a cryptographic solution to those applications, we consider the "delegation chain" by which the power of signing/authentication is inherited from one entity to another. A straightforward approach would be to make a chain of ordinary signatures where each public key is authorized by certificate authorities. This is a trivial solution so we construct a more efficient scheme both in communication complexity and computational complexity.

Our approach is derived from an observation of the interesting similarity between identity based signature schemes and the delegation of the power of signing. Identity-based authentication/signature was first introduced by Shamir in [15]. A key issuing authority issues a private key to a user based on the user's identity and the authority's private key. One can verify a signature issued by the user by using the user's identity and the public key of the authority. If verification is successful, it says something more than the validity of the signature; the signer has been authorized by the key issuing authority to use the private key. This establishes a kind of delegation system without certificates.

Our Results: The notion and definition of the delegation chain is introduced. We then propose an efficient and provably secure delegation chain scheme based on the Schnorr signature scheme [14]. We prove that our scheme is secure against an adaptive chosen message attack in the random oracle model. We clarify that the existing reduction method for proving the security guarantees security only up to the constant length of the chain. Regarding efficiency, our construction reduces the amount of communication compared to certificate-based schemes because certificates are never transmitted. Moreover, the scheme allows verifiers to efficiently compute a predicate that guarantees the integrity of the chain from the origin entity to the signing entity.

In addition to agent systems, we introduce other applications and variants of our scheme. We show that a threshold scheme can be easily integrated to our scheme. Another application is a secure identity-based signature scheme; this is the first efficient construction based on the discrete logarithm problem.

Related Works: A few investigations on delegation are known. In [5] Brands introduced "secret key certificates" and pointed out their usability for delegation, though the main application is electronic cash systems. In [10] Mambo et al. introduced the notion of "proxy signature" and constructed a scheme based on a Schnorr signature variant. Although their scheme was claimed to be secure under the intractability assumption of the discrete logarithm problem, no security proof against an adaptive chosen message attack was given. Moreover, both works consider only one step of delegation and did not discuss a generic structure for delegation and its adversary model as is done in this paper.

In [4], Blaze et al. presented another notion of proxy signatures where anyone can transform someone's signature to someone else's if those two players issue a kind of public key that is used for conversion. Such an approach may be usable even for the delegation chain. However, their construction requires the parties to mutually reveal their private keys. This is impractical in a delegation chain, since all parties in the chain would get the other member's private keys and all parties would have to change their private keys when the delegation authority expired.

The Fiat-Shamir identification scheme in [8] is a provably secure identity based scheme based on the factoring problem. Unfortunately their idea, which is to use the user's identity as a public key, is not directly usable because the user's private key issued by the authority has different form from the authority's, so he can not function as an authority to the next user. Regarding discrete logarithm based identity-based signature schemes, Beth presented a scheme that was a mixture of ElGamal and Schnorr signature [3], and Okamoto presented a scheme based only on the Schnorr signature scheme [12]. To the best of our knowledge however, the complexity of these schemes is comparable to that of ordinary certificate-based schemes. No efficient construction based on the discrete logarithm problem was known. Hence our construction is the first efficient and provably secure identity based signature scheme to be based on the discrete logarithm problem.

Organization: This paper is organized as follows. Section 2 defines delegation chain schemes and their security. We then construct a scheme in section 3. The proof of its security and an efficiency analysis are given in section 4. Some variants and other usages of the scheme are shown in section 5. We conclude with some remarks in section 6.

2 Definitions and the Security Model

2.1 Definitions

Let us first describe the players in our scheme. Some of the terms in the following came from agent systems mentioned as an example in the previous section. A *user* delegates the power of signing to an *agent* he or she trusts. Once an agent acquires the power of a user, we call it a *user agent*. Although a user agent may delegate the power to another agent again, he should be able to use it if necessary. A *verifier* confirms the signatures issued by user agents or users. All players are computationally bounded by a polynomial of a security parameter k.

An entity who wants to delegate his power first prepares a *warrant description* that specifies the identity of the agent being assigned the power and other necessary conditions like validity or permitted actions. The warrant description is assumed to be taken from warrant description space $\mathcal{W}$ which is disjoint with message space $\mathcal{M}$ from where signing messages are taken.

We now define a *delegation chain scheme* as follows.

Definition 1. A delegation chain scheme is a five-tuple of algorithms $(\mathcal{G}, \mathcal{D}, \mathcal{A}, \mathcal{S}, \mathcal{V})$ that satisfies the following:

- $\mathcal{G}$, the key generation algorithm, is a probabilistic polynomial time algorithm which on input $1^k (k \in N)$ outputs a pair (x_0, y_0) of a secret key and a corresponding public key: $\mathcal{G}(1^k) = (x_0, y_0)$.
- $\mathcal{D}$, the delegation algorithm, is a probabilistic polynomial time algorithm in k which produces a pair (x_{i+1}, v_{i+1}) with respect to a given secret key x_i, a warrant description W_{i+1} and a sequence of i pairs $(v_1, W_1), \ldots, (v_i, W_i)$ where each pair is of length k and each W_j taken from $\mathcal{W}$: $\mathcal{D}(x_i, v_1, \ldots, v_i, W_1, \ldots, W_i, W_{i+1}) = (x_{i+1}, v_{i+1})$.
- $\mathcal{A}$, the assignment verification algorithm, is a polynomial time algorithm whose inputs are a secret key x_{i+1}, a public key y_0, and a sequence of $i + 1$ pairs $(v_1, W_1), \ldots, (v_{i+1}, W_{i+1})$. The algorithm outputs 1 if (x_{j+1}, v_{j+1}) is in the range of $\mathcal{D}(x_j, v_1, \ldots, v_j, W_1, \ldots, W_{j+1})$ for $j = i, \ldots, 1$, and x_0 is a secret key that corresponds to y_0 generated by $\mathcal{G}(1^k)$, otherwise 0: $\mathcal{A}(x_{i+1}, y_0, v_1, \ldots, v_{i+1}, W_1, \ldots, W_{i+1}) \in \{0, 1\}$.
- $\mathcal{S}$, the signing algorithm, is a polynomial time probabilistic algorithm that produces signatures σ with respect to a given sequence of i pairs $(v_1, W_1), \ldots, (v_i, W_i)$ and a secret key x_i that satisfy $\mathcal{A}(x_i, y_0, v_1, W_1, \ldots, v_i, W_i) = 1$, for every message m of length k, taken from $\mathcal{M}$: $\mathcal{S}(x_i, v_1, W_1, \ldots, v_i, W_i, m) = \sigma$.
- $\mathcal{V}$, the verification algorithm, is a polynomial time algorithm whose inputs are a public key y_0, i-array of pairs $(v_1, W_1), \ldots, (v_i, W_i)$, a message m, and a signature σ. For every $(x_i, y_0, v_1, W_1, \ldots, v_i, W_i)$ that satisfies $\mathcal{A}(x_i, y_0, v_1, W_1, \ldots, v_i, W_i) = 1$, and every $m \in M$ of length k, the algorithm outputs 1 if σ is in the range of $\mathcal{S}(x_i, v_1, W_1, \ldots, v_i, W_i, m)$, otherwise 0: $\mathcal{V}(y_0, v_0, W_0, \ldots, v_i, W_i, m) \in \{0, 1\}$.

Algorithms $\mathcal{D}$ and $\mathcal{S}$ do not have to be probabilistic. Similarly, the definitions for $\mathcal{A}$ and $\mathcal{V}$ can be modified to cover probabilistic algorithms.

k-delegation chain refers to k triple $\{(x_i, v_i, W_i) | i = 1, \ldots, k\}$ and (x_0, y_0). If no confusion is expected, the term "delegation chain" is also used to describe a group of k entities which forms a valid k-delegation chain.

2.2 Security Model

Let A_i for $i = 0, \ldots, n$ form a delegation chain where A_0 is a user. It is assumed that A_0 is not concerned about what his agents does, as long as the action conforms to the warrant. On the other hand, A_0 must be *fully responsible* for what his agents will do in accordance with the warrant description. In our framework, accordingly, it is substantially meaningless that A_i frames A_{i+1} or further agents.

A simple attack is to forge a triple $(x_{i+1}, v_{i+1}, W_{i+1})$ issued by A_i by colluding with a subsequent agent A_j, $j > i$. A more serious attack would be to try to forge an entire chain starting from a specific user A_0. As a signature scheme, a delegation chain scheme should be existentially unforgeable against an *adaptive chosen message attack.* An adaptive chosen message attack against a delegation chain scheme is captured as follows: Let Adv be an adversary. The adversary is allowed to invoke S with any A_i to get a signature of arbitrary message in M at any moment. Let $(m_j, \sigma_j)_{A_i}$ be a message and signature pair created during the j-th invocation of S with A_i by Adv. Let S be a set that contains $(m_j, \sigma_j)_{A_i}$ for $i = 0, \ldots, n$ and for $j = 1, \ldots, q_i$, where q_i is the number of invocations of S with A_i by Adv. After obtaining S, Adv eventually produces a message and signature pair $(m, \sigma) \notin S$.

In the random oracle model [1,2], each A_i is assumed to invoke random oracle R_i. Adv also accesses R_i up to Q_i times. This paper assumes that Adv is a probabilistic polynomial time algorithm with regard to security parameter k, and the number of invocations to S and R_i should be polynomially bounded, i.e., q_i and Q_i are polynomial in k. The success probability ϵ of the forgery by Adv is taken over the coin flips of Adv, all A_i, and all R_i. Adv is allowed to collude with (or collapse) an agent, A_i. Assume that Adv collapses A_i in a delegation chain, $(A_0, A_1, \ldots, A_i, \ldots, A_n)$. Adv then can produce valid signatures of A_j for all $j \geq i$ for any messages without launching any additional attacks.

In considering the security of a delegation chain, $(A_0, A_1, \ldots, A_n)$, Adv is also allowed to perform any attack (e.g., to collapse an agent) on any agent outside the chain. For example, A_0 may produce several chains such as $(A_0, A_1, \ldots, A_n)$ and $(A_0, A_1', \ldots, A_l')$, and A_i may produce several sub-chains like $(A_0, A_1, \ldots, A_i, A_{i+1}, \ldots, A_n)$ and $(A_0, A_1, \ldots, A_i, A_{i+1}'', \ldots, A_n'')$. Therefore, the general structure of the delegation chains generated by A_0 is a tree rooted by A_0.

We now define the security of a delegation chain scheme as follows:

Definition 2. A delegation chain scheme is secure (*unforgeable* and *endorsed*), if and only if the following condition holds: For any $i \in [0, n]$, any $n \in N$ and any adversary, Adv, given $(y_0, (v_1, W_1), \ldots, (v_i, W_i))$, cannot generate $A_{(n)}$'s signature σ of arbitrary message m along with $(v_{i+1}, W_{i+1}), \ldots, (v_n, W_n)$ with non-negligible probability (over the coin flips of $\mathcal{G}_i$, Adv, actively attacked A_l and random oracle R_i), unless Adv collapses A_j and obtains x_j for $j \in \{0, 1, \ldots, i\}$. If Adv collapses A_j, then Adv can generate $A_{(j)}$'s signature ($j \leq i$). Here, $A_{(n)}$ is any agent who is in the n-th position of any delegation chain, and $\mathcal{G}_i$ is a probabilistic polynomial time algorithm (chain instance generator) which on input 1^k outputs $(y_0, (v_1, W_1), \ldots, (v_i, W_i))$ along with $(x_0, x_1, \ldots, x_i)$, where $i \in \{0, \ldots, n\}$. The distribution of the output of $\mathcal{G}_i$ is the same as that from the honest delegation chain $(A_0, \ldots, A_i)$ with $\mathcal{G}$.

To confirm that the above definition is reasonable, we consider intuitive requirements for delegation schemes as follows.

- User A_0 (A_j) should endorse any signature of an agent, A_i ($i \geq j$), in the delegation chain including A_0 (A_j). In other words, without the involvement of A_0 (A_j), no signatures of an agent, A_i ($i \geq j \geq 0$), can be produced.
- A_j can generate a signature of an agent, A_i ($i > j$), but A_i cannot generate a signature of A_j.

– The signatures of A_i $(i = 0, \ldots, n)$ should be existentially unforgeable against adversaries who can launch adaptive chosen message attacks against $A_0, \ldots, A_i$, and collapse the other agents.

The security definition captures these requirements of the delegation chain systems.

When $i = 0$ in the definition, y_0 is given to Adv, i.e., y_0 is determined (registered) as the public-key of A_0. In order for Adv to generate a signature of an agent of a chain started by A_0, Adv should collapse A_0 and obtain x_0. This property guarantees that no signature of an agent of a chain started by A_0 can be produced without the involvement of A_0. We can extend this property to A_j $(j > 0)$. If $(y_0, \ldots, (v_j, W_j))$ is given to Adv, i.e., $(y_0, \ldots, (v_j, W_j))$ is (directly/indirectly) endorsed (certified) by A_0, then A_j's power of signing is endorsed by A_0. It is then guaranteed that no signature of an agent lower than A_j can be produced without the involvement of A_j being endorsed by A_0. If and only if Adv generates a signature of A_i, should Adv be able to collapse A_j $(j \leq i)$, from the definition. This implies that no collusion of A_l $(l > i)$ can output a signature of A_i.

The third requirement is also implied by the definition, since Adv is, when Adv tries to forge a signature of A_i, allowed to perform adaptive chosen message attacks against $A_0, \ldots, A_i$, and collapse the other agents.

3 The proposed scheme

Let p and q be large primes that satisfy $q | p - 1$. By G_q we denote a multiplicative subgroup of order q in Z_p. Let $\mathcal{H} : \{0, 1\}^* \to Z_q^*$ denote a collision intractable hash function. We assume the use of a discrete logarithm instance generator that takes a security parameter k, and outputs (p, q, g), such that given $y := g^x \bmod p$ for randomly chosen x, it is infeasible to compute x in time polynomial in k. The generator is once executed and p, q, g are published as common system parameters. In the following, all arithmetic operations are done over Z_p unless otherwise stated. A user, say A_0 selects $x_0 \in_U Z_q$ and publishes $y_0 := g^{x_0}$ to all possible verifiers.

For delegation, the user composes a warrant description W_1 and then executes the following protocol with her intended agent A_1. The protocol must be performed via a secure channel.

[Primary Delegation Protocol]

D-1: A_0 selects $r_1 \in_U Z_q$ and computes $v_1 := g^{r_1}$, $c_1 := \mathcal{H}(v_1 \| W_1)$ and $x_1 := c_1 r_1 + x_0 \pmod{q}$. A_0 then sends (x_1, v_1, W_1) to A_1.

D-2: A_1 computes $c_1 := \mathcal{H}(v_1 \| W_1)$ and checks if $g^{x_1} \stackrel{?}{=} v_1^{c_1} y_0$.

[End]

Figure 1 illustrates the delegation protocol between two lower agents A_i and A_{i+1}.

A_i
$(x_i, v_i, \ldots, v_1, W_i, \ldots, W_1)$

A_{i+1}
(y_0)

$r_{i+1} \in_U Z_q^*$
$v_{i+1} := g^{r_{i+1}}$
$c_{i+1} := \mathcal{H}(v_1\|\ldots\|v_{i+1}\|W_1\|\ldots\|W_{i+1})$
$x_{i+1} := c_{i+1}r_{i+1} + x_i \pmod{q}$

$$x_{i+1}, (v_{i+1}, W_{i+1}), \ldots, (v_0, W_0) \longrightarrow$$

for $j = 1$ to $i+1$
$c_j := \mathcal{H}(v_1\|\ldots\|v_j\|W_1\|\ldots\|W_j)$

$$g^{x_{i+1}} \stackrel{?}{=} v_{i+1}^{c_{i+1}} \ldots v_1^{c_1} y_0$$

(OK/NG)

Fig. 1. The delegation protocol. A_{i+1} stores $x_{i+1}, (v_{i+1}, W_{i+1}), \ldots, (v_0, W_0)$.

To sign a document $m \in M$, signing agent A_n performs the following protocol with verifier V.

[Signing Protocol]

S-1: A_n checks if m is in M. A_n then selects $r_{s,n} \in_U Z_q$ and computes $v_{s,n} := g^{r_{s,n}}$ and $\sigma_n := r_{s,n} + \mathcal{H}(v_1\|\ldots\|v_n\|v_{s,n}\|W_1\|\ldots\|W_n\|m)x_n \pmod{q}$, and sends $(\sigma_n, v_{s,n}, v_n, W_n, \ldots, v_1, W_1)$ to V

S-2: V computes $c_j := \mathcal{H}(v_1\|\ldots\|v_j\|W_1\|\ldots\|W_j)$ for $j = 1, \ldots, n$, and $c_{s,n} := \mathcal{H}(v_1\|\ldots\|v_n\|v_{s,n}\|W_1\|\ldots\|W_n\|m)$. Then it checks $g^{\sigma_n} \stackrel{?}{=} v_{s,n}^{c_{s,n}} v_n^{c_n} \ldots v_1^{c_1} y_0$. If it holds, V examines all W_j to see if the agents have deviated from the original permission.

[End]

A signature of the end agent A_n on a document $m \in \mathcal{M}$ is $(\sigma_n, v_{s,n}, v_n, W_n, \ldots, v_1, W_1)$.

4   Analysis

4.1   Security

Hereafter we assume that an adversary, (Adv_1, Adv_2 etc.), is an active adversary, who is allowed to make adaptive chosen message attacks and to collapse agents (see the detailed adversary description in Section 2.2).

Let $\mathcal{G}_i$ be a chain instance generator which on input 1^k outputs $(y_0, (v_1, W_1), \ldots, (v_i, W_i))$ and $(x_0, x_1, \ldots, x_i)$ for i in $\{0, \ldots, n\}$. The output satisfies $g^{x_j} = v_j^{c_j} \ldots v_1^{c_1} y_0$ and $c_j := \mathcal{H}(v_1 \| \ldots v_j \| W_1 \ldots \| W_j)$ for $j = 0, 1, 2, \ldots, n$ (for $j = 0$, $g^{x_0} = y_0$). The distribution of the output is the same as that of the honest delegation chain $(A_0, \ldots, A_i)$ with $\mathcal{G}$.

Theorem 1. Let $n - i = O(1)$. If an adversary, Adv_1, given $(y_0, (v_1, W_1), \ldots, (v_i, W_i))$, can generate $A_{(n)}$'s signature $(\sigma_n, v_{s,n}, m)$ along with $(v_{i+1}, W_{i+1}), \ldots, (v_n, W_n)$ with non-negligible probability (over the coin flips of $\mathcal{G}_i$, Adv_1, actively attacked A_l and random oracle $\mathcal{H}$) then there exists a polynomial time algorithm (extractor M_1) that, using Adv as a black-box, $M_1^{Adv_1}$, can output $(x_i, x_{i+1}, \ldots, x_n)$ with overwhelming probability (over the coin flips of $\mathcal{G}_i$, Adv_1, M_1 and $\mathcal{H}$).

Proof (Sketch): The mathematical structure of our protocol is similar to that of the multi-signature Type II in [11] [1]. Therefore, we can apply their result to the proof of this theorem. The strategy of reducing Adv_1 to computing $(x_i, x_{i+1}, \ldots, x_n)$ is almost the same as that in [11]: First we convert Adv_1 to a prover, P, for a multi-round zero-knowledge interactive proof against an honest verifier. In this conversion, signers' signatures of adaptive chosen-message attacks by Adv_1 are simulated. We then reduce P to computing $(x_i, x_{i+1}, \ldots, x_n)$. Since we can use the analysis in [11], we omit the description of the detailed procedure and analysis of the reduction cost.

As a result, we can obtain the following: Suppose Adv_1 succeeds in generating A_n's signature within running time t, with probability of at least ϵ, using q_{sig} chosen message attacks, and q_H queries to the random oracle. $M_1^{Adv_1}$ can then compute $(x_i, x_{i+1}, \ldots, x_n)$ within running time t^* and with probability of at least ϵ^*, where

$$t^* = \frac{t'}{3\epsilon''}\left(2^{2(n-i)+1} + 1\right) + \Phi_3 \quad \text{and}$$

$$\epsilon^* = \left(\frac{1}{2}\right)^{(2(n-i)+1)}\left(1 - \frac{1}{e}\right)^{2^{n-i+1}} > \left(\frac{1}{2}\right)^{(2(n-i)+1)}\left(\frac{3}{5}\right)^{2^{n-i+1}}.$$

Here,

$$\epsilon_0 = \epsilon,$$

$$\epsilon_j = \frac{\epsilon_{j-1} - \frac{1}{q}}{q_H} \quad \text{for } 1 \leq j \leq n - i + 1,$$

$$\epsilon' = \epsilon_{n-i+1},$$

---

[1] The scheme in [11] is an ordinary multi-signature generated by several independent signers, while in our scheme, hierarchically dependent signers, delegation chain, produce a signature. Accordingly, only our scheme has the novel properties shown in Theorems 1 and 2.

$$\epsilon'' = \epsilon' - \frac{q_{sig}}{q} \geq \frac{2^{(n-i)}}{q^{n-i+1}}, \text{ and}$$

$$t' = t + \Phi_1 + \Phi_2.$$

Φ_1 is the verification time of the identification protocol, Φ_2 is the simulation time of q_{sig} signatures, and Φ_3 is the calculation time of $(x_i, x_{i+1}, \ldots, x_n)$ in the final stage of the reduction.

Since the power of Adv is polynomial time bounded, t, q_{sig} and q_H are polynomial in k. In addition, from the assumption, ϵ is non-negligible, and $n-i = O(1)$. Therefore, t^* should be polynomial in k, and ϵ^* is constant. By repeating this procedure polynomially times, we can raise the success probability to overwhelming. $\square$

Definition 3. Let G_{DL} be an instance generator such that on input 1^k outputs (p, g, q, x, y), where p, q : primes, $q|p - 1$, $x \in Z_q^*$, and $y := g^x \bmod p$. The "*discrete logarithm problem*" is, given (p, g, q, y) by G_{DL}, to find x. The discrete logarithm problem is *intractable*, if for any probabilistic polynomial time machine T, for any constant c, for sufficiently large k,

$$\Pr[T(1^k, n) = x] < 1/k^c.$$

The probability is taken over the coin flips of G_{DL} and T.

Theorem 2. Suppose the discrete logarithm problem is intractable. Let $n - i = O(1)$. The proposed delegation chain scheme is thus secure (unforgeable and endorsed) in the random oracle model.

Proof (Sketch): Suppose Adv_2 succeeds in generating A_n's signature with non-negligible probability. We will then construct a probabilistic polynomial time algorithm, M_2, using Adv_2 as a black-box to break the intractability assumption of the discrete logarithm.

Let (p, g, q, y^*) be a problem given by G_{DL}, and $y^* = g^{x^*}$. M_2 runs G_i to output $(y_0, (v_1, W_1), \ldots, (v_i, W_i))$ along with $(x_0, x_1, \ldots, x_i)$, where $i \in \{0, \ldots, n\}$. M_2 then replaces v_i by y^*, and inputs $((y_0, (v_1, W_1), \ldots, (v_{i-1}, W_{i-1}), (y^*, W_i))$ to Adv_2. From the assumption, there exists M such that $M_1^{Adv_2}$ can output $(x_i^*, x_{i+1}, \ldots, x_n)$ with overwhelming probability. Here $x_i^* = c^* x^* + x_{i-1} \bmod q$, and $c^* = \mathcal{H}(v_1 \| \ldots v_{i-1} \| y^* \| W_1 \ldots \| W_j)$. Finally M_2 calculates $x^* := (x_i - x_{i-1})/c^* \bmod q$.

This contradicts the intractability assumption of the discrete logarithm. $\square$

Corollary 1. Let $i = O(1)$. If an adversary, Adv_3, given $(y_0, (v_1, W_1), \ldots, (v_i, W_i))$, can calculate A_i's secret key x_i with non-negligible probability (over the coin flips of G_i, Adv_3, actively attacked A_l and random oracle $\mathcal{H}$) then there exists a polynomial time algorithm (extractor M_3) that, using Adv_3 as a black-box, $M_3^{Adv_3}$, can output $(x_0, x_1, \ldots, x_{i-1})$ with overwhelming probability (over the coin flips of G_i, Adv_3, M_3 and $\mathcal{H}$).

Proof (Sketch): Since A_i's secret key has the same form of A_{i-1}'s signature, we can apply the proof technique of Theorem 1 to this proof. When applying Theorem 1 with i and n, set $i \leftarrow 0$ and $n \leftarrow i - 1$. □

We then obtain the following corollary immediately from the previous corollaries.

Corollary 2. Suppose the discrete logarithm problem is intractable. Let $i = O(1)$.

For all $i = 0, \ldots, n$, and all $n = 0, 1, \ldots, O(1)$, an adversary, Adv_4, given $(y_0, (v_1, W_1), \ldots, (v_i, W_i))$, cannot generate A_i's secret key x_i with non-negligible probability (over the coin flips of $\mathcal{G}_i$, Adv_3, actively attacked A_l and random oracle $\mathcal{H}$), unless Adv collapses A_j and obtains x_j for $j \in \{0, 1, \ldots, i\}$.

4.2 Efficiency

We can employ a standard technique to reduce the computation and communication overheads. Instead of sending v_s, the signing agent sends $c_{s,n}$. The verification predicate in this case is as follows.

S'-2: For $j = 1, \ldots, n$, let $c_j := \mathcal{H}(v_1\|\ldots\|v_j\|W_1\|\ldots\|W_j)$. Check if

$$c_{s,n} \stackrel{?}{=} \mathcal{H}(v_1\|\ldots\|v_n\|(g^{\sigma_n}v_n^{-c_n}\cdots v_1^{-c_1}y_0^{-1})^{1/c_{s,n}}\|m) \pmod{q}$$

holds.

This saves one exponentiation in Z_p, and transmission of $|v_s| - |c_s|$ bits compared to **S-2**.

To discuss efficiency further, we briefly describe a certificate-based scheme that uses Schnorr signatures for both certification (delegation) and signing. Each agent A_i has his own public key y_i which is certified by A_{i-1} with A_{i-1}'s signature $(\sigma_i, c_i, W_i, y_{i-1})$ that satisfies $c_i = \mathcal{H}(g^{\sigma_i}y_{i-1}^{c_i}\|W_i\|y_i)$.

Verifying an n-delegation chain (or equivalently, $n - 1$-delegation chain and a signature) that contains those certificates costs n modular exponentiations of $g^{\sigma_i}y_{i-1}^{c_i}$. Such double-base modular exponentiation can be computed at the cost of $|q| - 1$ modular squaring and approximately $\frac{3}{4}|q|$ modular multiplications in Z_p where $|q|$ is the bit length of q. On the other hand, we can compute the $n+2$-base exponentiation $(g^{\sigma}v_n^{-c_n}\cdots v_1^{-c_1}y_0^{-1})^{1/c_s}$ of step **S'-2** using $|q| - 1$ squaring and approximately $\frac{n+2}{2}|q| + n + 2$ multiplications. For instance, at $n = 6$, assuming the cost for squaring and multiplication are comparable, our scheme achieves about $5|q|$ while the certificate-based scheme needs $10.5|q|$.

Regarding communication complexity, we simply saved $n|q|$ bits due to the elimination of certificate transmission.

5 Applications

This section shows three other applications of our scheme.

Identity-based Signatures: As we mentioned in section 1, a delegation chain scheme that allows only one link can be used as an identity-based signature scheme. Suppose an authority has key pair (x_0, y_0) where $y_0 = g_0^x$. The public key of the authority is the tuple (y_0, g, p, q) and is distributed to users beforehand. The authority and a user execute the delegation protocol using the user's identity I instead of W_0. The user obtains (x_1, v_0) at the end. To sign document m, the user follows step **S-1** and sends triple (σ, c_1, v_0) to the verifier. Such a signature can be verified by checking the predicate in **S'-2** with obvious modification.

Threshold Delegation: When a user wants to give limited authority to an agent, the threshold scheme is useful to distribute the power among a group of agents. It only needs sharing of x_{i+1}, see Figure 1, by using a verifiable secret sharing scheme [13,7]. Each agent of the assigned group then individually checks the validity of the share in step **D-2**. The assignment verification predicate can be computed easily in collaboration. Modification of the protocols for a group to group and group to single agent is also possible. Construction is straightforward if interaction among all group members is allowed.

Use with Individual Keys: Since we assumed that a user must be fully responsible for what his agents do, our scheme allows the user do everything that his agents are permitted to do. Depending on the application, such potential framing by the users may not be acceptable. In such cases, we give A_i an individual public and private key pair, say (x_{ai}, y_{ai}), which is independently generated by the discrete logarithm instance generator. A_i then uses the individual key together with a given secret key. This idea was first used in the secret key certificate issuing protocol in [5].

The delegation protocol is modified to use two independent hash functions $\mathcal{H}_1$ and $\mathcal{H}_2$ as follows.

D'-1: A_i selects $r_{i+1}, t_{i+1} \in_U Z_q$. He then computes

$$v_{i+1} := g^{r_{i+1}},$$
$$u_{i+1} := g^{t_{i+1}},$$
$$c_{i+1} := \mathcal{H}_1(v_1\|\ldots\|v_{i+1}\|W_1\|\ldots\|W_{i+1}),$$
$$e_{i+1} := \mathcal{H}_2(v_1\|\ldots\|v_{i+1}\|W_1\|\ldots\|W_{i+1}) \quad \text{and}$$
$$x_{i+1} := c_{i+1}r_{i+1} + x_i + u_i + e_{i+1}x_{ai} \pmod{q}.$$

Then he sends $x_{i+1}, (v_{i+1}, u_{i+1}, W_{i+1}), \ldots, (v_2, u_2, W_2), (v_1, W_1)$ to A_1.

In the above modification, we assume that each W_i contains y_{ai}. The verification step and signing protocol are also modified accordingly. Although the use of two hash functions makes the verification predicate complicated, such modification is needed if the security is to be provable. With this form, we can construct a knowledge extractor for the individual keys, which implies that an adversary who is successful in forging a signature knows x_i and x_{ai} independently.

6 Conclusion

We have introduced "delegation chains" where a user's power of signing or authentication is delegated to a series of agents that work on behalf of the user. The model and its security were defined, and security against adaptive chosen message attacks was proved in the random oracle model. Our proof guarantees that the n-delegation chain is secure, i.e., unforgeable and endorsed, as long as n is up to $O(1)$.

Many signature schemes can be used for constructing a delegation chain scheme following our approach. Obvious variations are to use other forms of the Schnorr signature scheme, e.g., [12]. It is also easy to construct a scheme based on the Guillou-Quisquater signature scheme [9] by analogy with the Schnorr signature scheme. The proposed scheme and its variants can be used for authentication schemes as well.

To construct a delegation chain scheme that is provably secure with up to $O(\log k)$ or more agents remains an open problem.

References

1. M. Bellare and P. Rogaway. Random oracles are practical: a paradigm for designing efficient protocols. In *First ACM Conference on Computer and Communication Security*, pages 62–73. Association for Computing Machinery, 1993.
2. M. Bellare and P. Rogaway. Optimal asymmetric encryption. In *Advances in Cryptology — EUROCRYPT '94*, LNCS 950, pages 92–111. Springer-Verlag, 1995.
3. T. Beth. Efficient zero-knowledge identification scheme for smart cards. In C. G. Günther, editor, *Advances in Cryptology — EUROCRYPT '88*, volume 330 of *Lecture Notes in Computer Science*, pages 77–86. Springer-Verlag, 1988.
4. M. Blaze, G. Bleumer, and M. Strauss. Divertible protocols and atomic proxy cryptography. In *Advances in Cryptology — EUROCRYPT '98*, LNCS 1403, pages 127–144. Springer-Verlag, 1998.
5. S. Brands. Secret-key certificates. Technical Report CS-R9510, CWI, 1995.
6. M. Ellison, B. Frantz, B. Lampson, R. Rivest, B. M. Thomas, and T. Ylonen. Simple public key certificate. Internet-Draft, 1998.
7. P. Feldman. A practical scheme for non-interactive verifiable secret sharing. 28th FOCS, pages 427–437, 1987.
8. A. Fiat and A. Shamir. How to prove yourself: Practical solutions to identification and signature problems. In *Advances in Cryptology — CRYPTO '86*, LNCS 263, pages 186–199. Springer-Verlag, 1986.
9. L. C. Guillow and J.-J. Quisquater. A practical zero-knowledge protocol fitted to security microprocessor minimizing both transmission and memory. In *Advances in Cryptology — EUROCRYPT '88*, LNCS 330, pages 123–128. Springer-Verlag, 1988.
10. M. Mambo, K. Usuda, and E. Okamoto. Proxy signatures: delegation of the power to sign messages. *IEICE Transaction of Fundamentals of electronic Communications and Computer Science*, 9(E79-A):1338–1354, 1996.
11. K. Ohta and T. Okamoto. On concrete security treatment of signatures derived from identification. In *Advances in Cryptology — CRYPTO '98*, LNCS 1462, pages 354–369. Springer-Verlag, 1998.

12. T. Okamoto. Provably secure and practical identification schemes and corresponding signature schemes. In *Advances in Cryptology — CRYPTO '92*, LNCS 740, pages 31–53. Springer-Verlag, 1993.
13. T. P. Pedersen. Non-interactive and information-theoretic secure verifiable secret sharing. In *Advances in Cryptology — CRYPTO '91*, LNCS 576, pages 129–140. Springer-Verlag, 1992.
14. C. P. Schnorr. Efficient signature generation for smart cards. *Journal of Cryptology*, 4(3):239–252, 1991.
15. A. Shamir. Identity-based cryptosystems and signature schemes. In *Advances in Cryptology — CRYPTO '84*, LNCS 196, pages 47–53. Springer-Verlag, 1985.

Optimal Construction of Unconditionally Secure ID-Based Key Sharing Scheme for Large-Scale Networks[*]

Goichiro Hanaoka[1][**], Tsuyoshi Nishioka[2], Yuliang Zheng[3] and Hideki Imai[1]

[1] The 3rd Department, Institute of Industrial Science, the University of Tokyo
7-22-1 Roppongi, Minato-ku, Tokyo 106-8558, JAPAN
Phone & Fax: +81-3-3402-7365
E-Mail:hanaoka@imailab.iis.u-tokyo.ac.jp
imai@iis.u-tokyo.ac.jp
[2] Information Technology R&D Center, Mitsubishi Electric Corporation
5-1-1 Ofuna, Kamakura, 247-8501, JAPAN
Phone: +81-467-41-2181 & Fax: +81-467-41-2185
E-Mail:nishioka@iss.isl.melco.co.jp
[3] The Peninsula School of Computing and Information Technology
Monash University, McMahons Road, Frankston
Melbourne, VIC 3199, Australia
Email: yzheng@fcit.monash.edu.au
URL: http://www-pscit.fcit.monash.edu.au/~yuliang/
Phone: +61 3 9904 4196, Fax: +61 3 9904 4124

Abstract. Efficient ID-based key sharing schemes are desired world-widely for secure communications on the Internet and other networks. The Key Predistiribution Systems (KPS) are a large class of such key sharing schemes. The remarkable property of KPS is that in order to share the key, a participant should only input its partner's identifier to the secret KPS-algorithm. Although KPS has a lot of advantages in terms of efficiency, to achieve unconditional security a large amount of memory is required. While conventional KPS establishes communication links between any pair of entities in a communication system, in many practical communication systems such as broadcasting, not all links are required. In this article, we show the optimal method to remove the unnecessary communication links. In our scheme, the required memory for each entity is just proportional to the number of its partners, while that in conventional KPS it is proportional to the number of entities in the whole system. E.g., if an entity communicates only with $1/r$ of others, the required memory is reduced to $1/r$ of conventional KPS. Furthermore, it is proven that this memory size is optimal. Our scheme provides a more efficient way for secure communication especially in large-scale networks.

[*] A part of this work was performed in part of Research for the Future Program (RFTF) supported by Japan Society for the Promotion of Science (JSPS) under contact no. JSPS-RETF 96P00604.
[**] A Research Fellow of JSPS

1 Introduction

For information security, ID-based key distribution technologies are very important. The concept of ID-based key cryptosystems was originally proposed by Shamir[3, 4]. Maurer and Yacobi presents an ID-based key distribution scheme following Shamir's concept [5, 6]. However, their scheme requires a huge computational power. Okamoto and Tanaka[7] also proposed a key-distribution scheme based on a user's identifier, but it requires prior communications between a sender and a receiver to share the employed key. Although Tsujii and others proposed several ID-based key-distribution schemes[8, 9], almost all of them have been broken[10]. Thus, the performance of these schemes is unsatisfactory. However, Blom's ID-based key-distribution scheme[2], which is generalized by Matsumoto and Imai[1], has very good properties in terms of computational complexity and non-interactivity. Many useful schemes based on Blom's scheme have been proposed[1, 11, 12, 13, 14, 15, 16, 17], and they are called Key Predistribution Systems (KPS).

In a KPS, no previous communication is required and its key-distribution procedure consists of simple calculations. Furthermore in order to share the key, a participant should only input its partner's identifier to its secret KPS-algorithm. Blundo et al.[14, 15, 16], Kurosawa et al.[18] showed a lower bound of memory size of users' secret algorithms and developed KPS for a conference-key distribution. Moreover Fiat and Naor[17], Kurosawa et al.[19] applied a KPS for a broadcasting encryption system.

Although KPS has many desired properties, the following problem exists: When a number of users, which exceeds a certain threshold, cooperate they can calculate the central authority's secret information. Thus, to achieve perfect security the collusion threshold is determined to be larger than the number of entities in the network. Setting up such a high collusion threshold in this scheme requires large amounts of memory in the center as well as for the users. Solving this problem will make KPS much more attractive for ID-based key-distribution.

Although KPS provides common keys for all possible communication links among entities, in practical communication systems most of them are not necessary. By removing such unnecessary communication links, we can significantly reduce the required memory[4]. In our scheme, the required memory for each entity is proportional to the number of its partners, while that in conventional KPS it is proportional to the number of entities in the whole system. E.g., if an entity communicates only with $1/r$ of others, the required memory is reduced for a factor $1/r$ in comparison with conventional KPS. Furthermore, this memory size is proven optimal. In this work, we also propose an optimal *asymmetric t-conference key distribution scheme*. Since this scheme has good properties, it is considered to be utilized effectively in other applications.

Section 2 gives a brief review of the KPS. Then, in section 3, straight-forward implementation of our scheme and the corresponding problem are described.

---

[4] We have shown an implementation of this concept for broadcasting in [20]. Our new scheme in this paper can be regarded as a generalized version of the above scheme.

Section 4 explains an asymmetric t-conference key scheme as the solution of straight-forward implementation problem. Afterwards, an optimal construction based on asymmetric t-conference key distribution scheme is described in section 5. This is followed by the evaluation and discussion of the security of our scheme in section 6. Finally, section 7 closes the paper with some concluding remarks.

2 A brief overview of KPS

A KPS consists of two kinds of entities: One entity is the KPS center, the other entities are the users who want to share a common key. The KPS center possesses the KPS-center algorithm by which it can generate an individual secret algorithm for each user. These individual algorithms are (pre-) distributed by the center to their users and allow each user to calculate a common key from the ID of his communication partner. This section explains how the users' secret KPS-algorithms are generated and how users share a common key.

Let a symmetric function $G(x, y)$ be the KPS-center algorithm. Then, each entity u_i $(i = 1, 2, \cdots, N)$ is given the secret algorithm $U_{u_i}(x)(= G(x, u_i))$ $(i = 1, 2, \cdots, N)$, respectively. In order to share the communication key between u_i and u_j, they should simply input u_j and u_i to their secret algorithms, respectively. Since $G(x, y)$ is a symmetric function, they both obtain $k_{u_i u_j} = U_{u_i}(u_j) = U_{u_j}(u_i) = G(u_i, u_j)$.

KPS has three noteworthy properties. First, there is no need to send messages for the key distribution between entities who want to establish a cryptographic communication channel. Second, its key-distribution procedure consists of simple calculations so that its computational costs are quite small. Finally, in order to share the key, a participant has only to input its partner's identifier to its secret algorithm. Thus, KPS is well applicable to one-pass or quick-response transactions, e.g. mail systems, broadcasting systems, electronic toll collection systems, and so on.

However, KPS has a certain collusion threshold; when more users cooperate they can calculate the KPS-center algorithm $G(x, y)$. Thus, to achieve perfect security the collusion threshold should be determined to be larger than the number of entities in the network. Accordingly, the required memory for users' secret algorithms are increased due to the collusion threshold. In the following subsection, the relationship between the memory size and the collusion threshold is discussed in more detail.

2.1 A lower bound of $|U_{u_i}(x)|$

For a random variable X, $H(X)$ denotes the entropy of X. Generally,

$$0 \leq H(X) \leq \log_2 |X|, \text{ where } X = \{x \mid \Pr(X = x) > 0\}. \tag{1}$$

In particular, $H(X) = \log_2 |X|$ iff X is uniformly distributed.

Blundo et al.[14] showed the lower bound of required memory size for users. Suppose that for each $P \subseteq \{u_1, u_2, \cdots, u_N\}$ such that $|P| = t$, there is a key k_P associated with P. Each user $u_i \in P$ can compute k_P, and if $F \subseteq \{u_1, u_2, \cdots, u_N\}, |F| \leq \omega$ and $|F \cap P| = 0$, then F cannot obtain any information about k_P. Accordingly, the lower bound of the amount of users' secret algorithm U_{u_i} is estimated as follows:

$$\log_2 |U_{u_i}| \geq \binom{t + \omega - 1}{t - 1} H(K), \tag{2}$$

where $k_P \in K$ for any P. In order to achieve perfect security, $\omega + t$ must be equal to the number of entities in the whole network. For $t = 2$, Eq. 2 becomes $\log_2 |U_{u_i}| \geq (\omega + 1)H(K)$.

2.2 Optimal schemes

Blundo et al.[14] presented a KPS which achieves the optimal memory size. In this scheme, the center chooses a random symmetric polynomial in t variables over $GF(q)$ in which the degree of any variable is at most ω, that is, a polynomial

$$f(x_1, \cdots, x_t) = \sum_{i_1=0}^{\omega} \cdots \sum_{i_t=0}^{\omega} a_{i_1 \cdots i_t} x_1^{i_1} \cdots x_t^{i_t}, \tag{3}$$

where $a_{i_1 \cdots i_t} = a_{\sigma(i_1 \cdots i_t)}$ for any permutation σ on $(i_1, \cdots, i_t)$. The center computes $U_{u_i} = f(u_i, x_2, \cdots, x_t)$ and gives U_{u_i} $(i = 1, \cdots, N)$ to u_i $(i = 1, \cdots, N)$, respectively. Then, they can share their communication keys by inputting $t - 1$ partners' identifiers. For $t = 2$, Blom's scheme[2], Matsumoto-Imai scheme[1] and some others are also known as being optimal schemes. Although these schemes achieve the optimal memory size: $\log_2 |U_{u_i}| = (\omega + 1)H(K)$, the amount of memory is still large (For perfect security, ω must be equal to $N - 2$). Especially, in large-scale networks required memory size is enlarged according to the high collusion threshold. Furthermore, on a smart card since its size of storage is strictly limited, the collusion threshold cannot be set up high enough to avoid strong collusion attacks by huge number of entities. "KPSL1 card"[21, 22], where the key length is 64bits. The secret-algorithm itself then consumes 64-KBytes of memory size in each IC card. Therefore KPS was considered to be somewhat expensive for real IC card systems at that time. By introducing 128~256bits symmetric key cryptosystems (namely, $H(K) = 128 \sim 256$bits), this problem will be more serious.

3 Straight-forward method for removing unnecessary functions

As already mentioned, some KPSs are proven to be optimal, and it is impossible to reduce the required memory size providing all the communication links. However, the required memory size of these schemes are still high. Although a possibility to reduce the required memory is to reduce the collusion threshold or the key length, the security is also reduced considerably. So, we pay attention to the unnecessary communication links in networks. In order to reduce the memory size maintaining the same security level. In this section, we show a basic concept to attain the specified goal and we consider its requirements.

3.1 Unnecessary communication links

In a large-scale network, there are many pairs of entities that do not communicate with each other at all. As an illustration we point out the following two reasons. Firstly, to avoid the illegal access, some users are not allowed to access specific resources. Secondly, there exist many resources which perform only to some specific client computers. Also, pairs of entities, which are not related with each other, do not need to communicate with each other.

Although there are many unnecessary communication links, conventional KPSs cannot deal with them efficiently. Namely, in conventional KPSs it seems to be impossible to remove only unnecessary communication links.

3.2 Straight-forward implementation and its problem

A possible solution to reduce memory size by removing unnecessary communication links is to construct a whole large network by using small KPSs. Namely, if small KPSs are provided only for necessary communication links, then the unnecessary ones can be removed. However, straight-forward implementation of this approach has a serious problem. The required memory can be even more than that of the conventional KPS. We explain this problem in more details below.

To construct a large network by using small KPSs, we first divide the set of entities Υ to N_Υ subsets $\{\Upsilon_1, \Upsilon_2, \cdots, \Upsilon_{N_\Upsilon}\}$ ($|\Upsilon_i \cap \Upsilon_j| = 0$). Each Υ_i ($i = 1, \cdots, N_\Upsilon$) fullfills following condition: all $v_i \in \Upsilon_i$ communicates with all $\bar{v}_i \in \bar{\Upsilon}_i$, where $\bar{\Upsilon}_i = \{\Upsilon_j \mid j \in P_i\}$, $P_i \subseteq \{1, 2, \cdots, N_\Upsilon\}$, and for $i_1, i_2 \in \{1, 2, \cdots, N_\Upsilon\}$, if $i_1 \in P_{i_2}$, then $i_2 \in P_{i_1}$. For convenience, if $i_1 \in P_{i_2}(i_2 \in P_{i_1})$, we say $< i_1, i_2 >= 1$, otherwise, $< i_1, i_2 >= 0$.

Then, if a whole network is constructed by small KPSs straight-forwardly, the critical problems will appear in following two situations:

Case1: $< i_1, i_2 >= 1$, $< i_1, i_1 >= 0$ $(i_1 \neq i_2)$
For the communication $< i_1, i_2 >= 1$, a small KPS is provided. The collusion threshold of this KPS is determined to be equal to $|\Upsilon_{i_1}| + |\Upsilon_{i_2}| - 2$. From Eq.2, the required memory size for this communication is also proportional to $|\Upsilon_{i_1}| + |\Upsilon_{i_2}| - 1$. However, since $< i_1, i_1 >= 0$, even $v_{i_1} \in \Upsilon_{i_1}$ communicates only with $|\Upsilon_{i_2}|$ entities using the above memory.

Case2: $< i_1, i_2 >= 1$, $< i_1, i_3 >= 1$, $< i_2, i_3 >= 0$ $(i_1 \neq i_2, i_1 \neq i_3, i_2 \neq i_3)$
For these communications, we consider 2 kinds of construction of small KPSs: a KPS for $\{\Upsilon_{i_1}, \Upsilon_{i_2}\}$ and a KPS for $\{\Upsilon_{i_1}, \Upsilon_{i_3}\}$ are set up, or a KPS for $\{\Upsilon_{i_1}, \Upsilon_{i_2}, \Upsilon_{i_3}\}$ is set up. For the first construction, the collusion threshold of 2 KPSs are $|\Upsilon_{i_1}| + |\Upsilon_{i_2}| - 2$ and $|\Upsilon_{i_1}| + |\Upsilon_{i_3}| - 2$, respectively. Thus, the

required memory size of $v_{i_1} \in \Upsilon_{i_1}$ for these communications is proportional to $2|\Upsilon_{i_1}| + |\Upsilon_{i_2}| + |\Upsilon_{i_3}| - 2$. On the other hand, for the second construction the collusion threshold of the KPS is $|\Upsilon_{i_1}| + |\Upsilon_{i_2}| + |\Upsilon_{i_3}| - 2$. Hence, the amount of memory of $v_{i_1} \in \Upsilon_{i_1}$ is proportional to $|\Upsilon_{i_1}| + |\Upsilon_{i_2}| + |\Upsilon_{i_3}| - 1$. Further, the required memory size for $v_{i_2} \in \Upsilon_{i_2}$ and $v_{i_3} \in \Upsilon_{i_3}$ are also proportional to $|\Upsilon_{i_1}| + |\Upsilon_{i_2}| + |\Upsilon_{i_3}| - 1$. Anyway, in both constructions since $v_{i_1} \in \Upsilon_{i_1}$ communicate only with $|\Upsilon_{i_2}| + |\Upsilon_{i_3}|$ entities by these KPS(s), the required memory size for v_{i_1} is large comparing with the amount of memory. Moreover, in the second construction $v_{i_2} \in \Upsilon_{i_2}$ and $v_{i_3} \in \Upsilon_{i_3}$ communicate only with $|\Upsilon_{i_1}|$ entities. Hence, their required memory size is also large.

In the worst case, the required memory size for an entity is almost 2 times of that in conventional KPS. Namely, if $< i_1, i > = 1$ $(i = 1, \cdots, N_\Upsilon, \ i \neq i_1)$ and $< j, k > = 0$ $(j, k = 1, \cdots, N_\Upsilon, \ j, k \neq i_1)$, then, the required memory size is $\left(\sum_{i \in \{1,2,\cdots,N_\Upsilon\}, i \neq i_1} (|\Upsilon_i| + |\Upsilon_{i_1}| - 1) \right) H(K)$. On the other hand, in conventional KPS the required memory size is $\left(\sum_{i \in \{1,2,\cdots,N_\Upsilon\}, i \neq i_1} |\Upsilon_i| - 1 \right) H(K)$. Hence, straight-forward implementation of constructing a whole network by small KPSs is inefficient. Note that **Case1** and **Case2** can not be considered as rare. These cases cannot occur iff Υ can be divided to hold the following condition:

$$< i, j > = \begin{cases} 0 & (i \neq j) \\ 1 & (i = j) \end{cases} \qquad \forall i, \forall j \in \{1, 2, \cdots, N_\Upsilon\}.$$

4 Optimal primitive

In order to construct a large-scale network by small KPSs efficiently, better primitives for **Case1** and **Case2** are required than the normal KPSs. In this section, the required property for the security primitive for **Case1** and **Case2** is discussed. Afterwards, we show an example which fulfills the requirements for **Case1** and **Case2**.

4.1 Generalization of Case1 and Case2

As already mentioned, **Case1** and **Case2** increase the memory size for entities. This problem is generalized by following **Lemma 1**.

Lemma 1: *The required memory size for entities can be optimal if a whole network is constructed by normal KPSs and other primitives whose memory size for $< i_1, i_2 > = 1, < i_1, i_1 > = 0, < i_2, i_2 > = 0$ is optimal.*

Proof: When $< i_1, i_2 > = 1$ $(i_1 \neq i_2)$ is realized by a KPS, $< i_1, i_1 > = 1$ and $< i_2, i_2 > = 1$ are always realized simultaneously even if they are not desired. In both **Case1** and **Case2**, such undesired functions bring the inefficiency. Thus, if an optimal primitive for $< i_1, i_2 > = 1, < i_1, i_1 > = 0, < i_2, i_2 > = 0$ is provided, **Case1** and **Case2** are dealt with optimally. If a function for communication

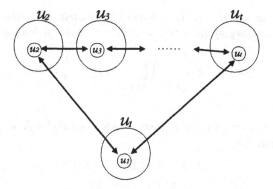

Fig. 1. An asymmetric t-conference key distribution.

$< i_1, i_1 >= 1 (\text{or} < i_2, i_2 >= 1)$ is required, a normal KPS is added as the optimal primitive for it. Hence, currently use of normal KPSs and optimal primitives for $< i_1, i_2 >= 1, < i_1, i_1 >= 0, < i_2, i_2 >= 0$ realizes optimal memory size for constructing large-scale networks by small key-sharing systems. □

4.2 Optimal primitive for $< i_1, i_2 >= 1, < i_1, i_1 >= 0, < i_2, i_2 >= 0$ ($i_1 \neq i_2$)

In this subsection, a lower bound of the memory size of a security primitive for $< i_1, i_2 >= 1, < i_1, i_1 >= 0, < i_2, i_2 >= 0$ ($i_1 \neq i_2$) is shown. For other applications, we further generalize the primitive and call it *asymmetric t-conference key distribution*. Asymmetric t-conference key distribution is defined as follows:

Definition: Let $\mathcal{U}$ be a set of entities and $\mathcal{U}$ is divided to t subsets $\mathcal{U} = \{\mathcal{U}_1, \mathcal{U}_2, \cdots, \mathcal{U}_t\}$. A key-sharing scheme for $\mathcal{U}$ is called asymmetric t-conference key distribution if

1. *$u_1 \in \mathcal{U}_1, u_2 \in \mathcal{U}_2, \cdots, u_t \in \mathcal{U}_t$ can compute their common key among them non-interactively (note that common keys among a same subset are not required).*

2. *Collusion thresholds $\psi_1, \psi_2, \cdots, \psi_t$ are independently set up for each of $\mathcal{U}_1, \mathcal{U}_2, \cdots, \mathcal{U}_t$. And unless a group of colluders $\mathcal{F}_i \in \mathcal{U}_i$ holds $|\mathcal{F}_i| > \psi_i$, any information of a common key is not exposed to entities who should not have it.*

Figure 1 illustrates an example of an asymmetric t-conference key distribution. Note that a security primitive for $< i_1, i_2 >= 1, < i_1, i_1 >= 0, < i_2, i_2 >= 0$ ($i_1 \neq i_2$) is an asymmetric 2-conference key distribution.

Lemma 2: *In an asymmetric t-conference key distribution system, for $u_i \in \mathcal{U}_i$ the amount of memory of the secret algorithm U_i holds following lower bound:*

$$\log_2 |U_i| \geq \left(\prod_{j \in \{1, \cdots, t\}, j \neq i} (\psi_j + 1) \right) H(K). \tag{4}$$

Proof: The mutual information between random valuables X and Y fulfills following two equations:

$$I(X;Y) = H(X) - H(X|Y), \tag{5}$$

$$I(X;Y) = I(Y;X). \tag{6}$$

From Eq.5 and Eq.6, following equation is obtained:

$$H(X) = H(Y) - H(Y|X) + H(X|Y). \tag{7}$$

Here, let $\mathcal{K}_{u_i}$ be the set of all u_i's common keys which are shared with all $u_1' \in \mathcal{U}_1' \subseteq \mathcal{U}_1$, all $u_2' \in \mathcal{U}_2' \subseteq \mathcal{U}_2, \cdots,$ all $u_t' \in \mathcal{U}_t' \subseteq \mathcal{U}_t$ (excluding all $u_i' \in \mathcal{U}_i' \subseteq \mathcal{U}_i$) by asymmetric t-conference key distribution, where each $\mathcal{U}_j'$ $(j = 1, \cdots, t)$ holds $|\mathcal{U}_j'| = \psi_j$.

From Eq.7, the entropy of U_i is described as follows:

$$H(U_i) = H(\mathcal{K}_{u_i}) - H(\mathcal{K}_{u_i}|U_i) + H(U_i|\mathcal{K}_{u_i}). \tag{8}$$

Since in an asymmetric t-conference key distribution scheme, all of $\mathcal{K}_{u_i}$ can be computed by using U_i,

$$H(\mathcal{K}_{u_i}|U_i) = 0. \tag{9}$$

Then, from Eq.8 and Eq.9 the following inequality is obtained:

$$H(U_{v_i}) = H(\mathcal{K}_{u_i}) + H(U_i|\mathcal{K}_{u_i}) \geq H(\mathcal{K}_{u_i}). \tag{10}$$

In a secure asymmetric t-conference key distribution system, mutual information between any pair of keys which belong to $\mathcal{K}_{u_i}$ must be 0. Thus, for $H(\mathcal{K}_{u_i})$ we have

$$H(\mathcal{K}_{u_i}) = \sum_{K \in \mathcal{K}_{u_i}} H(K) = \left(\prod_{j \in \{1, \cdots, t\}, j \neq i} (\psi_j + 1) \right) H(K). \tag{11}$$

From Eq.10 and Eq.11, we obtain

$$H(U_i) \geq \left(\prod_{j \in \{1, \cdots, t\}, j \neq i} (\psi_j + 1) \right) H(K). \tag{12}$$

Hence, from Eq.1 Eq.12 becomes Eq.4. □

For unconditional security, each ψ_i $(i = 1, \cdots, t)$ must be equal to $|\mathcal{U}_i| - 1$. By introducing this collusion threshold, Eq.4 becomes

$$\log_2 |U_i| \geq \left(\prod_{j \in \{1, \cdots, t\}, j \neq i} |\mathcal{U}_j| \right) H(K). \tag{13}$$

4.3 An example of optimal asymmetric t-conference key distribution schemes

In this subsection an optimal asymmetric t-conference key distribution scheme is shown. In this scheme, the symmetric polynomial which is the KPS-center algorithm in Blundo et al.'s scheme is replaced with an asymmetric polynomial in t variables $x_1, x_2, \cdots, x_t$ over $GF(q)$ in which the degree of each variable is $\psi_1, \psi_2, \cdots, \psi_t$, respectively, that is, a polynomial

$$f(x_1, \cdots, x_t) = \sum_{i_1=0}^{\psi_1} \cdots \sum_{i_t=0}^{\psi_t} a_{i_1 \cdots i_t} x_1^{i_1} \cdots x_t^{i_t}. \tag{14}$$

Note that $a_{i_1 \cdots i_t}$ is not necessary to be equal to $a_{\sigma(i_1 \cdots i_t)}$ for any permutation σ on $(i_1, \cdots, i_t)$. The center computes $U_i = f(x_1, x_2, \cdots, x_t)|_{x_i=u_i}$ and gives each U_i to u_i, respectively. When u_i communicates with $u_1 \in \mathcal{U}_1, u_2 \in \mathcal{U}_2, \cdots, u_t \in \mathcal{U}_t$ (excluding $u_i \in \mathcal{U}_i$), u_i computes their communication keys by $U_i|_{x_1=u_1, x_2=u_2, \cdots, x_t=u_t}$ (excluding $x_i=u_i$). By this procedure, an asymmetric t-conference key distribution is exactly realized.

In this scheme, the required memory size for U_i is estimated as follows:

$$\log_2 |U_i| = \left(\prod_{j \in \{1, \cdots, t\}, j \neq i} (\psi_j + 1) \right) H(K). \tag{15}$$

Thus, this scheme is optimal since we have already shown the lower bound on $|U_i|$ by Eq.4. For $t = 2$, we then obtain an optimal security primitive for $< i_1, i_2 >= 1, < i_1, i_1 >= 0, < i_2, i_2 >= 0$ ($i_1 \neq i_2$). When we apply this, the required memory for $v_{i_1} \in \Upsilon_{i_1}$ is $|\Upsilon_{i_2}| H(K)$, which is much less than that of normal KPSs as shown in 3.2.

5 Optimal construction by normal KPSs and asymmetric 2-conference key distribution schemes

As already mentioned in 4.1, if KPSs and asymmetric 2-conference key distribution schemes are applied, the required memory size can be optimal. In this section, we show an optimal construction of a large-scale network by removing unnecessary communication links.

Procedure of the center: The center provides center algorithms of normal KPSs and asymmetric t-conference key distribution systems. For each $< i, i >= 1$ ($i = 1, 2, \cdots, N_\Upsilon$), a normal KPS is applied. And for each $< i, j >= 1$ ($i \neq j$), an optimal asymmetric 2-conference key distribution scheme is applied. $G^i(x, y)$ and $G^{ij}(x, y)$ ($i, j \in \{1, 2, \cdots, N_\Upsilon\}$, $i \neq j$) denotes the center algorithms for normal KPSs and asymmetric t-conference key distribution systems, respectively

$(G^i(x,y)$ and $G^{ij}(x,y)$ holds $G^i(x,y) = G^i(y,x)$ and $G^{ij}(x,y) = G^{ji}(y,x)$, respectively). Then, the center gives the following secret algorithm U_{v_i} to $v_i \in \Upsilon_i$:

$$U_{v_i} = \{U_{v_i}^{ij} | \text{For} <i,j> \geq 1, U_{v_i}^{ij}(y) = G^i(v_i,y) \ (i = j), U_{v_i}^{ij}(y) = G^{ij}(v_i,y) \ (i \neq j)\} \tag{16}$$

Since applied KPSs and asymmetric 2-conference key distribution schemes are optimal, required memory size for each small key sharing system is estimated as follows:

$$\log_2 |U_{v_i}^{ij}(y)| = \begin{cases} (|\Upsilon_j| - 1)H(K) & (i = j : \text{ KPS}) \\ |\Upsilon_j|H(K) & (i \neq j : \text{ asymmetric 2-conference key distribution}). \end{cases} \tag{17}$$

As an optimal KPS, Blundo et al.'s scheme, Matsumoto-Imai scheme and Blom's scheme are available. On the other hand, as an optimal asymmetric 2-conference key distribution scheme, our scheme presented in 4.3 is available.

Procedure of the entities: v_i computes the common key with $v_j \in \Upsilon_j$ as follows:

$$\begin{aligned} v_i &: k_{v_i,v_j} = U_{v_i}^{ij}(v_j), \\ v_j &: k_{v_i,v_j} = U_{v_i}^{ij}(v_j) = U_{v_j}^{ji}(v_i). \end{aligned} \tag{18}$$

6 Evaluation

6.1 Memory size for entities

By our construction, the required memory size for $v_i \in \Upsilon_i$ is estimated as follows:

$$\log_2 |U_{v_i}| = \left(\sum_{j \in \{1,\cdots,N_T\}} (<i,j> |\Upsilon_j|) - <i,i> \right) H(K). \tag{19}$$

Here let $\log_2 |U'|$ be the required memory size for an entity when the a whole network is constructed only by one normal KPS. Then, we obtain following equation:

$$\log_2 |U_{v_i}| = \frac{\sum_{j \in \{1,\cdots,N_T\}} (<i,j> |\Upsilon_j|) - <i,i>}{\left(\sum_{j \in \{1,\cdots,N_T\}} |\Upsilon_j| \right) - 1} \log_2 |U'|. \tag{20}$$

Since $\sum_{j \in \{1,\cdots,N_T\}} (<i,j> |\Upsilon_j|) - <i,i>$ is equal to the number of partners of v_i, and $\sum_{j \in \{1,\cdots,N_T\}} |\Upsilon_j|$ is equal to the number of entities in the whole system, Eq.20 becomes

$$\log_2 |U_{v_i}| = \frac{\text{the number of partners}}{\text{the number of entities} - 1} \log_2 |U'|. \tag{21}$$

Namely, by using our scheme the memory size for an entity is reduced to be almost same as (# of partners)/(# of entities). Moreover, our scheme is optimal due to the discussion in Section 4.

6.2 Security

Our scheme is perfectly secure since any subset of entities have no information on a key they should not know. When we reduce the collusion threshold for reduction of memory size, the security becomes non-perfect. Namely, if the collusion threshold is less than the given one, by a collusion attack colluders can compute common keys of a victim. However, to success this attack a huge number of colluders is required and it seems still impossible in the real world.

7 Conclusion

In this paper, an optimal construction of ID-based key sharing scheme for large-scale networks is proposed. It has been pointed out that in order to achieve perfect security a huge amount of memory is required in conventional KPS, and it has been shown how KPS can be improved for practical communication systems. By removing communication links that are not required in a practical communication system. This approach yields that the amount of memory is reduced significantly. In our scheme, the required memory for each entity is (the number of partners) × (the length of a common key), while that in conventional KPS is (the number of entities −1) × (the length of a common key). So, if an entity communicates only with $1/r$ of others, the required memory is reduced almost $1/r$ of that required in conventional KPS. Furthermore, our scheme is proven to be optimal. The previous makes our scheme attractive for various applications like broadcasting or E-commerce in the Internet. In this paper, we also propose an optimal asymmetric t-conference key distribution scheme. Since this scheme has good properties, it could be utilized effectively in other applications. Additionally, since public-key cryptosystems have no advantages over KPS in terms of computational costs, ID-basedness, and so on, the efficient combination of a public-key cryptosystem and our scheme will yield a more efficient and secure communication system in comparison with single use of a public-key cryptosystem.

References

1. T. Matsumoto and H. Imai, "On the KEY PREDISTRIBUTION SYSTEM: A Practical Solution to the Key Distribution Problem," Proc. of CRYPTO'87, LNCS 293, Springer-Verlag, pp.185-193, 1987.
2. R. Blom, "Non-public Key Distribution," Proc. of CRYPTO'82, Plenum Press, pp.231-236, 1983.
3. A. Fiat and A. Shamir, "How to Prove Yourself: Practical Solutions to Identification and Signature Problems," Proc. of CRYPTO'86, LNCS 263, Springer-Verlag, pp.186-194, 1986
4. A. Shamir, "Identity-Based Cryptosystems and Signature Schemes," Proc. of CRYPTO'84, LNCS 196, Springer-Verlag, pp.47-53, 1985.
5. U. Maurer and Y. Yacobi, "Non-interactive Public-Key Cryptography," Proc. of Eurocrypt'91, LNCS 547, Springer-Verlag, pp.498-407, 1992.

6. U. Maurer and Y. Yacobi, "A Remark on a Non-interactive Public-Key Distribution System," Proc. of Eurocrypt'92, LNCS 658, Springer-Verlag, pp.458-460, 1993.

7. E. Okamoto and K. Tanaka, "Identity-Based Information Security management System for Personal Comuputer Networks," IEEE J. on Selected Areas in Commun., 7, 2, pp.290-294, 1989.

8. H. Tanaka, "A Realization Scheme of the Identity-Based Cryptosystems," Proc. of CRYPTO'87, LNCS 293, Springer-Verlag, pp.340-349, 1988.

9. S. Tsujii and J. Chao, "A New ID-Based Key Sharing System," Proc. of CRYPTO'91, LNCS 576, Springer-Verlag, pp.288-299, 1992.

10. D. Coppersmith, "Attack on the Cryptographica Scheme NIKS-TAS," Proc. of CRYPTO'94, LNCS 839, Springer-Varlag, pp.40-49, 1994.

11. L. Gong and D. J. Wheeler, "A Matrix Key-Distribution Scheme," Journal of Cryptology, vol. 2, pp.51-59, Springer-Verlag, 1993.

12. W. A. Jackson, K. M. Martin, and C. M. O'Keefe, "Multisecret Threshold Schemes, " Proc. of CRYPTO'93, LNCS 773, pp.126-135, Springer-Verlag, 1994.

13. Y. Desmedt and V. Viswanathan, "Unconditionally Secure Dynamic Conference Key Distribution," IEEE, ISIT'98, 1998.

14. C. Blundo, A. De Santis, A. Herzberg, S. Kutten, U. Vaccaro and M. Yung, "Perfectly Secure Key Distribution for Dynamic Conferences," Proc. of CRYPTO '92, LNCS 740, Springer-Verlag, pp.471-486, 1993.

15. C. Blundo, L.A. Frota Mattos and D.R. Stinson, "Trade-offs between Communication and Strage in Unconditionally Secure Schemes for Broadcast Encryption and Interactive Key Distribution," Proc. of CRYPTO '96, LNCS 1109, Springer-Verlag, pp.387-400, 1996.

16. C. Blundo and A. Cresti, "Space Requirements for Broadcast Encryption," Proc. of Eurocrypt '94, LNCS 950, Springer-Verlag, pp.287-298, 1995.

17. A. Fiat and M. Naor, "Broadcast Encryption," Proc. of CRYPTO '93, LNCS 773, Springer-Verlag, pp.480-491, 1994.

18. K. Kurosawa, K. Okada and H. Saido, "New Combimatorial Bounds for Authentication Codes and Key Predistribution Schemes," Designs, Codes and Cryptography, 15, pp.87-100, 1998.

19. K. Kurosawa, T. Yoshida, Y. Desmedt and M. Burmester, "Some Bounds and a Construction for Secure Broadcast Encryption," Proc. of ASIACRYPT '98, LNCS 1514, Springer-Verlag, pp.420-433, 1998.

20. G. Hanaoka, T. Nishioka, Y. Zheng and H. Imai, "An Efficient Hierarchical Identity-based Key-Sharing Method Resistant against Collusion-Attacks," Proc. of ASIACRYPT '99, Springer-Verlag, to appear.

21. T. Matsumoto, Y. Takashima, H. Imai, M. Sasaki, H. Yoshikawa, and S. Watanabe, "A Prototype KPS and Its Application - IC Card Based Key Sharing and Cryptographic Communication -," Trans. of IEICE Vol. E 73, No. 7, July 1990, pp.1111-1119, 1990.

22. T. Matsumoto, Y. Takashima, H. Imai, M. Sasaki, H. Yoshikawa, and S. Watanabe, "THE KPS CARD, IC Card for Cryptographic Communication Based on the Key Predistribution System," Proc. of SMART CARD 2000, IC Cards and Applications, Today and Tomorrow, Amsterdam, Oct., 1989.

Enhancing the Resistance of a Provably Secure Key Agreement Protocol to a Denial-of-Service Attack

Shouichi Hirose[1] and Kanta Matsuura[2]

[1] Graduate School of Informatics, Kyoto University, Kyoto, 606–8501 JAPAN
E-mail: hirose@i.kyoto-u.ac.jp
[2] Institute of Industrial Science, University of Tokyo, Tokyo, 106–8558 JAPAN
E-mail: kanta@imailab.iis.u-tokyo.ac.jp

Abstract. In this manuscript, two key agreement protocols which are resistant to a denial-of-service attack are constructed from a key agreement protocol in [9] provably secure against passive and active attacks. The denial-of-service attack considered is the resource-exhaustion attack on a responder. By the resource-exhaustion attack, a malicious initiator executes a key agreement protocol simultaneously as many times as possible to exhaust the responder's resources and to disturb executions of it between honest initiators and the responder. The resources are the storage and the CPU. The proposed protocols are the first protocols resistant to both the storage-exhaustion attack and the CPU-exhaustion attack. The techniques used in the construction are stateless connection, weak key confirmation, and enforcement of heavy computation. The stateless connection is effective to enhancing the resistance to the storage-exhaustion attack. The weak key confirmation and the enforcement of heavy computation are effective to enhancing the resistance to the CPU-exhaustion attack.

1 Introduction

In using private key encryption for secure communication, it is necessary for the participants to share a common key securely in advance with some key agreement scheme. Key agreement schemes can be classified into two types: non-interactive schemes and interactive schemes. In this manuscript, the interactive schemes are discussed, which are called the key agreement protocols (KAP's).

A three-pass KAP was proposed in [9] which is based on the Diffie-Hellman KAP [4]. This protocol is provably secure against passive and active attacks in the random oracle model [2] on the assumptions that the Diffie-Hellman problem is intractable and that the secret pieces of information of users are selected at random and independently of each other. This protocol also provides key confirmation, and the method for providing it is different from those of the other protocols such as in [5, 10, 3, 12]. This protocol provides key confirmation without using the shared key explicitly. Each participant only shows the other participant that he can compute the shared key. This kind of key confirmation is called weak key confirmation.

In this manuscript, two KAP's which are resistant to a denial-of-service attack are constructed from the KAP in [9]. The denial-of-service (DoS) attack considered is a resource-exhaustion attack on a responder; a malicious initiator launches as many (bogus) requests as possible one after another without establishing connection with the target responder in order to exhaust the responder's resource and to disturb executions with honest initiators. The resources are the storage and the CPU. The amount of storage of a responder determines the upper limit of the number of initiators who are able to execute a KAP with the responder simultaneously. The computation load on the CPU of a responder determines the throughput. An anti-clogging token called Cookie [11] is an efficient way to avoid DoS with IP-address spoofing. But when it is used in public-key based authenticated protocols like corresponding modes of IKE (the Internet Key Exchange) [8], the responder must perform expensive public-key based operation *before* he/she becomes sure of the initiator's identity. By contrast, the proposed protocols are the first authenticated protocols that are resistant both to the storage-exhaustion attack and to the CPU-exhaustion attack.

The techniques used in our transformation are stateless connection [1], weak key confirmation, and enforcement of heavy computation. The stateless connection is effective to enhancing the resistance to the storage-exhaustion attack. It releases responders from keeping connection states, which is kept by initiators. Once released, the responders are confronted with a risk of replay-flooding attack. In a comparison of the replay-flooding attack against stateless protocols with the storage-exhaustion attack against stateful ones, however, the stateless protocols are shown to provide a better performance [1].

The weak key confirmation and the enforcement of heavy computation are effective to enhancing the resistance to the CPU-exhaustion attack. In the proposed KAP's, the most time-consuming computation is a modular exponentiation. The weak key confirmation enables a responder to compute the shared key, which needs a modular exponentiation, only if the received messages are valid.

The validity verification starts with a light step of hashing before main heavy computation on the responder's side. Since the light step can tell whether the initiator has really completed the exponentiation, a malicious initiator will fear a "falling-together" nightmare: he/she has to compute modular exponentiations in order to make the responder compute modular exponentiations for checking the validity of the received messages. The basic concept of this trick is "attackers must pay a lot for the attack". A *pricing function* by Dwork and Naor [6] is based on the same concept. The pricing function was, however, introduced for combatting junk mails in one-pass protocols *without* authentication mechanisms. On the other hand, we studied three-pass protocols equipped *with* authentication.

The proposed KAP's are still practical in computation although they require the participants to exchange the connection state and additional authenticators; they cost encryption/decryption of a private-key cipher and one-way hashing, which are computationally inexpensive in comparison with modular exponentiations, the main cost of the prototype KAP. In addition, depending on the operation mode, the required number of modular exponentiations in effect can be reduced by the use of simultaneous multiple exponentiation. The proposed KAP's also remain secure and one of them almost inherits the provable security, except that they become vulnerable to the replay attack.

This paper is constructed as follows. Section 2 first overviews the KAP in [9]. DoS-resistant KAP's are then proposed in Section 3, with a discussion on the resistance. Section 4 includes the performance analysis of the proposed protocols. Conclusion is given in Section 5.

2 The base key agreement protocol

In this section, the KAP in [9] is reviewed, which is the base of the DoS-resistant KAP's proposed in the next section. It is an authenticated KAP based on the Diffie-Hellman KAP. In this manuscript, it is called the base KAP.

Let p and q be large primes such that $q \mid p - 1$ and $g \in GF(p)$ of order q. Let $h : \{0,1\}^* \to Z_q$ be a collision-free hash function. p, q, g, h are public and shared among all of the users. Let $s_i \in Z_q$ be a secret piece of information of the user i and $v_i = g^{-s_i} \bmod p$ be a public piece of information of the user i. Let I_i be the ID of the user i. It is assumed that the users know the public pieces of information of the other users or that the public piece of information of each user is contained in his ID.

The base KAP

1. The initiator A randomly selects $k_A \in Z_q$ and computes $u_A = g^{-k_A} \bmod p$.
2. A sends u_A, I_A to B.
3. The responder B randomly selects $k_B \in Z_q$ and computes $u_B = g^{-k_B} \bmod p$. B also selects $r_B \in Z_q$ at random and computes $x_B = g^{r_B} \bmod p$. B computes $e_B = h(x_B, u_B, u_A)$ and $w_B = r_B + e_B k_B + e_B^2 s_B \bmod q$.
4. B sends u_B, e_B, w_B, I_B to A.
5. A computes $z_B = g^{w_B} u_B^{e_B} v_B^{e_B^2} \bmod p$ and checks if $e_B = h(z_B, u_B, u_A)$. If it does not hold, then A terminates the execution. Otherwise, A randomly selects $r_A \in Z_q$ and computes $x_A = g^{r_A} \bmod p$. Then, A computes $e_A = h(x_A, u_A, u_B)$ and $w_A = r_A + e_A k_A + e_A^2 s_A \bmod q$.
6. A sends e_A, w_A to B.
7. A computes $K_A = u_B^{-k_A} \bmod p$.

   B computes $z_A = g^{w_A} u_A^{e_A} v_A^{e_A^2} \bmod p$ and checks if $e_A = h(z_A, u_A, u_B)$. If it does not hold, then B terminates the execution. Otherwise, B computes $K_B = u_A^{-k_B} \bmod p$.

This protocol provides key confirmation without using the shared key explicitly. Each participant only shows the other participant that he can compute the shared key. This kind of key confirmation is called the weak key confirmation.

A signature scheme called the redundant signature scheme is used for providing the weak key confirmation. It is an extension of Schnorr's signature scheme [13]. For example, (u_B, e_B, w_B) is able to be regarded as a signature for u_A, which is used as a nonce. By checking the validity of (u_B, e_B, w_B), that is, by computing z_B and checking that $e_B = h(z_B, u_B, u_A)$ in Step 5, A is able to make sure that B received u_A, that B sent u_B in response to the receipt of u_A and that B knows k_B. Thus, A is able to make sure that B is able to compute K_B which is equal to K_A. B does not use K_B in computing (u_B, e_B, w_B).

This protocol is provably secure against passive and active attacks in the random oracle model [2] on the assumptions that the Diffie-Hellman problem is intractable and that the secret pieces of information of users are selected at random and independently of each other [9]. All of these attacks are assumed to be known-key attacks; it is provably secure even if all the previously shared keys are disclosed to the attacker. In addition, this protocol provides forward secrecy; the security against passive eavesdropping is proved on the assumption that the attacker knows the secret pieces of information of the participants.

3 Key agreement protocols resistant to a denial-of-service attack

3.1 Denial-of-service attack

The DoS attack considered is the resource-exhaustion attack on a responder by a malicious initiator. The resources are the storage and the CPU. KAP's usually require the storage which keeps connection states. The amount of storage of a responder determines the upper limit of the number of initiators who are able to execute a KAP with the responder simultaneously. On the other hand, the computation load on the CPU of a responder determines the throughput.

By the DoS attack, a malicious initiator tries to disturb executions of the KAP between honest initiators and a target responder. The malicious initiator launches as many (bogus) requests as possible one after another without establishing connections with the responder by using arbitrary ID's. In the worst case, the storage of the responder is exhausted and new requests are refused. Even if the storage is not exhausted, the throughput is degraded by the computation load on the responder. Thus, it is preferable that the amount of the storage and the computation load required by each execution are as small as possible.

The base KAP is vulnerable to the DoS attack. In the DoS attack against the base KAP, a malicious initiator X uses arbitrary ID I_X, and sends u_X, I_X to the target responder B. Then, X receives u_B, e_B, w_B, I_B from B and stops sending e_X, w_X or postpones it until just before a certain expiration time. X repeats this procedure as many times as possible. For each request of this attack, B has to keep (I_X, u_X, k_B, u_B) until the expiration or until he receives e_X, w_X. Moreover, he has to compute two modular exponentiations in Step 3, that is, $u_B = g^{-k_B} \bmod p$ and $x_B = g^{r_B} \bmod p$. B also has to compute modular exponentiations in Step 6, that is, $z_X = g^{w_X} u_X^{e_X} v_X^{e_X^2} \bmod p$, if he receives e_X, w_X to check the validity of u_X, e_X, w_X. On the other hand, X need not compute any modular exponentiation during the attack. X need not check the validity of u_B, e_B, w_B nor compute valid u_X, e_X, w_X. His only purpose is to make the responder be over-loaded, and he achieves it with much less computation than B.

In the next subsection, the base KAP is transformed to the protocols more resistant to the DoS attack. The techniques used in the transformation are the stateless connection [1], the weak key confirmation, and the enforcement of heavy computation.

3.2 Protocols

Two DoS resistant protocols are presented in this subsection.

The first DoS resistant protocol can be constructed by applying the idea of the stateless connection directly to the base KAP. The protocol is shown below. Let E_B (D_B) be an encryption (a decryption) function of B. These are assumed to be the functions of some symmetric key encryption scheme.

DoS-resistant KAP1

0. (Precomputation)
   The initiator A randomly selects $k_A, r_A \in \mathbf{Z}_q$ and computes $u_A = g^{-k_A} \bmod p$ and $x_A = g^{r_A} \bmod p$.
   The responder B randomly selects $k_B, r_B \in \mathbf{Z}_q$ and computes $u_B = g^{-k_B} \bmod p$ and $x_B = g^{r_B} \bmod p$.
1. A sends u_A, I_A to B.
2. B computes $e_B = h(x_B, u_B, u_A)$ and $w_B = r_B + e_B k_B + e_B^2 s_B \bmod q$. B also computes $c_B = E_B(k_B, x_B)$.
3. B sends u_B, e_B, w_B, c_B, I_B to A.
4. A computes $z_B = g^{w_B} u_B^{e_B} v_B^{e_B^2} \bmod p$ and checks if $e_B = h(z_B, u_B, u_A)$. If it does not hold, then A terminates the execution. Otherwise, A computes $a_A = h(z_B, u_A, u_B)$. A also computes $e_A = h(x_A, u_A, u_B)$ and $w_A = r_A + e_A k_A + e_A^2 s_A \bmod q$.
5. A sends $u_A, e_A, w_A, a_A, u_B, e_B, c_B, I_A$ to B.
6. A computes $K_A = u_B^{-k_A} \bmod p$.
   B recovers $(k_B, x_B) = D_B(c_B)$ and checks if $a_A = h(x_B, u_A, u_B)$ and $e_B = h(x_B, u_B, u_A)$. If they do not hold, then B terminates the execution. Otherwise, B computes $z_A = g^{w_A} u_A^{e_A} v_A^{e_A^2} \bmod p$ and checks if $e_A = h(z_A, u_A, u_B)$. If it does not hold, then B terminates the execution. Otherwise, B computes $K_B = u_A^{-k_B} \bmod p$.

This protocol is also shown in Fig. 1. It requires the participants to pass the connection state and its authenticators back and forth, which requires additional bandwidth. The computational costs of the authenticators are an encryption/decryption of a private-key cipher and a few one-way hashings, which are computationally inexpensive.

Precisely speaking, for DoS-resistant KAP1, the provable confidentiality of the shared key of the base KAP is ruined in general, because k_B is encrypted with E_B and the ciphertext is transformed over an insecure channel. DoS-resistant

Step	A		B
0	$k_A, r_A \in_R \mathbf{Z}_q$ $u_A = g^{-k_A} \bmod p$ $x_A = g^{r_A} \bmod p$		$k_B, r_B \in_R \mathbf{Z}_q$ $u_B = g^{-k_B} \bmod p$ $x_B = g^{r_B} \bmod p$
1		$\Longrightarrow (u_A, I_A) \Longrightarrow$	
2			$e_B = h(x_B, u_B, u_A)$ $w_B = r_B + e_B k_B + e_B^2 s_B \bmod q$ $c_B = E_B(k_B, x_B)$
3		$\Longleftarrow (u_B, e_B, w_B, c_B, I_B) \Longleftarrow$	
4	$z_B = g^{w_B} u_B^{e_B} v_B^{e_B^2} \bmod p$ $e_B = h(z_B, u_B, u_A)$? $a_A = h(z_B, u_A, u_B)$ $e_A = h(x_A, u_A, u_B)$ $w_A = r_A + e_A k_A + e_A^2 s_A \bmod q$		
5		$\Longrightarrow (u_A, e_A, w_A, a_A, u_B, e_B, c_B, I_A) \Longrightarrow$	
6	$K_A = u_B^{-k_A} \bmod p$		$(k_B, x_B) = D_B(c_B)$ $a_A = h(x_B, u_A, u_B)$? $e_B = h(x_B, u_B, u_A)$? $z_A = g^{w_A} u_A^{e_A} v_A^{e_A^2} \bmod p$ $e_A = h(z_A, u_A, u_B)$? $K_B = u_A^{-k_B} \bmod p$

Fig. 1. DoS-resistant KAP1

KAP2 in Fig. 2 solves this problem. This protocol assumes that the responder B makes use of the same u_B in many executions. This assumption does not ruin the provable security of the base KAP. In this case, B can use the same k_B to compute shared keys in many executions of the protocol and he can keep k_B in his own storage. In this protocol, x_B is used as a nonce for A instead of u_B, and $e_A = h(x_A, u_A, u_B, x_B)$.

3.3 Resistance to the DoS attack

In the following, resistance to the DoS attack is discussed only for the DoS-resistant KAP1. The discussion for the DoS-resistant KAP2 is similar to it.

First, let us consider the storage-exhaustion attack. By the stateless connection for the responder B, the connection state (I_A, u_A, k_B, u_B) need not be kept by B during an execution of the protocol, and the storage-exhaustion attack has no effect on DoS-resistant KAP1. e_B and c_B guarantee the integrity and confidentiality of the state information. In Step 6 of DoS-resistant KAP1, B can confirm that u_A and u_B are used together for computing the shared key if $e_B = h(x_B, u_B, u_A)$. B can recover k_B by decrypting c_B. Since $c_B = E_B(k_B, x_B)$ and $e_B = h(x_B, u_B, u_A)$, x_B implies to B that $-k_B$ is the discrete logarithm of u_B. Hence, B need not check whether $u_B = g^{-k_B} \bmod p$.

Step	A		B
0	$k_A, r_A \in_R \mathbf{Z}_q$ $u_A = g^{-k_A} \bmod p$ $x_A = g^{r_A} \bmod p$		$(k_B,) r_B \in_R \mathbf{Z}_q$ $(u_B = g^{-k_B} \bmod p)$ $x_B = g^{r_B} \bmod p$
1		$\Longrightarrow (u_A, I_A) \Longrightarrow$	
2			$e_B = h(x_B, u_B, u_A)$ $w_B = r_B + e_B k_B + e_B^2 s_B \bmod q$ $c_B = E_B(u_B, x_B)$
3		$\Longleftarrow (u_B, e_B, w_B, c_B, I_B) \Longleftarrow$	
4	$z_B = g^{w_B} u_B^{e_B} v_B^{e_B^2} \bmod p$ $e_B = h(z_B, u_B, u_A)$? $a_A = h(z_B, u_A, u_B)$ $e_A = h(x_A, u_A, u_B, z_B)$ $w_A = r_A + e_A k_A + e_A^2 s_A \bmod q$		
5		$\Longrightarrow (u_A, e_A, w_A, a_A, u_B, e_B, c_B, I_A) \Longrightarrow$	
6	$K_A = u_B^{-k_A} \bmod p$		$(u_B, x_B) = D_B(c_B)$ $a_A = h(x_B, u_A, u_B)$? $e_B = h(x_B, u_B, u_A)$? $z_A = g^{w_A} u_A^{e_A} v_A^{e_A^2} \bmod p$ $e_A = h(z_A, u_A, u_B, x_B)$? $K_B = u_A^{-k_B} \bmod p$

Fig. 2. DoS-resistant KAP2. B can use the same u_B and k_B in many executions; B updates k_B (and consequently u_B) in not every execution of the protocol.

Next, let us consider the CPU-exhaustion attack. The computation of modular exponentiations are mainly considered, because it is the most time-consuming computation. $u_B = g^{-k_B} \bmod p$ and $x_B = g^{r_B} \bmod p$ are able to be computed off-line; Step 0 can be implemented as a precomputation step. The shared key need be computed only if a_A, e_B and u_A, e_A, w_A are valid because this protocol provides the weak key confirmation. Hence, B need not compute any modular exponentiation in Step 2 though this protocol is based on the Diffie-Hellman KAP. In this protocol, B computes some modular exponentiations on-line, only if $a_A = h(x_B, u_A, u_B)$ and $e_B = h(x_B, u_B, u_A)$. Before the malicious initiator makes B compute $z_A = g^{w_A} u_A^{e_A} v_A^{e_A^2} \bmod p$ in Step 6, he has to pay quite similar computational cost in Step 4 for $z_B = g^{w_B} u_B^{e_B} v_B^{e_B^2} \bmod p$ and $a_A = h(z_B, u_A, u_B)$ due to the enforcement of computation; heavy computation of z_B is required for providing the correct extra hash a_A which will be inexpensively checked by the responder at the beginning of Step 6.

The KAP with stateless connection such as those proposed in this manuscript is intrinsically vulnerable to the replay-flooding attack. Looking at a comparison of the replay-flooding attack against stateless protocols with the storage-exhaustion attack against stateful ones, however, we can find that the stateless protocols perform better [1].

4 Performance analysis

In an actual network, one application may differ from another in requirements on the efficiency of a key agreement protocol. For example, a network-layer application may probably have much more stringent requirements than an upper-layer application. This section analyzes the computational costs of the proposed versions of KAP.

4.1 Basic analysis

Since the main source of the computational cost is modular exponentiation, Step 4 and Step 6 have to be analyzed both in DoS-resistant KAP1 and in KAP2; specifically, the computation of z_B and z_A, and the key establishment (K_A on the initiator's side and K_B on the responder's side).

The cost of the key establishment is trivial: one modular exponentiation on each side. The exponentiation is carried out by using one-time exponent and one-time base; none of (u_A, u_B, k_A, k_B) is a long-term value. The rest of this section is devoted to the analysis of the cost of z_B and z_A.

A simple implementation may pay three exponentiations for computing z_B. This, however, can be reduced to 2.17 exponentiations by the use of the simultaneous multiple exponentiation technique which was attributed by ElGamal [7] to Shamir[3]. $z_B = g^{w_B} v_B^{e_B^2} u_B^{e_B} \bmod p$, where the first two bases g and v_B are known in advance. Since we can use Shamir's trick to $g^{w_B} v_B^{e_B^2} \bmod p$, the total cost is given by

$$\frac{2}{3}\left\{2 - \left(\frac{1}{2}\right)^2\right\} + 1 = 2.17$$

modular exponentiations.

On the responder's side, fully conforming to the DoS consideration, computation of z_A costs three modular exponentiations.

4.2 Optional saving

One approach for saving in computational cost on the responder's side is a DoS-resistance-dependent use of three different modes.

The lightest mode requires the responder to keep the state information carried by the message in Step 1. This makes the protocol less resistant against the storage-exhaustion attack by the storage of (u_A, I_A). Since both u_A and v_A can be regarded as a pre-known value in this case, the computation of z_A in Step 6

---

[3] This algorithm is summarized in Appendix A.

Table 1. Comparison of the three modes on the responder's side. The cost is measured by the equivalent number of modular exponentiations; the cost in Step 4 on the initiator's side is 2.17. After precomputation of simultaneous multiple exponentiation, Light Mode requires additional storage of four precomputed results, while Moderate Mode one.

Mode	Additional storage	Additional precomputation	Cost in Step 6
Light Mode	(u_A, I_A)	4 multiplications	1.25
Moderate Mode	I_A	1 multiplication	2.17
Simple Mode	0	0	3

can use the simultaneous multiple exponentiation for three bases: g, u_A, and v_A. Thus the cost in Step 6 can be reduced to

$$\frac{2}{3}\left\{2 - \left(\frac{1}{2}\right)^3\right\} = 1.25$$

exponentiations. It should be noted that the precomputation for simultaneous multiple exponentiation here costs only four modular multiplications. We refer to this mode as "Light Mode".

The second one requires the responder to keep the state information I_A. Since v_A can be regarded as a pre-known value in this case, the computation of z_A in Step 6 can use the simultaneous multiple exponentiation for two bases: g and v_A. Thus the cost in Step 6 can be reduced to

$$\frac{2}{3}\left\{2 - \left(\frac{1}{2}\right)^2\right\} + 1 = 2.17$$

exponentiations. It should be noted that the precomputation for simultaneous multiple exponentiation here costs only one modular multiplication. We refer to this mode as "Moderate Mode".

The third one is the simple mode with the cost of three exponentiations mentioned in the previous subsection. We refer to this mode as "Simple Mode".

These three modes are summarized in Table 1. Depending on the real-time situation of the required DoS care[4] or traffic condition, the mode can be switched from one to another.

4.3 DoS-resistance evaluation

The DoS-resistance of KAP1 is compared with that of the other protocols with key confirmation. These protocols are

---

[4] Precomputed values require additional storage as state information.

STS: Station-to-station protocol [5] with Schnorr's signature scheme [13],
JV: Protocol IIA of Just and Vaudenay [10],
BJM: Protocol 2 of Blake-Wilson, Johnson and Menezes [3],
LMQSV: Protocol 3 of Law, Menezes, Qu, Solinas and Vanstone [12].
BJM are provably secure in the random oracle model on the assumptions that
the Diffie-Hellman problem is intractable and that there exists a secure MAC
(Message Authentication Code). On the other hand, STS, JV and LMQSV are
not provably secure. All protocols are assumed to be executed in Light Mode.

Stateless connection is a general technique which is applicable to other protocols. Thus, in this subsection, only the CPU-exhaustion attack is considered,
and the number of modular exponentiations required to a responder under the
CPU-exhaustion attack is evaluated. First, the number of all modular exponentiations is considered. Then, the number of on-line modular exponentiations is
considered.

Three types of requests to a responder are considered in this subsection. One
is a valid request, another is an invalid request with a last message and the other
is an invalid request without a last message. The last two types of requests are
from malicious initiators who try to make the CPU-exhaustion attack. Malicious
initiators tend to make invalid requests without last messages, because he is able
to forget his past requests and concentrate on making new invalid requests. Thus,
in this manuscript, an invalid request with a last message is called a minor bad
request and an invalid request without a last message is called a major bad
request. A valid request is called a good request.

Table 2 shows the numbers of all modular exponentiations for the three types
of requests. For BJM, the computational load of a MAC involved in it does not
considered here. KAP1 is the least efficient for a good request. However, weak
key confirmation of KAP1 reduces the number of exponentiations of KAP1 for
a bad request. For a minor bad request, KAP1 is more efficient than STS and
JV, but still less efficient than BJM and LMQSV. For a major bad request, a
responder need not check the authenticity of an initiator, and KAP1 is the most
efficient protocol.

Notice that the enforcement of heavy computation of KAP1 makes minor
bad requests less applicable to it. The number of exponentiations for a minor
bad request is equal to that for a major bad one unless a responder receives
valid a_A, which requires a (malicious) initiator to compute 1.25 exponentiations,
$z_B = g^{w_B} u_B^{e_B} v_B^{e_B^2} \bmod p$.

Table 3 shows the lower bounds of the ratios of bad requests to all requests
for the average number of modular exponentiations of KAP1 to be smaller than
those of the other protocols. These are evaluated based on Table 2. Good ∨
MinorBad represents each request is a good request or a minor bad request.
Good ∨ MajorBad represents each request is a good request or a major bad
request. In the case of Good ∨ MajorBad, KAP1 can be more efficient than the
other protocols. For example, if each request is a good or major bad request
and 7.4% or more of all requests are the latter one, then the average number
of exponentiations of KAP1 is smaller than that of STS. In the case of Good

∨ MinorBad, the average number of exponentiations of KAP1 is always larger than those of BJM and LMQSV. Notice that the computational load of MAC in BJM is not considered here and that LMQSV provides no provable security. The computational load of BJM gets much larger if its MAC is implemented based on discrete logarithm.

Table 4 shows the numbers of on-line modular exponentiations for the three types of requests. For a bad request, KAP1 is the most efficient protocol in terms of on-line computation. Especially, for a major bad request, KAP1 requires no on-line exponentiations to a responder.

The average number of on-line modular exponentiations of KAP1 can be smaller than those of the other protocols when some of the requests are bad. Table 5 shows the lower bounds of the ratios of bad requests to all requests for the average number of exponentiations of KAP1 to be smaller than those of the other protocols. These are evaluated based on Table 4.

Table 2. The number of modular exponentiations required to a responder for a good request, a minor bad request and a major bad request. STS is the station-to-station protocol [5] with Schnorr's signature scheme [13], JV is Protocol IIA of Just and Vaudenay [10], BJM is Protocol 2 of Blake-Wilson, Johnson and Menezes [3], and LMQSV is Protocol 3 of Law, Menezes, Qu, Solinas and Vanstone [12]. All protocols are in Light Mode. For BJM, the computational load of MAC involved in it is not considered here.

request \ protocol	KAP1	STS	JV	BJM	LMQSV
good request	4.25	4.17	4	3	2.17
minor bad request	3.25	4.17	4	3	2.17
major bad request	2	3	3	3	2.17

Table 3. The lower bounds of the ratios(%) of bad requests to all requests for the average number of modular exponentiations of KAP1 to be smaller than those of protocols compared. These are evaluated based on Table 2. Good ∨ MinorBad represents each request is a good request or a minor bad request. Good ∨ MajorBad represents each request is a good request or a major bad request. "xxx" means that the average number of modular exponentiations of KAP1 is always larger.

request \ protocol	STS	JV	BJM	LMQSV
Good ∨ MinorBad	8.0	25.0	xxx	xxx
Good ∨ MajorBad	7.4	20.0	55.6	92.4

Table 4. The number of on-line modular exponentiations required to a responder for a good request, a minor bad request and a major bad request. For BJM, the computational load of MAC involved in it is not considered here.

request \ protocol	KAP1	STS	JV	BJM	LMQSV
good request	2.25	2.17	3	2	1.17
minor bad request	1	2.17	3	2	1.17
major bad request	0	1	2	2	1.17

Table 5. The lower bounds of the ratios(%) of bad requests to all requests for the average number of on-line modular exponentiations of KAP1 to be smaller than those of protocols compared. These are evaluated based on Table 4.

request \ protocol	STS	JV	BJM	LMQSV
Good ∨ MinorBad	6.4	0	20.0	86.4
Good ∨ MajorBad	7.4	0	11.1	48.0

5 Conclusion

Two key agreement protocols which are resistant to a DoS attack have been proposed in this manuscript. The denial-of-service attack considered is the resource-exhaustion attack on a responder, where the resources are the storage and the CPU. The proposed KAP's are the first protocols resistant to both the storage-exhaustion attack and the CPU-exhaustion attack. This Dos-resistance may simplify DoS-related bandwidth-control mechanisms in different layers.

Acknowledgement

The authors would like to thank Dr. P. Nikander for giving them the motivation and for fruitful discussions.

References

1. T. Aura and P. Nikander. Stateless connections. In *Information and Communications Security (ICICS'97)*, pages 87–97, 1997. Lecture Notes in Computer Science 1334.
2. M. Bellare and P. Rogaway. Random oracles are practical: A paradigm for designing efficient protocols. In *Proceedings of the 1st ACM Conference on Computer and Communications Security*, pages 62–73, 1993.

3. S. Blake-Wilson, D. Johnson, and A. Menezes. Key agreement protocols and their security analysis. In *Cryptography and Coding (IMA'97)*, pages 30–45, 1997. Lecture Notes in Computer Science 1355.

4. W. Diffie and M. E. Hellman. New directions in cryptography. *IEEE Trans. Infor. Theory*, IT-22:644–654, 1976.

5. W. Diffie, P. C. van Oorschot, and M. J. Wiener. Authentication and authenticated key exchanges. *Designs, Codes and Cryptography*, 2(2):107–125, 1992.

6. C. Dwork and M. Naor. Pricing via processing or combatting junk mail. In *CRYPTO'92*, pages 139–147. Springer-Verlag, 1993. Lecture Notes in Computer Science 740.

7. T. ElGamal. A public key cryptosystem and a signature scheme based on discrete logarithms. *IEEE Transactions on Information Theory*, IT-31(4):469–472, 1985.

8. D. Harkins and D. Carrel. The internet key exchange (IKE). RFC2409, 1998.

9. S. Hirose and S. Yoshida. An authenticated Diffie-Hellman key agreement protocol secure against active attacks. In *PKC'98*, pages 135–148, 1998. Lecture Notes in Computer Science 1431.

10. M. Just and S. Vaudenay. Authenticated multi-party key agreement. In *ASIACRYPT'96*, pages 36–49, 1996. Lecture Notes in Computer Science 1163.

11. P. Karn and W. Simpson. Photuris: Session-key management protocol. RFC2522, 1999.

12. L. Law, A. Menezes, M. Qu, J. Solinas, and S. Vanstone. An efficient protocol for authenticated key agreement. Technical Report CORR98-05, Department of C&O, University of Waterloo, 1998.

13. C. P. Schnorr. Efficient identification and signatures for smart cards. In *CRYPTO'89*, pages 239–252, 1990. Lecture Notes in Computer Science 435.

A Simultaneous Multiple Exponentiation

The followings are a brief description of a simultaneous multiple exponentiation algorithm, which was attributed by ElGamal [7] to Shamir.

Input: group elements $g_0, g_1, \cdots, g_{k-1}$ and non-negative t-bit integers $e_0, e_1, \cdots, e_{k-1}$.

Output: $g_0^{e_0} g_1^{e_1} \cdots g_{k-1}^{e_{k-1}}$.

Assumption: $g_0, g_1, \cdots, g_{k-1}$ are known in advance and thus can be used in precomputation.

1. For $i = 1$ to $\left(2^k - 1\right)$, $G_i := \prod_{j=0}^{k-1} g_j^{i_j}$ where $i = (i_{k-1} \cdots i_0)_2$ (*Precomputation*).

2. Let I_i be the i-th column of an exponent array (binary representation).

3. $A := 1$.

4. For $i = 1$ to t, $A := A \cdot A$ and then $A := A \cdot G_{I_i}$. If $I_i = 0$ (and hence $G_{I_i} = 1$), the latter multiplication is trivial.

5. Return(A).

Expected Cost: $\left\{2 - \left(\frac{1}{2}\right)^k\right\} t$ non-trivial multiplications when the exponents are independently and randomly generated. Assuming that the cost of "one modular exponentiation" means the expected number of non-trivial modular multiplications in the case of "square and multiply" method, the simultaneous multiple exponentiation reduces the cost from k exponentiations down to

$$\frac{\left\{2 - \left(\frac{1}{2}\right)^k\right\} t}{\frac{3}{2}t} = \frac{2}{3}\left\{2 - \left(\frac{1}{2}\right)^k\right\}. \tag{1}$$

Example: Let us consider an example of $k = 3$ and $t = 5$:
$e_0 = 30 = (11110)_2$, $e_1 = 10 = (01010)_2$, $e_2 = 24 = (11000)_2$.
The exponent array is

	I_1	I_2	I_3	I_4	I_5
e_0	1	1	1	1	0
e_1	0	1	0	1	0
e_2	1	1	0	0	0

The precomputation (step 2) gives

i	0	1	2	3	4	5	6	7
G_i	1	g_0	g_1	$g_0 g_1$	g_2	$g_0 g_2$	$g_1 g_2$	$g_0 g_1 g_2$

Since $G_0 = 1$, $G_1 = g_0$, $G_2 = g_1$, and $G_4 = g_2$ are trivial and G_7 can be computed as $G_7 = (g_0 g_1)g_2 = G_3 g_2$, this precomputation costs four modular multiplications.

Finally, each iteration gives

i	1	2	3	4	5
A	$g_0 g_2$	$g_0^3 g_1 g_2^3$	$g_0^7 g_1^2 g_2^6$	$g_0^{15} g_1^5 g_2^{12}$	$g_0^{30} g_1^{10} g_2^{24}$

An Extended Logic for Analyzing Timed-Release Public-Key Protocols

Michiharu Kudo[1] and Anish Mathuria[2]

[1] kudo@jp.ibm.com
[2] anish@trl.ibm.co.jp
IBM Japan Ltd., Tokyo Research Laboratory
1623-14, Shimotsuruma, Yamato-shi
Kanagawa-ken 242-0001
JAPAN

Abstract. A logic is presented for analyzing public key protocols which provide time-dependent confidentiality using a trusted party. The logic is developed as an extension to an existing cryptographic modal logic with time due to Coffey and Saidha. The extension is designed to help capture aspects of timed-release public key protocols that are not captured in the Coffey-Saidha logic. The explicit use of time in the logic is shown to facilitate reasoning about the correctness of an example protocol.

1 Introduction

The universality of time makes time-dependent security especially important in practice. Time-dependent confidentiality is a familiar protection requirement. For example, many office documents are marked "confidential until month/day". We typically see that product releases should be published only after a certain time on a certain day. Another motivating example is the following. Suppose Alice wants to send a confidential letter to Bob by e-mail now, but she does not want him to open the (electronically sealed) envelope until a week later. She does not want to wait a week before sending it, because she wants him to know now that she has sent the letter. Moreover, she wants to send it only once, not twice, because she will not available at a later date. The type of time-dependent confidentiality required in the previous examples can be solved using the notion of timed-release crypto described in Rivest, Shamir and Wagner [RSW96]. According to them, the essential goal of timed-release crypto is to "send information into the future". They propose two different approaches to implementing timed-release crypto, one approach based on computational puzzles that require a precise amount of time to solve, and another approach based on a trusted agent who does not reveal certain information until a specific time. The former approach is less practical and requires much more computational resources than the latter.

In this paper we present a timed-release protocol using public-key cryptography, and a logic for verifying such protocols. The idea behind the protocol is similar as in an off-line protocol sketched in Rivest *et al.*, namely the use of a trusted agent who maintains asymmetric key pairs for each future time. A principal motivation for our work is to capture reasoning about such protocols in a formal yet intuitively appealing way. The work described here arose from our attempts to explore the use of existing BAN-like logics to help us analyze our protocol. We found the explicit use of time unavoidable in reasoning about the correctness of the protocol. For this reason we could not directly use the BAN logic [BAN90] and many other related logics that avoid the explicit use of time. To mention just a few of the many logics published in the literature which extend BAN but which do not require the explicit use of time: Gong, Needham and Yahalom [GNY90], Abadi and Tuttle [AT91], Mao and Boyd [MB93], Syverson and van Oorschot [SVO94], and Wedel and Kessler [WK96]. The paper describes how we extended a logic for verifying public-key protocols, introduced by Coffey and Saidha [CS97], to analyze our protocol. Their logic allows time to be explicitly associated with constructs of the logic, as in some existing BAN-like logics with time (see, for example, Syverson [S93], and Stubblebine and Wright [SW96]). Apart from the explicit use of time, however, the Coffey-Saidha logic provides axioms that help express and prove some secrecy aspects of protocols. In their logic, there is a direct way of saying that an agent cannot learn the decryption of some message. Thus their logic differs in an important way from BAN-like logics, which generally do not address the issue of secrecy. The limitations of BAN-like logics in regards to analysis of secrecy properties have previously been noted in the literature (see Gong and Syverson [GS95] for example). Coffey and Saidha's logic is closely based on a logic of communication, knowledge and time, CKT5, proposed by Bieber [B90]. The major difference between their logic and CKT5 is that the former is designed for analysis of public key protocols, whereas the latter is designed for analysis of shared key protocols. Like CKT5, the Coffey-Saidha logic makes use of a knowledge operator and negation to analyze confidentiality requirements. This feature of the logic, when combined with the explicit use of time, makes the logic fruitful for our purposes.

The rest of the paper is organized as follows. In Section 2 we describe our protocol and give an informal statement of the protocol goals. In Section 3 we review the logic proposed by Coffey and Saidha [CS97], hereafter referred to as the CS logic. In Section 4 we present our modifications to their logic, extending its applicability to timed-release public-key protocols. In Section 5 we apply the extended logic to analyze the protocol. We close the paper in Section 6 with our conclusions.

2 A Timed-Release Protocol

The protocol has three participants: Alice, Bob, and Trent. Alice is a party who wishes to send a time-confidential message into the future. Bob is a receiver who wishes to decrypt this message. We assume that neither Alice nor Bob can tell the correct time on their own. The third participant, Trent, is assumed to be capable of knowing the correct time, generating asymmetric key pairs for use in a suitable public key cryptosystem, and binding a generated key pair to a specific time instant. The message exchanges comprising the protocol are illustrated in Figure 1. We use the following notations to represent cryptographic operations:

- $\{M\}_K$ - Message M encrypted under a public key K.
- $\{M\}_{K^{-1}}$ - Message M signed under a private key K^{-1}. This is implemented by concatenating M and the signature of hash of M with K^{-1}.

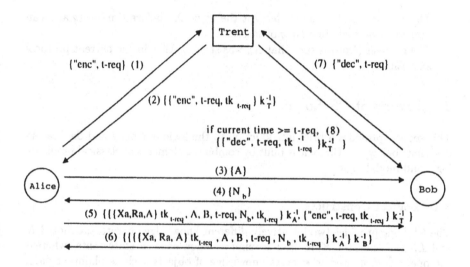

Fig. 1. Timed-Release Protocol

First Alice sends a message requesting Trent to generate a time-key pair for some time $t\text{-}req$ in the future (message 1). Upon receiving this message, Trent generates a time-key pair and responds with a signed message containing the time-key for encryption, $tk_{t\text{-}req}$, to Alice (message 2). Trent holds the corresponding decryption key, $tk_{t\text{-}req}^{-1}$, secret until the time $t\text{-}req$. Alice verifies Trent's signature and sends a message containing her name to Bob as the request for a challenge (message 3). Bob responds with a random nonce challenge N_b (message 4). Next Alice generates a random nonce value R_a and encrypts it together with time-confidential data X_a and the name A, under the time key for encryption received from Trent earlier. The purpose of R_a is to ensure that the resulting ciphertext differs from all previous and future ones. Alice appends the encrypted value with the names A and B, nonce value N_b, the time-key for encryption $tk_{t\text{-}req}$, and the release time $t\text{-}req$. She signs the resulting message and sends it to Bob, along with message 2 (message 5). The nonce N_b is included in message 5 as protection against replay attacks. Bob verifies Alice's signature and acknowledges receipt of the previous message by sending a signed message to Alice (message 6). When Bob wants to decrypt the time-confidential data sent by Alice, he requests the time-key for decryption from Trent (message 7). If Trent's clock indicates that the current time is greater or equal to *t-req*, then Trent returns a message containing the time-key for decryption (message 8); otherwise he does not respond. When the time *t-req* finally arrives Bob is able to decrypt X_a. It is worth noting that the protocol aims to ensure confidentiality of X_a with respect to a specific time rather than the receiver. If the release time has passed, anyone can possibly ask Trent for the decryption time key and learn the value X_a.

The protocol goals can be informally specified as follows:

- The receiver (Bob) cannot decrypt the value X_a before time *t-req* and can decrypt it only at time *t-req* or later.
- The receiver (Bob) is sure that X_a was sent by Alice in the current protocol execution.

3 Review of CS Logic

This section provides a brief introduction to the logic of Coffey and Saidha. As described below, the CS logic is built by combining elements of classical predicate logic and modal logic.

3.1 Logical Constructs

The CS logic provides two intuitively different knowledge operators, denoted K and L. The operator K is used to express knowledge of statements, whereas the operator L is used to express knowledge of objects such as plaintext data, ciphertext data, keys, etc. Both these operators are indexed with time:

$K_{\Sigma,t}\phi$ Entity Σ knows statement ϕ at time t
$L_{\Sigma,t}x$ Entity Σ knows and can reproduce message x at time t

The notion of belief is represented using another operator, denoted B, which is intuitively different from the operator K. The belief operator too is indexed with time:

$B_{\Sigma,t}\phi$ Entity Σ believes statement ϕ at time t

A public-key of entity Σ is denoted k_Σ, and the corresponding private-key is denoted k_Σ^{-1}. The notations $e()$ and $d()$ are used to represent encryption and decryption functions, respectively:

$e(x, k_\Sigma)$ The encryption of x under a public-key k_Σ
$d(x, k_\Sigma^{-1})$ The decryption of x under the corresponding private-key k_Σ^{-1}

The operators for expressing the sending and receiving of a message are denoted S and R, respectively. Both these operators are indexed with time, like the modal operators for knowledge and belief:

$S(\Sigma, t, x)$ Entity Σ sends a message x at time t
$R(\Sigma, t, x)$ Entity Σ receives a message x at time t

The logic also provides an operator, denoted C, to express the intuitive concept of a message being contained, directly or indirectly, in another message:

$C(x, y)$ The message x contains a message y. y may be cleartext or ciphertext in x.

Finally, the logic includes standard propositional connectives: '$\wedge$' for conjunction, '$\vee$' for disjunction, '$\neg$' for negation, and '$\rightarrow$' for implication, as well as standard quantifiers: '$\exists$' for existential quantification, and '$\forall$' for universal quantification.

3.2 Axiomatization

For ease of description it is convenient to divide the axioms of the logic into several groups. In what follows, the symbols $\in$ and $\setminus$ are used to denote set membership and set difference, respectively. The set of all entities in the system is denoted ENT. The first group of axioms consists of some standard modal axioms for knowledge and belief:

$$A1(a). \quad K_{\Sigma,t}p \wedge K_{\Sigma,t}(p \rightarrow q) \rightarrow K_{\Sigma,t}q$$
$$(b). \quad B_{\Sigma,t}p \wedge B_{\Sigma,t}(p \rightarrow q) \rightarrow B_{\Sigma,t}q$$
$$A2. \quad K_{\Sigma,t}p \rightarrow p$$

(Axiom A2 essentially distinguishes the K operator from the B operator.) The second group consists of axioms capturing monotonicity of knowledge and belief:

$$A3(a). \quad L_{i,t}x \rightarrow \forall t' \geq t \; L_{i,t'}x$$
$$(b). \quad K_{i,t}x \rightarrow \forall t' \geq t \; K_{i,t'}x$$
$$(c). \quad B_{i,t}x \rightarrow \forall t' \geq t \; B_{i,t'}x$$

The third group consists of a single axiom for the C operator. The axiom below captures the intuition that if a piece of data is constructed from other pieces, then each contributory piece must be known to some entity:

$$A4. \quad L_{i,t}y \wedge C(y, x) \rightarrow \exists j \in ENT \; L_{j,t}x$$

The fourth group consists of axioms capturing relationships between the S, L and R operators:

$$A5. \quad S(\Sigma, t, x) \rightarrow L_{\Sigma,t}x \wedge \exists i \in ENT \setminus \{\Sigma\} \; \exists t' > t \; R(i, t', x)$$
$$A6. \quad R(\Sigma, t, x) \rightarrow L_{\Sigma,t}x \wedge \exists i \in ENT \setminus \{\Sigma\} \; \exists t' < t \; S(i, t', x)$$

The fifth group consists of axioms capturing the ability to encrypt and decrypt a message under the knowledge of a public key and private key, respectively:

$$A7(a). \quad L_{i,t}x \wedge L_{i,t}k_{\Sigma} \rightarrow L_{i,t}(e(x, k_{\Sigma}))$$
$$(b). \quad L_{i,t}x \wedge L_{i,t}k_{\Sigma}^{-1} \rightarrow L_{i,t}(d(x, k_{\Sigma}^{-1}))$$

The sixth group consists of axioms capturing the impossibility of encrypting or decrypting a message without knowledge of the correct key:

A8(a). $\neg L_{i,t}k_{\Sigma} \wedge \forall t' < t\ \neg L_{i,t'}(e(x,k_{\Sigma})) \wedge \neg(\exists y(R(i,t,y) \wedge C(y,e(x,k_{\Sigma}))))$
$\rightarrow \neg L_{i,t}(e(x,k_{\Sigma}))$

(b). $\neg L_{i,t}k_{\Sigma}^{-1} \wedge \forall t' < t\ \neg L_{i,t'}(d(x,k_{\Sigma}^{-1})) \wedge \neg(\exists y(R(i,t,y) \wedge C(y,d(x,k_{\Sigma}^{-1}))))$
$\rightarrow \neg L_{i,t}(d(x,k_{\Sigma}^{-1}))$

The seventh group consists of axioms capturing properties of private keys:

$$\text{A9.}\ \ L_{i,t}k_i^{-1} \wedge \forall j \in ENT \setminus \{i\}\ \neg L_{j,t}k_i^{-1}$$
$$\text{A10.}\ \ L_{i,t}(d(x,k_{\Sigma}^{-1})) \rightarrow L_{\Sigma,t}x$$

Axiom A9 is called the key secrecy axiom; it logically characterizes private keys as secrets known only to their owners. Axiom A10 essentially captures the intuition that if a piece of data is constructed as a decryption under a private key, then the private key owner must be involved in the construction.

We will discuss some of the axioms presented here in greater detail in the next section.

The notion of *theorem* is defined as usual and expressed by the symbol $\vdash$. The logic incorporates several rules of inference, including:

R1. From $\vdash p$ and $\vdash p \rightarrow q$ infer $\vdash q$ (Modus ponens)
R2(a). From $\vdash p$ infer $\vdash K_{\Sigma,t}p$ (Generalization)
(b). From $\vdash p$ infer $\vdash B_{\Sigma,t}p$ (Generalization)

The remaining rules are those for standard natural deduction systems. A number of desirable properties can be derived as theorems of the logic, for example:

K1. $K_{\Sigma,t}(p \wedge q) \rightarrow K_{\Sigma,t}p \wedge K_{\Sigma,t}q$
K2. $K_{\Sigma,t}p \wedge K_{\Sigma,t}q \rightarrow K_{\Sigma,t}(p \wedge q)$

4 Extension to CS Logic

This section motivates some modifications to the CS logic that are required for analysis of the protocol shown in Section 2. The modifications carried out here can broadly be described as follows. First, we extend the logical syntax to provide distinct notations for time keys. Second, we modify several existing axioms and introduce new axioms to increase the scope of the logic. Section 5 below shows how these modifications make the analysis of our proposed protocol possible. We emphasize that although the modifications are designed with analysis of this protocol in mind, it is hoped that they will also be useful in analysis of other less specialized public key protocols.

4.1 Modeling impossibility of decrypting (encrypting) without the correct key

The construct $C(x, y)$ of the CS logic expresses the statement "the object x contains the object y". To grasp the intended meaning of this statement, let us look at axiom A4:

$$L_{i,t}y \wedge C(y, x) \to \exists j \in ENT\ L_{j,t}x$$

Observe that neither this axiom nor any other axiom of the CS logic allows us to conclude that if y contains x, then i can reproduce x whenever i can reproduce y. This observation suggests the following informal meaning for $C(y, x)$: x is potentially recoverable from y. It does not mean that any entity who knows y necessarily obtains x.

The treatment of the C operator in the original logic is rather incomplete. It lacks some intuitively appealing axioms for this operator. We propose the following additional axioms:

A4(b). $C(x, x)$

(c). $C(x, y) \wedge C(y, z) \to C(x, z)$

(d). $C(e(x, k_\Sigma), x) \wedge C(d(x, k_\Sigma^{-1}), x)$

We hereafter refer to axiom A4 of the original logic as A4(a). The axioms A4(b) and A4(c) say that 'C' is reflexive and transitive, respectively. The axiom A4(d) says that x is contained in any message formed by encrypting (respectively, decrypting) x under a public key (respectively, private key). We don't adopt $C(e(x, k_\Sigma), k_\Sigma) \wedge C(d(x, k_\Sigma^{-1}), k_\Sigma^{-1})$ as an axiom. This is because the definition of the C operator, as given in Coffey and Saidha, seemingly precludes this axiom. Their definition says that $C(x, y)$ holds in the case when y is a plaintext or ciphertext in x. But it says nothing about the case when y is a key that encrypts some plaintext in x.

Recall that axioms A8(a)-(b) of the the CS logic specify the conditions under which we can deduce that an entity doesn't know the result of a particular encryption or decryption. We repeat them below for convenience:

A8(a). $\neg L_{i,t}k_\Sigma \wedge \forall t' < t\ \neg L_{i,t'}(e(x, k_\Sigma)) \wedge \neg(\exists y(R(i, t, y) \wedge C(y, e(x, k_\Sigma))))$
$\to \neg L_{i,t}(e(x, k_\Sigma))$

(b). $\neg L_{i,t}k_\Sigma^{-1} \wedge \forall t' < t\ \neg L_{i,t'}(d(x, k_\Sigma^{-1})) \wedge \neg(\exists y(R(i, t, y) \wedge C(y, d(x, k_\Sigma^{-1}))))$
$\to \neg L_{i,t}(d(x, k_\Sigma^{-1}))$

We discuss axiom A8(b) in some detail below. It should be noted that a similar discussion applies to axiom A8(a). Consider the antecedent of A8(b). The condition captured in the antecedent is for most part self-explanatory, but the conjunct, $\neg(\exists y(R(i, t, y) \wedge C(y, d(x, k_\Sigma^{-1}))))$, requires some thought. It is intended to capture the statement that i doesn't obtain $d(x, k_\Sigma^{-1})$ from any message that it receives at time t. However, the statement expressed by the formula, $\neg(\exists y(R(i, t, y) \wedge C(y, d(x, k_\Sigma^{-1}))))$, is in fact stronger than that required. To see why, suppose this formula is falsified. That is, suppose that i receives a message y at t such that $C(y, d(x, k_{\Sigma}^{-1}))$ holds. Further suppose that i can't obtain $d(x, k_{\Sigma}^{-1})$ from y. For example, consider the following situation: y is $e(y', k_{\psi})$, $C(y', d(x, k_{\Sigma}^{-1}))$, and $\neg L_{i,t} k_{\psi}^{-1}$, where k_{ψ} is the public key of some entity ψ other than i, and k_{ψ}^{-1} is the corresponding private key. In the above situation, it makes sense to conclude that the succedent, $\neg L_{i,t}(d(x, k_{\Sigma}^{-1}))$, should hold, although we have a message y for which $R(i, t, y) \wedge C(y, d(x, k_{\Sigma}^{-1}))$ holds. Using axioms A4(d) and (c), it is easy to check that $C(y, d(x, k_{\Sigma}^{-1}))$ holds given the assumption that $C(y', d(x, k_{\Sigma}^{-1}))$ holds. It is possible to support the desired conclusion, $\neg L_{i,t}(d(x, k_{\Sigma}^{-1}))$, by reformulating the axiom. To do so, we define the construct $\sigma_{i,t}(x, y)$ to express the statement "i can obtain y from x at time t". The intended meaning of this construct can be grasped from the following axiom:

$$\text{A11. } L_{i,t} y \wedge \sigma_{i,t}(y, x) \to L_{i,t} x$$

Some other straightforward axioms come to mind; for example, $\sigma_{i,t}(x, y) \wedge \sigma_{i,t}(y, z) \to \sigma_{i,t}(x, z)$, which captures transitivity. However, we do not give such basic axioms simply because we don't need them here. Instead we focus on axioms that help in connecting the σ operator with other logical operators, like axiom A11 above. We will later give some more axioms of this kind.

Returning to axioms A8(a) and (b), we modify them as follows:

A8(a). $\neg L_{i,t} k_{\Sigma} \wedge \forall t' < t \ \neg L_{i,t'}(e(x, k_{\Sigma})) \wedge$
$\neg(\exists y (R(i, t, y) \wedge C(y, e(x, k_{\Sigma})) \wedge \sigma_{i,t}(y, e(x, k_{\Sigma}))))$
$\to \neg L_{i,t}(e(x, k_{\Sigma}))$

(b). $\neg L_{i,t} k_{\Sigma}^{-1} \wedge \forall t' < t \ \neg L_{i,t'}(d(x, k_{\Sigma}^{-1})) \wedge$
$\neg(\exists y (R(i, t, y) \wedge C(y, d(x, k_{\Sigma}^{-1})) \wedge \sigma_{i,t}(y, d(x, k_{\Sigma}^{-1}))))$
$\to \neg L_{i,t}(d(x, k_{\Sigma}^{-1}))$

4.2 Modeling message reception

The CS logic has an axiom which essentially relates the reception of a message by some entity to the emission of the *same* message by another entity, and imposes a constraint between the time of reception and the time of emission. Namely, axiom A6, which we hereafter refer to as A6(a):

$$R(\Sigma, t, x) \to L_{\Sigma,t} x \wedge \exists i \in ENT \setminus \{\Sigma\} \ \exists t' < t \ S(i, t', x)$$

This axiom says nothing about a message y that Σ may receive indirectly as a result of receiving x. We propose the following more general axiom:

A6(b). $R(j, t, x) \wedge C(x, y) \wedge \sigma_{j,t}(x, y) \to$
$\exists i \in ENT \ \exists t' < t \ \exists z (S(i, t', z) \wedge C(z, y) \wedge L_{i,t'} y \wedge \sigma_{j,t}(x, z) \wedge \sigma_{j,t}(z, y))$

This axiom is predicated on a number of intuitive properties about the system:

(P1) if a message y is indirectly received by j as a result of receiving x at time t, then y must be part of some message z which was sent at a time earlier than t,

(P2) the entity who originally sends a message containing y must have created y and therefore known y, and

(P3) j must have obtained y from x via z.

The intuition behind (P1) is that if j receives x and y is a part of x, then y need not have been sent by itself. However, it must at least be a part of some message sent earlier, possibly different from x. The intuition behind (P2) is that y must be known to at least one entity in the system who sends a message containing it. Therefore, if t' is the earliest time at which any entity i sends a message containing y, then i must know y at t'. Notice that by (P1) we already know that $S(i, t', z) \wedge C(z, y)$ holds for some i and some $t' < t$ and some z. We can therefore reason the additional fact, $L_{i,t'}y$, using (P2). We remark that this fact cannot be captured by using axioms A5 and A4(a), since their application would yield a conclusion that is weak for our purposes here: some entity i', *possibly different from i*, knows y at t'. We therefore build the desired fact into axiom A6(b). The intuition behind (P3) is that although z was sent by i at time t', by the time it reaches j it is subsumed under x. So, j must use x to obtain y from z.

4.3 Modeling time keys and their properties

The CS logic treats cryptographic keys with the usual working of public-key cryptography in mind. In particular, it assumes that a private key corresponding to a public key is known only to its owner. This assumption holds for all times, as reflected by the key secrecy axiom A9 of the original logic:

$$L_{i,t}k_i^{-1} \wedge \forall j \in ENT \setminus \{i\} \ \neg L_{j,t}k_i^{-1}$$

Clearly, this axiom does not hold for time-keys since a private time-key is made public (on demand) once its time of release is reached. Let us introduce the notations tk_τ and tk_τ^{-1} to denote a public time key and the corresponding private time key, where τ is the desired release time for the private time key. Let the notation T denote a trusted time-key authority. We propose a modified version of axiom A9 to reflect that the secrecy assumption applies to a private time key until its time of release:

$$\text{TA1.} \ \forall t < \tau \ L_{T,t}tk_\tau^{-1} \wedge \forall i \in ENT \setminus \{T\} \ \neg L_{i,t}tk_\tau^{-1}$$

This axiom says that the private time-keys used within the system are known only to the time-key authority prior to their release times.

Since we introduced distinct notations for time-keys, we need to provide additional axioms similar in form to axioms A7 and A8:

$$\text{TA2(a).} \ L_{i,t}x \wedge L_{i,t}tk_\tau \rightarrow L_{i,t}(e(x, tk_\tau))$$

(b). $L_{i,t}x \wedge L_{i,t}tk_\tau^{-1} \to L_{i,t}(d(x, tk_\tau^{-1}))$

TA3(a). $\neg L_{i,t}tk_\tau \wedge \forall t' < t \ \neg L_{i,t'}(e(x, tk_\tau)) \wedge$
$\qquad \neg(\exists y(R(i, t, y) \wedge C(y, e(x, tk_\tau)) \wedge \sigma_{i,t}(y, e(x, tk_\tau))))$
$\qquad \to \neg L_{i,t}(e(x, tk_\tau))$

(b). $\neg L_{i,t}tk_\tau^{-1} \wedge \forall t' < t \ \neg L_{i,t'}(d(x, tk_\tau^{-1})) \wedge$
$\qquad \neg(\exists y(R(i, t, y) \wedge C(y, d(x, tk_\tau^{-1})) \wedge \sigma_{i,t}(y, d(x, tk_\tau^{-1}))))$
$\qquad \to \neg L_{i,t}(d(x, tk_\tau^{-1}))$

An additional axiom refers to the fact that the time-key authority must know any public time-key which has been used to encrypt any data:

$$\text{TA4. } L_{i,t}(e(x, tk_\tau)) \to L_{T,t}tk_\tau$$

Finally, we propose an axiom which captures the fact that if a message is encrypted under a public time key for time τ, then no part of it can be recovered from the resulting ciphertext, prior to time τ, by any entity other than Trent:

$$\text{TA5. } \forall i \in ENT \setminus \{T\} \ \forall t < \tau \ L_{i,t}y \wedge y = e(y', tk_\tau) \wedge C(y', x) \to \neg\sigma_{i,t}(y, x)$$

5 The Protocol Analyzed

The first step in analyzing the protocol is to explicitly associate time with individual messages. We label each protocol message with two distinct times, distinguishing the instants at which the message is sent and received. If $P \to Q : m$ denotes a protocol step where P sends a message m to Q, then we affix the time instant when P sends m to the left of the $\to$ symbol, and the time instant when Q receives m to the right of this symbol. Figure 2 shows the protocol of Section 2 with times affixed to individual messages.

In what follows, we use t_0 to denote some time instant before the protocol starts. Thus $t_0 < t_1$. Let t_g denote the time when T generates the private key $tk_{t_8}^{-1}$. Let x denote the encrypted message $e(\{X_a, R_a, A\}, tk_{t_8})$.

5.1 Protocol Assumptions

To analyze the protocol, we first list the assumptions. The first group of assumptions concerns trust:

$$\forall t < t_8 \ \forall y \ S(A, t, y) \wedge C(y, d(x, tk_{t_8}^{-1})) \to \tag{1}$$
$$y = e(y', tk_{t_8}) \wedge C(y', d(x, tk_{t_8}^{-1}))$$
$$\forall t < t_8 \ \neg\exists y(S(T, t, y) \wedge C(y, d(x, tk_{t_8}^{-1}))) \tag{2}$$
$$t_g < t_8 \tag{3}$$

The first of these indicates that Alice is trusted to encrypt the time-confidential data correctly. The second indicates that Trent is trusted not to send decrypted

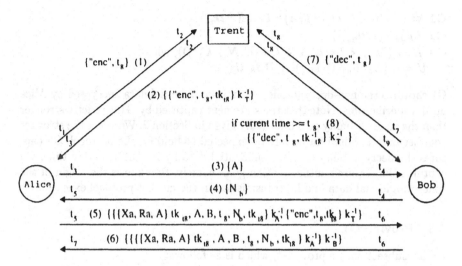

Fig. 2. Protocol with times affixed to messages

time-confidential data in any message prior to the release time. The third indicates that Trent generates the time-key pair, $(tk_{t_8}, tk_{t_8}^{-1})$, prior to the release time.

The next group of assumptions show that tk_{t_8} and $tk_{t_8}^{-1}$ have the properties of a nonce:

$$\forall t < t_g \ \forall i \in ENT \setminus \{A\} \ \neg L_{i,t} d(x, tk_{t_8}^{-1}) \tag{4}$$

$$\forall t < t_g \ \forall i \in ENT \ \neg L_{i,t} tk_{t_8} \tag{5}$$

The first of these refers to the impossibility of using a private time key prior to its generation. The second refers to the impossibility of anyone knowing the public time key prior to its generation.

Finally, we make assumptions relating to the initial state of Bob:

$$K_{B,t_0}(L_{B,t_0} k_A) \tag{6}$$

$$K_{B,t_0}(\forall t < t_0 \ \forall i \in ENT \ \neg L_{i,t} N_b) \tag{7}$$

The first of these says that Bob knows that it knows Alice's public key. The second of these says that Bob knows that no entity knows its nonce challenge before the protocol starts.

5.2 Protocol Goals

We formally express the protocol goals in the logic as follows:

G1 $\forall t < t_8 \; \forall i \in ENT \setminus \{T, A\} \; \neg L_{i,t}(d(x, tk_{t_8}^{-1}))$

G2 $L_{B,t_9}(d(x, tk_{t_8}^{-1}))$

G3 $K_{B,t_6}(\exists t \; t_0 < t < t_6 \; S(A, t, d(\{U, N_b\}, k_A^{-1})))$, where
$U = e(\{X_a, R_a, A\}, tk_{t_8}), A, B, t_8, tk_{t_8}$

G1 captures the non-decryptability of time-confidential data encrypted by Alice until a specific time. Note that the statement captured by G1 is strictly stronger than the confidentiality goal informally given in Section 2. We select the stronger goal because it should intuitively be expected to hold for the protocol. G2 captures the decryptability of time-confidential data by Bob at a specific time. G3 captures the required authentication goal, namely, Bob knows the origin of the time-confidential data and its transmission in the current protocol execution.

5.3 Proving G1

In this subsection, we prove G1, which is as follows:

$$\forall t < t_8 \; \forall i \in ENT \setminus \{T, A\} \; \neg L_{i,t}(d(x, tk_{t_8}^{-1}))$$

Let $t < t_8$ and $j \in ENT \setminus \{T, A\}$ be arbitrary but fixed. We show that $\neg L_{j,t}(d(x, tk_{t_8}^{-1}))$ by induction over time. We use the time when T generates the private key $tk_{t_8}^{-1}$, i.e. t_g, as the basis case of the induction.

1. (Basis) Let $t = t_g$. The required statement is: $\neg L_{j,t_g}(d(x, tk_{t_8}^{-1}))$. We prove it by contradiction. Suppose $L_{j,t_g}(d(x, tk_{t_8}^{-1}))$. By the contrapositive of axiom TA3(b) it follows that: $L_{j,t_g} tk_{t_8}^{-1} \vee \exists t < t_g \; L_{j,t}(d(x, tk_{t_8}^{-1})) \vee \exists y(R(j, t_g, y) \wedge C(y, d(x, tk_{t_8}^{-1})) \wedge \sigma_{j,t_g}(y, d(x, tk_{t_8}^{-1})))$.

   Case (i): $L_{j,t_g} tk_{t_8}^{-1}$.
   This is impossible by the time-key secrecy axiom TA1 and assumption (3).

   Case (ii): $\exists t < t_g \; L_{j,t}(d(x, tk_{t_8}^{-1}))$.
   This is impossible by assumption (4).

   Case (iii): $\exists y(R(j, t_g, y) \wedge C(y, d(x, tk_{t_8}^{-1})) \wedge \sigma_{j,t_g}(y, d(x, tk_{t_8}^{-1})))$.
   By axiom A6(b) we have:

   $$\exists i \in ENT \; \exists t' < t_g \; \exists z(S(i, t', z) \wedge C(z, d(x, tk_{t_8}^{-1})) \wedge L_{i,t'}(d(x, tk_{t_8}^{-1}))).$$

   Case (a): $i \in ENT \setminus \{A\}$.
   This is impossible since $L_{i,t'}(d(x, tk_{t_8}^{-1}))$ is false by assumption (4).

   Case (b): i is A.
   So, $\exists t' < t_g \; \exists z(S(A, t', z) \wedge C(z, d(x, tk_{t_8}^{-1})) \wedge L_{A,t'}(d(x, tk_{t_8}^{-1})))$. Hence, by assumption (1), $z = e(z', tk_{t_8}) \wedge C(z', d(x, tk_{t_8}^{-1}))$. By axiom A5 we have $L_{A,t'} z$, so $L_{A,t'}(e(z', tk_{t_8}))$. Hence, by axiom TA4, we must have $L_{T,t'} tk_{t_8}$, which is impossible by assumption (5) since $t' < t_g$.

2. (Induction) Let $t_g < t' < t$ be arbitrary. We assume the inductive hypothesis:

$$\forall t'' < t' \ \forall i \in ENT \setminus \{T, A\} \ \neg L_{i,t''}(d(x, tk_{t_8}^{-1})),$$

and show that this implies that $\neg L_{j,t'}(d(x, tk_{t_8}^{-1}))$. Suppose $L_{j,t'}(d(x, tk_{t_8}^{-1}))$.
By the contrapositive of axiom TA3(b) it follows that: $L_{j,t'} tk_{t_8}^{-1} \vee \exists t <$
$t' \ L_{j,t}(d(x, tk_{t_8}^{-1})) \vee \exists y (R(j, t', y) \wedge C(y, d(x, tk_{t_8}^{-1})) \wedge \sigma_{j,t'}(y, d(x, tk_{t_8}^{-1})))$.
Case (i): $L_{j,t'} tk_{t_8}^{-1}$.
    This is impossible by the time-key secrecy axiom TA1 since $t' < t_8$.
Case (ii): $\exists t < t' \ L_{j,t}(d(x, tk_{t_8}^{-1}))$.
    This is impossible by the inductive hypothesis.
Case (iii): $\exists y (R(j, t', y) \wedge C(y, d(x, tk_{t_8}^{-1})) \wedge \sigma_{j,t'}(y, d(x, tk_{t_8}^{-1})))$.
    By axiom A6(b) we have: $\exists i \in ENT \ \exists t'' < t' \ \exists z$ such that

$$S(i, t'', z) \wedge C(z, d(x, tk_{t_8}^{-1})) \wedge L_{i,t''}(d(x, tk_{t_8}^{-1})) \wedge \sigma_{j,t'}(y, z) \wedge \sigma_{j,t'}(z, d(x, tk_{t_8}^{-1}))$$

Case (i): i is T.
    We then have $\exists z(S(T, t'', z) \wedge C(z, d(x, tk_{t_8}^{-1})))$, which is impossible
    by assumption (2) since $t'' < t_8$.
Case (ii): $i \in ENT \setminus \{T, A\}$.
    We then have $L_{i,t''}(d(x, tk_{t_8}^{-1}))$, which is impossible by the inductive
    hypothesis since $t'' < t_8$.
Case (iii): i is A.
    By axiom A6(a) we have $L_{j,t'} y$. Since $\sigma_{j,t'}(y, z)$ it follows by axiom
    A11 that $L_{j,t'} z$. By assumption (1), $z = e(z', tk_{t_8}) \wedge C(z', d(x, tk_{t_8}^{-1}))$.
    So, by axiom TA5 we must have $\neg \sigma_{j,t'}(z, d(x, tk_{t_8}^{-1}))$, which gives a
    contradiction.

5.4 Proving G2

In this subsection, we prove G2, which is as follows:

$$K_{B,t_9}(d(e(\{X_a, R_a, A\}, tk_{t_8}), tk_{t_8}^{-1}))$$

If Bob receives the messages in protocol steps 5 and 8, the following statements
hold true:

$$K_{B,t_6}(R(B, t_6, d(\{e(\{X_a, R_a, A\}, tk_{t_8}), A, B, t_8, N_b, tk_{t_8}\}, k_A^{-1}))) \tag{8}$$
$$K_{B,t_9}(R(B, t_9, d(\{\text{``dec``}, t_8, tk_{t_8}^{-1}\}, k_T^{-1}))) \tag{9}$$

Since we assume that a signed message consists of a message concatenated with
the signature for that message, the following statement hold true:

$$K_{B,t_6}(R(B, t_6, e(\{X_a, R_a, A\}, tk_{t_8})) \tag{10}$$
$$K_{B,t_9}(R(B, t_9, tk_{t_8}^{-1})) \tag{11}$$

We remark that it may be possible to capture the above inference in the logic
itself. For example, we can have as axiom the formula, $R(\Sigma, t, (x_1, \ldots, x_n)) \rightarrow$

$R(\Sigma, t, x_i)$, which is similar in form to axiom A8 of the Stubblebine-Wright logic. For brevity, however, we do not treat this axiom explicitly in the logic. To continue with the proof, by (10), (11) and axiom A2, we have:

$$R(B, t_6, e(\{X_a, R_a, A\}, tk_{t_8})) \tag{12}$$

$$R(B, t_9, tk_{t_8}^{-1}) \tag{13}$$

By (12), (13) and axiom A6, we have:

$$L_{B,t_6}(e(\{X_a, R_a, A\}, tk_{t_8})) \tag{14}$$

$$L_{B,t_9}(tk_{t_8}^{-1}) \tag{15}$$

By (14) and axiom A3(a), we have:

$$L_{B,t_9}(e(\{X_a, R_a, A\}, tk_{t_8})) \tag{16}$$

Finally, by (15), (16) and axiom TA2(b), we obtain:

$$L_{B,t_9}(d(e(\{X_a, R_a, A\}, tk_{t_8}), tk_{t_8}^{-1})),$$

which is as required.

5.5 Proving G3

In this subsection, we prove G3, which is as follows:

$$K_{B,t_6}(\exists t \ t_0 < t < t_6 \ S(A, t, d(\{U, N_b\}, k_A^{-1}))), \text{where}$$
$$U = e(\{X_a, R_a, A\}, tk_{t_8}), A, B, t_8, tk_{t_8}$$

We first discuss an approach to proving G3 that doesn't seem to work. Note that G3 is proved if we prove that the following statements hold:

$$K_{B,t_6}(\exists i \in ENT \ \exists t \ t_0 < t < t_6 \ S(i, t, d(\{U, N_b\}, k_A^{-1}))) \tag{17}$$

$$K_{B,t_6}(\forall i \in ENT \setminus \{A\} \ \forall t < t_6 \ \neg S(i, t, d(\{U, N_b\}, k_A^{-1}))) \tag{18}$$

However, although (17) should intuitively be expected to hold, the same cannot be said of (18). Clearly, some entity other than Alice and Bob can intercept the message sent by Alice and *re-send* it, which violates the statement expressed in (18). For this reason we find it unsatisfying to attempt proving G3 in the above way. We choose to augment the logic with the following axiom to help us prove the desired authentication goal:

A12. $\forall i \in ENT \setminus \{\Sigma\} \ L_{i,t}d(x, k_\Sigma^{-1}) \wedge L_{i,t}k_\Sigma \rightarrow \exists t' < t \ S(\Sigma, t', d(x, k_\Sigma^{-1}))$

This axiom captures the origin of a signed message; it closely follows the BAN logic message meaning rule for public keys and some comparable axioms found in other logics (see, for example, axiom A8 of the Stubblebine-Wright logic). The main difference is that it explicitly captures a time constraint, which is generally not expressed in equivalent axioms found in BAN-like logics.

The proof of G3 is done in two parts. Using axiom A12, we first prove that:

$$K_{B,t_6}(\exists t \ t < t_6 \ S(A, t, d(\{U, N_b\}, k_A^{-1}))) \tag{19}$$

It then remains to show that:

$$K_{B,t_6}(\forall i \in ENT \ \forall t < t_0 \ \neg S(i, t, d(\{U, N_b\}, k_A^{-1}))) \tag{20}$$

holds. The proof of (19) is as follows. By (8), and axioms A6 and A1, we have:

$$K_{B,t_6}(L_{B,t_6}(d(\{U, N_b\}, k_A^{-1}))) \tag{21}$$

By assumption (6) and axioms A3(a), (b) and A1, we have:

$$K_{B,t_6}(L_{B,t_6}k_A) \tag{22}$$

It is now easy to see that (19) follows by (21), (22), theorem (K2) and axioms A12 and A1.

The proof of (20) is by contradiction. Suppose that the opposite holds, that is:

$$K_{B,t_6}(\exists i \in ENT \ \exists t < t_0 \ S(i, t, d(\{U, N_b\}, k_A^{-1}))) \tag{23}$$

By (23) and axioms A5 and A1, we have:

$$K_{B,t_6}(\exists i \in ENT \ \exists t < t_0 \ L_{i,t}(d(\{U, N_b\}, k_A^{-1}))) \tag{24}$$

By (24) and axioms A4(a), (d) and A1, we have:

$$K_{B,t_6}(\exists j \in ENT \ \exists t < t_0 \ L_{j,t}N_b), \tag{25}$$

which yields a contradiction by axiom A3(b) and assumption (7).

6 Conclusions

We presented a logic to reason about timed-release public-key protocols, which is built upon an existing logic due to Coffey and Saidha. We have argued that the explicit use of time is essential to proving the correctness of such protocols. Our logic helps us formally prove the correctness of an example protocol, and is especially useful in identifying a number of important assumptions that need to be made when proving the protocol correct. For example, using the contains operator one can formalize assumptions which require that Alice and Trent do not divulge time-confidential data prior to the release time, see assumptions (1) and (2).

There are two important directions for further work. Work needs to be done to define the formal semantics of the augmented logic and to prove the soundness of the new axioms. We also need to apply the logic to more protocols to illustrate its scope and applicability.

References

[AT91]   M. Abadi and M. Tuttle, "A Semantics for a Logic of Authentication," *Proc. 10th ACM Symposium on Principles of Distributed Computing*, pp. 201–216, Aug. 1991.

[B90]   P. Bieber, "A Logic of Communication in Hostile Environment," *Proc. IEEE Computer Security Foundations Workshop*, pp. 14–22, 1990.

[BAN90] M. Burrows, M. Abadi, and R. Needham, "A Logic of Authentication," *ACM Trans. on Computer Systems*, Vol. 8, No. 1, pp. 18–36, Feb. 1990.

[CS97]   T. Coffey and P. Saidha, "Logic for Verifying Public-Key Cryptographic Protocols," *IEE Proc. Computers and Digital Techniques*, Vol. 144, No. 1, pp. 28–32, Jan 1997.

[GNY90] L. Gong, R. Needham and R. Yahalom, "Reasoning about Belief in Cryptographic Protocols," *Proc. IEEE Symposium on Security and Privacy*, pp. 234–248, May 1990.

[GS95]   L. Gong and P. Syverson, "Fail-Stop Protocols: An Approach to Designing Secure Protocols", *Pre-proc. 5th IFIP Working Conference on Dependable Computing for Critical Applications*, pp. 44–55, Sep 1995.

[MB93]   W. Mao and C. Boyd, "Towards Formal Analysis of Security Protocols," *Proc. IEEE Computer Security Foundations Workshop*, pp. 147–158, 1993.

[RSW96] R. L. Rivest, A. Shamir, and D. A. Wagner, "Time-Lock Puzzles and Timed-Release Crypto," Technical Report, MIT Laboratory for Computer Science, 1996.

[SW96]   S. G. Stubblebine, and R. N. Wright, "An Authentication Logic Supporting Synchronization, Revocation, and Recency," *Proc. 3rd ACM Conference on Computer and Communications Security*, pp. 95–105, Mar. 1996.

[S93]   P. Syverson, "Adding Time to a Logic of Authentication," *Proc. 1st ACM Conference on Computer and Communications Security*, pp. 97–101, Nov 1993.

[SVO94] P. Syverson and P. van Oorschot, "On Unifying Some Cryptographic Protocol Logics," *Proc. IEEE Symposium on Security and Privacy*, pp. 14–28, May 1994.

[WK96]   G. Wedel and V. Kessler, "Formal Semantics for Authentication Logics," *Proc. European Symposium on Research in Security and Privacy*, pp. 219–241, 1996.

Bringing Together X.509 and EDIFACT Public Key Infrastructures: The DEDICA[1] Project

Montse Rubia, Juan Carlos Cruellas, and Manel Medina

Telematics Applications Group - Department of Computer Architecture
Universitat Politècnica de Catalunya
c / Jordi Girona 1-3, Mòdul D6
08034 - Barcelona (SPAIN)
E-mail : {montser, cruellas, medina}@ac.upc.es

Safelayer Secure Communications S.A.
Edificio World Trade Center (s4)
Moll de Barcelona s/n
08039 – Barcelona (SPAIN)
E-mail : montse@safelayer.com

Abstract. This paper introduces the barriers of interoperability that exist between the X.509 and EDIFACT Public Key Infrastructures (PKI), and proposes a solution to remove them. The solution goes through the DEDICA (Directory based EDI Certificate Access and management) Project. The main objective of this project is to define and to provide the means to make these two infrastructures inter-operable without increasing the amount of information to be managed by them. The proposed solution is a gateway tool interconnecting both PKIs. The main goal of this gateway is to act as a TTP that "translates" certificates issued by one PKI to the other's format, and then signs the translation to make it a new certificate. The gateway will, in fact, act as a proxy Certification Authority (CA) of the CAs of the other PKI, and will take the responsibility of the certified data authenticity, on the behalf of the original CA.

1. Introduction

The growth and expansion of the electronic means of communication involves the need for certain mechanisms providing services that make secure these communications. These services are mostly based on asymmetric cryptography, which requires an infrastructure (PKI) to make available the public keys.

---

[1] This project has been funded by the EU Telematics program and the Spanish CICYT, and has been selected as one of the pilot projects to promote the telematic applications by the SMEs by the G7.

A Public Key Infrastructure (PKI) is a system for publishing the public-key values used in public-key cryptography. Each type of PKI defines how the security objects are managed. A big amount of effort has been devoted to specify and develop several PKIs during last years. On the one hand, Electronic Data Interchange (EDI) has been used for electronic transactions within certain economical sectors. Security concerns related to the EDI infrastructure have lead to the emergence of security services based on electronic certificates and digital signatures, which have been defined by the corresponding standardisation bodies [2]. On the other hand, security concerns have also resulted in the emergence of similar security services in open systems and distributed applications. In this scenario, several initiatives around the world have caused the emergence of PKIs based on X.509 certificates[ITU-T X.509], as SET (Secure Electronic Transaction) [18] or PKIX (Internet Public Key Infrastructure) [2].

However these PKIs (the EDIFACT infrastructure based on EDIFACT certificates, and the infrastructure based on X.509 certificates) are not inter-operable, mainly due to the fact that certificates and messages use different encoding mechanisms (ASN.1 and DER are used for X.509 PKI, EDIFACT syntax for EDIFACT PKI).

1.1 X.509-based Public Key Infrastructure

X.509 is the authentication framework designed to support X.500 directory services. Both X.509 and X.500 are part of the X series of international standards proposed by the ISO and ITU[2].

The mechanism used by X.509 in order to manage the revocation issues is the Certificate Revocations Lists (CRLs).

The version 3 of the X.509 certificate introduced changes in the X.509 standard. The main change is the introduction of extensions in the certificate and CRL formats. The most meaningful X.509 extensions are those related to: the certificate policies and policy mapping, the alternative names for the subject and issuer, the constraints of the certification path, and the attributes for the directory entries related to the certificate subject. Extensions have also been standardised for CRLs, the most significant ones are: the CRL number, distribution points and reason codes, and the extensions related to the mechanisms for reducing size of the revocation lists.

Nowadays many companies around the world use products based on X.509 certificates, such as: Visa, MasterCard and Eurocard (the Secure Electronic Transaction standard is based on X.509), some WEB browsers (as the Netscape software), the IETF's S/MIME Internet mail [21], or PKIX (Internet Public Key Infrastructure) [12].

1.2. EDIFACT-based Public Key Infrastructure

Another PKI type is based on the EDIFACT certificate. Electronic Data Interchange (EDI) is the computer to computer transfer of commercial or administrative documents, using an agreed standard to structure the transaction and message data. In the EDI world the international acknowledged standard is EDIFACT (EDI For

---

[2] ISO: International Standard Organization, <http://www.iso.ch>, ITU: International Telecommunication Union, <http://www.itu.ch>

Administration, Commerce and Transport). Groups of experts on different areas have worked on the specification of EDIFACT compliant messages, producing UN/EDIFACT (a set of internationally agreed standards, and guidelines for the electronic interchange of structured data). EDIFACT syntax defines a way of structuring information from a basic level where one finds sequences of characters (representing numbers, names, codes, etc.) to the highest level (the interchange). The interchange is composed of sequences of messages (the electronic version of paper documents), which in turn are built of sequences of relevant pieces of information called segments (to represent date and time, for instance). The EDIFACT certificates are encoded in the EDIFACT syntax, and are formed by segment groups related to general certificate information, algorithm and key information and the CAs digital signature (USC-USA(3)-USR). Since EDIFACT certificates are structured and encoded using the EDIFACT syntax, they can be included inside EDIFACT messages.

Certificate management is done using the UN/EDIFACT KEYMAN message [1]. This message can be used to request, deliver, renew, replace or revoke certificates, and to deliver lists of revoked certificates and certification paths.

In the EDIFACT environment a Certificate Revocation List (CRL) is a KEYMAN message, digitally signed by the CA, containing the identifiers of certificates that have been revoked by the issuing CA. A receiver of a signed EDIFACT message with a certificate can retrieve the CRL from a publicly accessible repository to determine if that certificate is in the list of revoked certificates. Alternatively, a security domain could delegate in a trusted authoritative entity the facility for validating individual certificates. In this context, users wanting to validate a received certificate would request validation by sending a KEYMAN message to this trusted entity.

EDIFACT also defines revoked certificates. In this way, a certificate owner can request to the CA the generation of the revoked certificate. The CA modifies the certificate by setting a flag indicating the new certificate status, inserting the reason of the revocation, and signing again the certificate.

EDIFACT is also able to support the transmission of X.509 certificates. The EDIFACT syntax defines a structure, the *package*, that allows the transport of binary data (in the case of X.509 certificates, the DER-coded X.509 certificate). This structure is basically a bit string placed between one header and one trailer (UNO and UNP segments) where, among other things, the type of binary information in the package and the length of the bit string is specified. The package is placed in an interchange at the same level than a message. A KEYMAN message is used to indicate that a package containing a X.509 certificate is included in the interchange and to identify the certification service using that certificate.

1.3. DEDICA Project: the solution to the interoperability problems between the X.509-based PKI and the EDIFACT PKI

DEDICA (Directory based EDI Certificate Access and management) is a research and development project established by the European Commission under the Telematics Applications programme. Its main objective is to define and to provide the means to make the two above-mentioned infrastructures inter-operable without increasing the amount of information to be managed by them. The proposed solution involves the design and implementation of a gateway tool interconnecting both PKIs: the certification infrastructure, currently available, based on standards produced in the open systems world, and the existing EDI applications, which follow the UN/EDIFACT standards for certification and electronic signature mechanisms.

The main goal of the gateway proposed by DEDICA is to act as a Trusted Third Party (TTP) that "translates" certificates issued in one PKI to the other's format, and then signs the translation to make it a new certificate. In this way, any user certified, for instance, within an X.509 PKI could get an EDIFACT certificate from this gateway without having to register against an EDIFACT authority. The gateway will act, in fact, as a proxy CA of the CAs of the other PKI, taking the responsibility to certify the authenticity of the signed data.

The tools developed for the gateway can also be used in systems with a mixture of components, as they can allow CAs to behave as such in both PKIs, i.e. issuing certificates in the two formats. This gives a broader scope to the work done, as it can be the starting point of further specifications and developments leading to inter-operability among other currently arising PKIs (SPKI, for instance). This environment is also easier to manage, from the responsibility viewpoint, since both, CA and Gateway, run the same certification policy and belong to the same security domain.

The figure below shows the DEDICA gateway context. Each user is registered in his PKI and accesses the certification objects repository related to this PKI. The DEDICA gateway must be able to interact with the users of both PKIs in order to serve requests from them. It must also be able to access the security object stores of both PKIs, and to be certified in EDIFACT and X.509 CAs.

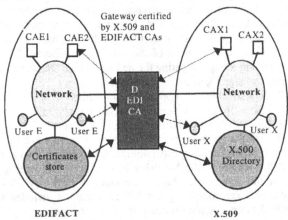

Fig. 1. DEDICA gateway context

2. Functionality of the gateway

The interoperability problem between the X.509 and EDIFACT PKIs was focused by the DEDICA project in two levels: the different certificate formats and the different messages interchanged by the PKIs entities.

Concerning the first one, the DEDICA consortium, after an in-depth study of the contents of both types of certificates, specified a set of mapping rules that make possible the two-way translation of both types of certificates [20]

The other problem that DEDICA has addressed is related to differences in messages and certification services: whereas in the EDIFACT world the UN/EDIFACT KEYMAN message is used to provide certification services, in the X.509 world a set of messages specified for each PKI (PKIX on the Internet, for instance) are used.

The DEDICA gateway assumes the role of a TTP for users of both infrastructures. The gateway accomplishes a process of certificates translation from EDIFACT to X.509 and conversely; however **this translation process is not strictly a mapping at level of certificate formats, since the gateway adds a digital signature to the mapped data**. In addition to that, in some cases, it is not possible just to move data from one certificate to the other, due to format restrictions (size, encoding). In these cases the gateway has to generate tagged data for the derived certificate, that will allow it to reproduce the original data, kept in internal records (e.g. names mapping table, see fig. 3). When the X.509 certificate has private extensions, the gateway will just ignore them, since they are assumed relevant only to other applications.

Full details of the mapping mechanism between both infrastructures may be found at: http://www.ac.upc.es/DEDICA/ and at DEDICA CEC-Deliberable WP03.DST3: *Final Specifications of CertMap Conversion Rules* [6].

The DEDICA gateway is able to offer four services:

1. Request of an EDIFACT certificate from an X.509 certificate generated by an X.509 CA.
2. Verification of an EDIFACT certificate generated by the DEDICA gateway (coming from the mapping of an X.509 certificate).
3. Request of an X.509 certificate from an EDIFACT certificate generated by an EDIFACT CA.
4. Verification of an X.509 certificate generated by the DEDICA gateway (coming from the mapping of an EDIFACT certificate).

Figure 1 shows the DEDICA gateway context. This context is as follows: An X.509 PKI with CAs, users and X.500 Directory access (the Directory operating as a repository of security objects), an EDIFACT PKI with CAs and users, and the DEDICA gateway, certified by CAs in both PKI and access to X.500 Directory.

The DEDICA gateway defines the terms *Initial certificates* and *Derived certificates* as follows:

Initial Certificate: any certificate supplied by a Certification Authority of any Public Key Infrastructure (EDIFACT or X.509).

Derived Certificate: certificate coming from an initial certificate, generated by the DEDICA gateway through the mapping rules defined by DEDICA. The derived certificate depends on the initial one; if the initial certificate is revoked, then the derived certificate will also be considered revoked.

2.1. Request of a derived certificate

In the scenario shown in Figure 2, an X.509 user (user X) that may want to send EDIFACT messages to an EDIFACT user (user E) using digital signatures or any security mechanism that involves the management of certificates. This user needs a certificate from the other Public Key Infrastructure (in this case, the EDIFACT PKI). He then sends an interchange to the gateway requesting the production of an EDIFACT certificate "equivalent" to its provided X.509 one. This interchange will contain a KEYMAN message (indicating a request for an EDIFACT certificate) and the X.509 certificate of this user in an EDIFACT package (EDIFACT structure capable of containing binary information).

The gateway will validate the X.509 certificate. If the certificate is valid (the signature is correct, it has not been revoked, and it has not expired), it will perform the mapping process, and will generate the new derived EDIFACT certificate. After that the gateway will send it to user X within a KEYMAN message.

Now user X can establish a communication with user E using security mechanisms that involve the use of electronic certificates through the derived EDIFACT certificate, sending him an EDIFACT interchange with this derived certificate.

2.2. Validation of a derived certificate

The DEDICA gateway also validates derived certificates upon user request. Following the process described in the previous section, user E, after receiving the interchange sent by user X, requests validation of the derived certificate by sending the corresponding KEYMAN message to the gateway.

The gateway determines whether the EDIFACT certificate has been generated by itself, and proceeds with the validation of the original X.509 certificate and the derived EDIFACT certificate. It will have to access the X.500 Distributed Directory to get both the original X.509 and the Certificate Revocation Lists (CRL) needed for the certificate validation process. The general process of validation of derived certificates is as follows:

1. It verifies the validity of the derived certificate. This requires checking of:
    (a) The correctness of signature, using the public key of the gateway.
    (b) Whether the certificate can be used, in view of the validity period.
2. The gateway accesses to the X.500 Distributed Directory, in order to get the original X.509 certificate and the necessary Certificate Revocation Lists.
3. It verifies the signature of the original certificate, and checks the validity period.
4. The gateway verifies the certification path related to the original X.509 certificate, and checks that its certificates have not been revoked.

If all these verifications are successful, then the derived EDIFACT certificate can be considered to be a valid one, and the gateway will send the positive validation response to the EDIFACT user within a KEYMAN message.

3. Gateway architecture

The DEDICA gateway has two main architectural blocks: the **CertMap** and the **MangMap** modules. The first is responsible for the certificate mapping, and the second converts the functionality of the KEYMAN message into equivalent operations of X.509 PKI (including X.500 access).

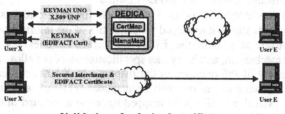

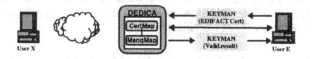

Validation of a derived certificate

Fig. 2. Functionality of the DEDICA gateway

3.1. CertMap module

The CertMap module is responsible for performing the certificate translations following the mapping rules specified by the DEDICA consortium [6].

Mapping between X.509 and EDIFACT certificates. The Certificate Mapping Rules developed in DEDICA were designed in such a way that the translated information was relayed as precisely as possible from the original certificate to the derived one. A number of issues had to be taken into account:

- Specification syntax and transfer syntax for the transmission. The EDIFACT certificates are specified following the EDIFACT syntax, and they are transmitted coded in printable characters. However, in the X.509 environment the ASN.1 Abstract Syntax and the DER rules are used.

- Naming System. In the X.509 world, the basic mechanism of identification is the Distinguished Name (DN) [7], which is associated with an entry in the DIT (Directory Information Tree) of the X.500 Distributed Directory. Furthermore, the X.509 certificate supports a variety of types of names in addition to DN, such as the RFC822 name, the URLs and even the EDI party names. On the other hand, the EDIFACT certificate supports both codes (i.e., identifiers assigned by authorities) and EDI party names. The new version of the EDIFACT certificate incorporates mechanisms that allow it to carry Distinguished Names in certain fields of this certificate. The DEDICA gateway performs a name mapping between the Distinguished Names and the EDI Names, according to guidelines defined in EDIRA (EDIRA Memorandum of Understanding) [8]. EDIRA proposes an identification mechanism compatible with the DN strategy in X.500. The DEDICA Deliverable WP03.DST2([5]) contains the specifications of the conversion rules that are used by the CertMap module to execute the mapping between Distinguished Names and EDI Names.

- Extension mechanism. Version 3 of the X.509 certificate has an extension mechanism that allows it to extend the semantics of the information that it carries out. However, at present the EDIFACT certificate does not have any the extension mechanism, and its syntax specification does not allow to specify such a wide variety of information. In the mapping of X.509 certificates version 3, only the following extensions will be mapped: *keyUsage* and *subjectAltName*. The other extensions will not be mapped since the associated information is not carried in the EDIFACT certificate.

- Digital signature. When the gateway finishes the mapping process, it automatically generates a new digital signature. In the certificate field identifying the issuer entity, the DEDICA gateway identifier will appear, instead of the original certificate issuer identification.

Internal structure of the CertMap module. The CertMap is composed of three main modules:

- **CM_Kernel** module (**CM_KE**). This module co-ordinates the operations performed by all the other CertMap modules. It receives an original EDIFACT as input, or a DER encoded X.509 certificate, and it returns the derived certificate to the MangMap.
  Four groups of information presentes in both certificates have been identified: Names, Algorithms, Time and Keys. For each one of these groups, a software module that implements the appropriate relevant translation process has been designed: the CM_Names, the CM_Algorithm, the CM_Time and the CM_Keys modules.

- **EDIFACT certificate coding/decoding** module (**CM_CE**). This module is able to extract all the information contained in a serialised EDIFACT certificate. This information will be stored in the format of an agreed structure data type to manipulate it during the mapping process. This module is also able to generate the character stream corresponding to a derived certificate.

- A set of **APIs** needed to allow the CM_KE to interact with external software tools. Two of these tools have been identified as necessary: an ASN.1 tool and a cryptographic tool. The APIs needed are the following ones:

  - **CM_KE:ASN1 API**. This API will provide the CM_KE with the means to extract the information from the DER coded X.509 initial certificate, and instructs the ASN.1 tool to obtain the DER stream corresponding to an X.509.

  - The **CM_KE:CRYPTOGRAPHIC API**. This API will provide the means for the CM_KE to interact with cryptographic tools that will allow the signing of the derived certificates.

In DEDICA, a table that contains mapping information was defined: **The Names Mapping Table**. This table will establish links between initial and derived certificates. These links are generated by the CertMap module.

The figure 3 shows the internal structure of the CertMap module. It also shows the sequence of operations that will take place inside the CertMap to generate an EDIFACT derived certificate from the initial X.509 one.

It can be seen how the different parts of the system take part in the generation of the derived certificate. Note that the effective mapping process is performed, as mentioned above, by modules CM_NM, CM_AL, CM_TM, and CM_PK.

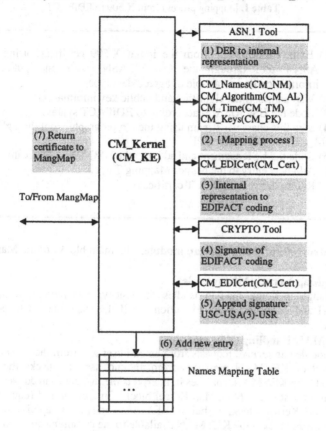

Fig. 3. Mapping process from X.509 to EDIFACT

Table 1 shows a high level description of the tasks that allow the generation of an EDIFACT derived certificate from an initial X.509 certificate.

3.2. MangMap module

In addition to the data elements, the DEDICA gateway has to convert certain operations of the KEYMAN message into equivalent operations (messages) in the X.509 PKI (including X.500 access). This is done by the MangMap module of the DEDICA gateway.

MangMap is also the general management module for DEDICA gateway. It receives all the requests sent to it and chooses which information has to be recovered from external repositories, what type of translation is needed, and what results must be generated and sent to the requesting entity.

Table 1. Mapping process from X.509 to EDIFACT

(1) Extract information from the initial X.509 certificate using the ASN.1 tool through the CM_KE:ASN1 API and put such information into a variable of agreed data type.
(2) Map names, time, algorithm and public key information.
(3) Code the USC-USA(3) according to EDIFACT syntax.
(4) Sign this character stream using the Cryptographic tool through the CM_KE:Cryptographic API.
(5) Code the USR segment and add it to the USC-USA (3) stream.
(6) Add new entry to the Names Mapping Table.
(7) Return derived EDIFACT certificate.

Internal structure of the MangMap module. The main blocks of the MangMap are as follows:

- **MangMap Kernel (MK) module**
  The MangMap Kernel module handles different types of requests from both KM and XH and co-ordinates the execution of all the steps needed to perform the request.
- **KEYMAN Handling (KH) module**
  This module can receive requests from an end user and from the kernel block. On reception of KEYMAN messages from an end user, it checks the protection applied to the KEYMAN, analyses it, interprets the message and converts it into an internal request to the MangMap Kernel block. On reception of requests from the MangMap Kernel block, it builds KEYMAN messages, applies the required protection and makes the KEYMAN available to the communication services.
- **X.509 Public Key Infrastructure Messages Handling (XH) module**
  On reception of relevant X.509 public key infrastructure messages from an end user, XH module checks the protection applied to the message, analyses it and converts the message into an internal request to the MK.
  XH is also able to access the X.500 Directory in order to get X.509 certificates, revocation lists and certification paths. XH will be able to send requests to X.500 and to obtain and interpret answers from it.
  On reception of requests from MK, it builds relevant X.509 public key infrastructure messages, applies the required protection and makes the messages available to the communication service.
  Figure 4 shows the building blocks of the MangMap module and its relationships with the CertMap module.

Information flow example: Derived EDIFACT certificate request. This section shows the information flow inside the building blocks of the MangMap module, when an X.509 certificated user requests a derived EDIFACT certificate. A slightly different flow occurs when validation of this certificate is required.

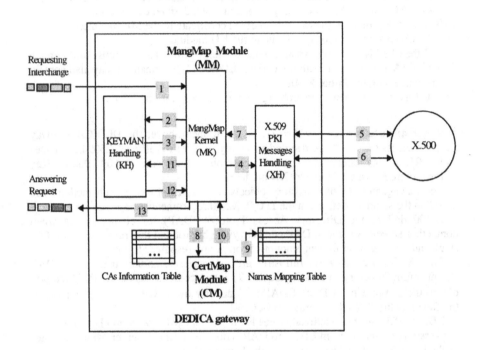

Fig. 4. Derived EDIFACT certificate request

The following list shows a high level description of the tasks performed by the building blocks in the gateway:

1. The requesting interchange arrives at the MangMap Kernel (MK) block in MangMap. This interchange will carry a KEYMAN message with a package containing the DER-coded X.509 certificate, or even the DER-coded Distinguished Name of the user.
2. MK identifies the request and forwardsforwards it to the KEYMAN Handling (KH) block.
3. KH analyses the KEYMAN message and sends back to the MK the information needed to provide the derived certificate.
4. MK instructs the X.509 PKI Messages Handling block (XH) to access the X.500 server to retrieve all the information needed to verify the validity of the initial X.509 certificate (the original X.509 certificate and CRLs).
5. XH fetches that information from the X.500 Directory and checks whether the initial certificate has been revoked.
6. XH fetches the Certification Path from the X.500 Directory and verifies the signatures of the involved CAs.

7. If the initial certificate has not been revoked and the Certification Path has been successfully verified, XH informs MK that the corresponding derived EDIFACT certificate can be generated.
8. MK then instructs the CertMap module to generate this EDIFACT certificate.
9. The CertMap generates the derived certificate and also creates a new entry in the Names Mapping Table associating both initial and derived certificates.
10. The CertMap module forwards the derived certificate to the MK module.
11. MK forwards the derived certificate to the KH module.
12. KH then builds the response and sends it back to MK. This response will contain the KEYMAN message with the derived EDIFACT certificate. It may also contain a package with the initial X.509.
13. MK forwards the answer interchange to the communication system.

X.500 Distributed Directory Access. As mentioned above, the DEDICA gateway needs to access the X.500 Distributed Directory in order to retrieve X.509 certificates and CRLs. This access is mainly accomplished by the X.509 Public Key Infrastructure Messages Handler (XH module).

The management of the directory objects and the employed search methods are based on the standard defined in the ITU-T X.500 recommendation.

X.500 defines the Directory Access Protocol (DAP) for clients to use when contacting Directory servers. DAP is a heavyweight protocol that runs over a full OSI stack and requires a significant amount of computing resources to run.

The X.509 Public Key Infrastructure Messages Handler uses the LDAP (Lightweight Directory Access Protocol) [9] interface to access the X.500 Directory, due to the complexity of DAP. LDAP runs directly over TCP and provides most of the functionality of DAP at a much lower cost.

LDAP offers all the functionality needed to interact with the X.500 Directory. The conversion of requests from LDAP to DAP is done by a LDAP server, which operates as the DUA (Directory User Agent) of the Directory.

4. Systems in which DEDICA is being integrated

The DEDICA consortium has developed both the gateway and client software that is being integrated within the existing EDI applications. Within the project several pilot schemes have been launched in the following fields: customs, EDI software providers, electronic payment in banking and public administration application forms for electronic commerce.

DEDICA is also being integrated as part of the TEDIC system. The TEDIC system has been developed by THOMSON C.S.F. / ISR / URACOM, and it offers a legal solution for EDI Trade transactions without any prior contact to establish a specific "interchange agreement" between the involved parties. The TEDIC system offers a set of security services to the different kinds of messages based on a range of security levels. It sets up a hierarchy based on a system of security policies, and allows dynamic and automatic negotiation of the transaction policy level. The integration of the DEDICA gateway in the TEDIC system means that the TEDIC users can use X.509 certificates as well as the EDIFACT certification. In this way all the users

registered in an X.509 PKI can become TEDIC users, and they do not need to register also in an EDIFACT CA.

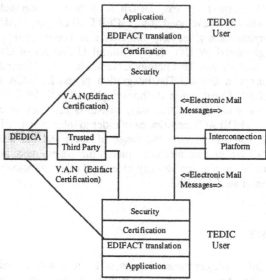

Fig. 5. TEDIC in a heterogeneous environment

Another system where DEDICA is also being integrated is the EDIINT project EDI over INTERNET), promoted by AECOC (the EAN Spanish representative)[3] and UPC (Polytechnical University of Catalonia). This working group is specifying methods and developing tools that allow secure EANCOM interchanges[4] and sends them using any single or multiple transport protocols over Internet. This approach makes possible the transparent transfer of EDI interchanges through any protocol gateway or proxy interposed between sender and recipient. These security services follow the standards developed by the UN/EDIFACT "Security Joint Working Group" (SJWG).

The figure below shows the scenario of this project. In this scenario a security services module will be added to the EDI stations. The EDI messages sender generates EDIFACT messages, applying the security services required. The interchanges may either be stored in an internal repository, or sent to the EDI message recipient. When the recipient receives the EDIFACT interchange, he can verify the validity of the applied security objects or store it in the interchange repository.

Because public key cryptography is used, a Certification Authority is needed in order to generate the public key certificates to the users. The users can request the CA for both the certification of a public key and a specific certificate.

In the described scenario only EDI items appear, although an EDI user could have an X.509 certificate generated by an X.509 CA. The DEDICA gateway allows them

---

[3] AECOC: Since 1977 AECOC has held the representation of EAN International in Spain. AECOC is responsible for the management and promotion of the bar-codes of products and services, as well as the Electronic Data Interchange (EDI); <http://www.aecoc.es>

[4] EANCOM is in charge of the management of profiles of EDIFACT messages standardised by EAN.

to use also the described scenario, without changing the tools and modules related to the EDIINT infrastructure.

In the EDIINT project context, DEDICA is being integrated into an X.509 Certification Authority developed by esCERT-UPC[5] for SAFELAYER[6]. The esCERT-UPC organisation gives help and advice on computer security and incident handling, it is sponsored by UPC (Polytechnical University of Catalonia), CICYT (Interministerial Science and Technology Commission), the Generalitat of Catalonia and the EU (European Union). The integration of the DEDICA gateway and the SAFELAYER X.509 Certification Authority[7] gives the X.509 CA the capability to manage EDI security objects. In this way, an X.509 user can have not only X.509 certificates but also EDIFACT certificates in order to interact with EDI users.

The DEDICA gateway extends the range of users and the infrastructures with which the EDIINT system can interact. It makes the system more heterogeneous and provides it with tools to manage security objects of other infrastructures, such as the X.509 certificates of an X.509 PKI

5. Future work

Part of the DEDICA project consortium is still under way in order to extend the functionality offered by the gateway, especially in the domain of validation and revocation management. In EDIFACT, a Certificate Revocation List is a KEYMAN message, digitally signed by the CA, containing the identifiers of certificates that have been revoked by the issuing CA. A receiver of a signed EDIFACT message with a certificate can retrieve the CRL from a publicly accessible repository to determine whether that certificate is in the list of revoked certificates. Alternatively, a security domain could delegate to a trusted authoritative entity the facility for validating individual certificates. In this context, users wanting to validate a received certificate would request validation by sending a KEYMAN message to this trusted entity.

At present, the revocation management is indirectly solved:

a) the user asks the gateway for the validation of a derived EDIFACT certificate;
b) then the gateway recovers the original X.509 certificate related to the derived one, and
c) it looks for this certificate in CRLs published by the issuing CA.

The revocation management of the DEDICA gateway could then go through the publication of EDIFACT Revocation Lists of derived certificates generated by the gateway. The DEDICA gateway could periodically check the status of the original X.509 certificates it has translated. When one of such certificates has been revoked, the derived one should also be flagged as revoked. The gateway would then generate the KEYMAN message with the corresponding revocation list of EDIFACT derived certificates, and it would publish it in the X.500 Distributed Directory.

When the DEDICA gateway receives a validation request of an EDIFACT derived certificate, it would look at this EDIFACT CRL published in the X.500. If this

---

[5] esCERT-UPC: Spanish Computer Emergency Response Team; <http://escert.upc.es>

[6] Safelayer Secure Communications S.A. is a company provider of PKI and SET software solutions <http://www.safelayer.com>

[7] A demonstration of this integration can be found at <http://ca.upc.es/dedica> URL

certificate is not is in this list, it would follow the previously explained normal process.

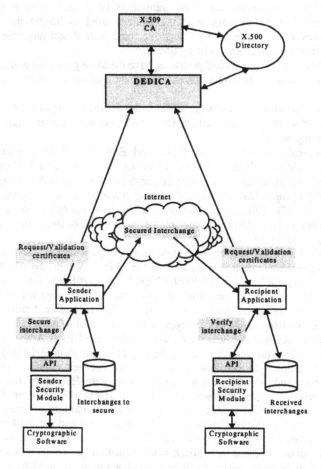

Fig. 6. AECOC project and DEDICA integration context

6. Conclusions

This work has proved the suitability to launch a TTP service to translate security objects between different protocol environments. The problems found in this project are of general nature, and the solutions adopted here may be extrapolated to any other pair of environments.

Other arising PKI services, like SPKI or UMTS (Universal Mobile Telephone System), or XML-EDI are potential candidates to use the results of this work. But it will be possible to extend the results of this work to other TTP services, like Time Stamping, Attribute certification, etc.

The data type conversion based on translation table may solve any format incompatibility, and the message mapping strategy used to handle the different certificate management strategies may also overcome with almost any mismatching services between the two protocols being linked.

As far as both, environments and protocols, have the same goals, the details of data and service elements not having a corresponding element on the other environment may either:

a) be just overridden because it is not useful in the destination application, or
b) be replaced by an equivalent data or service element with similar meaning in the destination protocol.

The interoperability between the X.509 and EDIFACT PKIs can be greatly enhanced by facilities such as the DEDICA gateway, which acts as a TTP capable of offering a basic set of certificate management services to users of both infrastructures.

The DEDICA project has set up a gateway to translate the security objects between X.509 and EDIFACT. This solution also provides interoperability between EDIFACT and all the other tools used in electronic commerce, since all of them authenticate the entities using X.509 certificates.

The DEDICA gateway is being integrated in several pilot schemes and projects in the context of electronic certification, such as the TEDIC system, the AECOC-UPC EDI over Internet project, or in the SAFELAYER X.509 Certification Authority.

The DEDICA service is interesting to both the large enterprises and SMEs, although this gateway is mostly interesting to SMEs. This is because it allows them to use security in the interchange of messages, without the need to pay registration fees in several infrastructures. This was the reason for which the DEDICA was selected as one of the G7 pilots projects to promote the use of information technology by the SMEs. The sharing of the certification infrastructure between e-mail or distributed application users and the EDI users will quickly satisfy the EDI users' requirements for global service provision. The integration of public key certificates from X.509 and EDI will provide the means to share common information and infrastructure between the most widely used telematic application (electronic mail), and the most interesting one from the economical point of view: EDI. The main advantage for the user will be the possibility to use the same authentication mechanism (digital signature, tools, etc.) in the various applications where it can be applied, avoiding the burden of having to register with various services in order to satisfy one single user requirement.

Moreover, the service has been quickly deployed and made available, thanks to the fact that no additional registration infrastructure is needed, due to its compatibility with the EDIFACT and X.509 infrastructures. This service will promote the use of Internet by EDI applications (since it will allow them to secure the interchanges which has been identified) in spite of the major barriers to the deployment of EDI over Internet in the past.

Other functional advantages of the DEDICA gateway are the independence of the communication protocols used by the applications, and its transparency to both types of PKI users. In this way, an EDI/X.509 user does not have to know that the certificate that he is managing is a certificate automatically generated by the DEDICA gateway, and coming from an X.509/EDIFACT certificate.

Within the project, several pilot schemes have been launched to demonstrate the system in the following fields: customs, electronic chambers of commerce, tourism, electronic products manufacturers, EDI software providers and electronic payment in banking and public administration.

References

1. Security Joint Working Group, Proposed Draft of a MIG Handbook UN/EDIFACT Message KEYMAN, 30. June 1995.
2. Security Joint Working Group: Committee Draft UN/EDIFACT CD 9735-5, Electronic Data Interchange for Administration, Commerce and Transport (EDIFACT) - Application Level Syntax Rules, Part 5: Security Rules for Batch EDI (Authenticity: Integrity and Non-Repudiation of Origin, Release 1, 14. December 1995.
3. United Nations, Economic and Social Council, Economic Commission for Europe, Committee on the Development of Trade: TRADE/WP.4/R.1026/Add.2, EDIFACT Security Implementation Guidelines, 22. February 1994.
4. DEDICA Consortium, CEC Deliverable WP03.DST1: Technical description of X.509 and UN/EDIFACT certificates, July 1996.
5. DEDICA Consortium, CEC Deliverable WP03.DST2: Naming Conversion Rules Specifications Requirements, July 1996.
6. DEDICA Consortium, CEC Deliverable WP03.DST3: Final Specifications of CertMap Conversion Rules, July 1996.
7. Network Working Group, RFC 1779: *A String Representation of Distinguished Names*, ISODE Consortium, 1995.
8. EDIRA - Memorandum of Understanding for the Operation of EDI Registration Authorities, Final Draft. November, 1993.
9. Network Working Group, RFC 1959: *An LDAP URL Format*, 1996.
10. Network Working Group, INTERNET-DRAFT: *Lightweight Directory Access Protocol (v3): Attribute Syntax Definitions*, 1997.
11. Network Working Group, INTERNET-DRAFT: *A Summary of the X.500(96) User Schema for use with LDAPv3*, 1997.
12. PKIX Working Group: Internet Public Key Infrastructure - Part III: *Certificate Management Protocols, Internet Draft*, June 1996.
13. PKIX Working Group, INTERNET-DRAFT: *Internet Public Key Infrastructure, X.509 Certificate and CRL Profile*, 1997.
14. PKIX Working Group, INTERNET-DRAFT: *Internet Public Key Infrastructure, Operational Protocols - LDAPv2*, 1997.
15. Fritz Bauspieß, Juan Carlos Cruellas, Montse Rubia, *DEDICA Directory based EDI Certificate Access and Management*, Digital Signature Conference, July 1996.
16. Juan Carlos Cruellas, Damián Rodriguez, Montse Rubia, Manel Medina, Isabel Gallego, WP07.DST2. *Final Specification of MangMap Conversion Rules*, DEDICA Project, 1996.
17. Juan Carlos Cruellas, Damián Rodriguez, Montse Rubia, Manel Medina, Isabel Gallego, WP07.DST1. *Final Specifications of MangMap*, DEDICA Project, 1996.
18. Secure Electronic Transaction (SET) Specification. Book 1: *Business Description*. Version 1.9, May 31, 1997.
19. Secure Electronic Transaction (SET) Specification. Book 2: *Programmer's Guide*. Version 1.9, May 31, 1997.
20. Montse Rubia, Juan Carlos Cruellas, Manel Medina. Removing Interoperability Barriers Between the X.509 and EDIFACT Public Key Infrastructures: The DEDICA Project. In *International wokshop on Practice an Theory in Public Key Cryptography (PKC '99)* (Lectures Notes in Computer Science 1560), pages 245-262, Kamakura, Japan, March 1999. Springer-Verlag.
21. P. Hoffman, *Enhanced Security Services for S/MIME*, Ed. June 1999 (RFC2634)

User Identification System Based on Biometrics for Keystroke

Kazumasa Omote[1] and Eiji Okamoto[2]

[1] School of Information Science
Japan Advanced Institute of Science and Technology,
Asahidai 1-1, Tatsunokuchi, Nomi, Ishikawa, 923-1292 JAPAN
E-mail: omote@jaist.ac.jp
[2] Center for Cryptography, Computer and Network Security
University of Wisconsin Milwaukee, WI 53201, USA
E-mail: okamoto@cs.uwn.edu

Abstract. In this paper, a new type of user identification system based on characteristics (biometrics) for keystroke is proposed. In this system, we do not require any special devices. We utilize two basic parameters: Press-Time, when a user presses a key, Release-Time, when a uses releases the key. Release-Time is a new parameter. By using Release-Time, a user can be identified much more clearly than using only Press-Time. Furthermore, even if a password is stolen, the attacker can hardly find how the owner types because of user's biometrics for keystroke. Hence this system can overcome the weak point of password checking schemes.

1 Introduction

Today computers are connected to others that may be linked to a world-wide network like Internet. So many people get information through such a network. Therefore users must be exactly identified when they get that information. Computer systems must distinguish legal users from illegal users who try to impersonate. That is the reason why user identification systems become more and more important.

Personal identification is a technique that a user registers his/her information like a password in advance, and then his/her input information identify the user who has the correct information. This information consists of objects, knowledge, action, physiology and so on.

The Electronic Commerce Promotion Council of Japan (ECOM) decided base evaluation of personal identification skills in electronic commerce business. This base evaluation is classified into five groups as means of personal identification.

- Biometrics 1 $\cdots$ fingerprints, retina pattern, face pattern
- Biometrics 2 $\cdots$ voice print, handprint
- Possessions $\cdots$ ID card, credit-card
- Secret Information 1 $\cdots$ password, password number
- Secret Information 2 $\cdots$ digital signature

Biometrics is related to biological characteristics of physiology. Biometrics 1 consists of characteristics that cannot be distorted by user's own will. On the other hand, Biometrics 2 uses changeable characteristics with user's own will.

When we introduce biometrics skill into user identification system, we must examine the following four items.

[Security] The acceptance rate should be high. Furthermore it should be difficult for an illegal user to forge or steal legal user's information.

[Cost] The system should realize high performance according to investing costs.

[Handiness] The system should be handy and easy to operate, and it must compute fast to identify a legal user.

[Social acceptance] We can use the system without incompatibility or resistance.

Performance of biometrics skill is decided not only by security but by all of the four items mentioned above. Even if a system lacks any one of the four items, it will not satisfy necessary level of users. So any of the four items should be high, and have good balance.

Nowadays user identification systems based on keyboard mainly adopt a password system. But one of the disadvantages is that somebody may steal his/her password at a glance. To overcome this problem, user identification systems based on biometrics for keystroke have been proposed. For each person, his/her typing rhythm is somewhat different from that of others. This difference shows personal characteristics as far as personal identification. These systems can identify a user by picking out his/her characteristics of keyboard typing.

These systems add biometrics of keystroke (Biometrics 2) to password (Secret Information 1) as base evaluation of personal identification skill. We do not use expensive special purpose terminal, but realize user identification system using keyboard only (standard device). When we use a personal computer or a network computer, we usually use a keyboard. This system can easily fulfill all of the previous items, that is, security, cost, handiness and social acceptance as mentioned above, because password checking schemes are already used.

We will improve these systems by not only a key press but also a key release.

2 Related Work

In 1990 Rick Joyce and Gopal Gupta[1] reported a work related to keystroke dynamics. In this system, new users need to register eight reference signatures by typing their {user name, password, first name, last name} eight times. In verifying a user, a test signature is compared with the mean reference signature.

The signature in this system consists of the press intervals between two successive typings.

In 1993 Kasukawa et al.[2] extended the method of [1]. The main difference between [1] and [2] is as follows: the signature of [1] consists of the successive press intervals, whereas the signature of [2] consists of the press intervals between all two typings in addition to the two successive typings.

In 1997 Fabian Monrose and Aviel Rubin[3] extended the basic research on keystroke dynamics previously studied. They examined the use of keystroke duration (i.e., length of time keys are depressed) in addition to keystroke latencies (i.e., time between successive keystrokes). Collecting users' data, their system uses "Free Style" (i.e., nonstructured) text, which is a few sentences from a list of available phrases. Moreover they applied clustering method to their system in order to reduce the search time from database. Therefor it partition the data into cluster domains.

In 1998 we[4] reported comparative experiments related to four kinds of characteristics which are undermentioned in section 4.3. These experiments say that Release-Press Interval(Characteristic 3) can identify a user the most clearly of the four. In this paper, however, we will examine the extraction of personal characteristics from not only Characteristic 3 but all of the four. Collecting users' data, users need to type their password many times unlike [3]. Each password was different from each other.

3 User Identification System Using Biometrics of Keystroke

3.1 Outline

This method can utilize the human characteristics because each person's typing rhythm is somewhat different. For each person, some key releases may be fast, other keeping presses may be a little longer. The system can distinguish a user from all other users with the aid of this difference. Hence even if your password is disclosed, your computer cannot be abused.

The flow of the system is following.

1. While a user types a password many times, the system measures the typing time.
2. The system preserves the typing time in the computer as a database.
3. When a user types a password, computer figures out the difference of the typing time between that input data and the average database.
4. If this difference is less than a certain threshold, the system regards a user as the legal one.

3.2 Problem Points

When we use this identification system, we need to pay attention as well as we do for an identification system based on handwritten signature or voice (Biometrics

2). The attention is that we need to decrease the difference of typing between registration and identification. We have four problems as follows:

1. We may not type as usual under different conditions.
2. We need to type a keyboard fluently in order to make your typing rhythm stable.
3. If you stop in the middle of typing, you have to retype a password from the beginning.
4. Personal characteristics may change as time goes by.

4 Propose System

4.1 Password

There are two kinds of passwords. One is a short password which can be typed rapidly at a time, and the other is the longer passphrase like a sentence. For example, Unix password is of the former type. It is impossible for a user to type passphrase rapidly at a time. It will impose a burden on a user.

On the other hand, it is possible for a user to type a short password rapidly at a time, and in this case all of the user's finger movement of typing is important. It is a premise that we are used to typing a password and make the typing stable. We type a short password more easily than a passphrase because you type a keyboard without thinking. Hence we use a short password "okamotoken" which is our laboratory's name because all of the users are usually used to typing it.

4.2 Basic Parameter

We think Press-Time and Release-Time as fundamental parameters of personal characteristics. For example, consider the password "word" of four characters as a simple example. If you think not only Press-Time but also Release-Time, you can regard the password as eight characters as follows.

$$w < w > o < o > r < r > d < d >$$

Note that the character in $<>$ represents releasing that key. We explain four kinds of personal characteristics in this section. The value of p_i stands for the ith Press-Time and the value of r_i stands for the ith Release-Time.

4.3 Personal Characteristics[4]

We have investigated the following four kinds of personal characteristics utilizing the two basic parameters, that is, Press-Time, when a user presses a key, Release-Time, when a user releases a key. The characteristics depend upon the Press/Release-Time. Four kinds of the characteristics are

1. Press Interval (Characteristic 1)

2. Release Interval (Characteristic 2)
3. Release-Press Interval (Characteristic 3)
4. Press-Release Interval (Characteristic 4).

These four kinds of characteristics are figured out from Press-Times and Release-Times. Now, we see the precise definitions for these characteristics in sequence.

Press Interval (Characteristic 1) A Press Interval is the interval between two typing. For example, if the number of password is L, $(L-1)$ Press Intervals are obtained from the password. Denoting Press Interval by P_i, P_i is defined as

$$P_i = p_{i+1} - p_i. \tag{1}$$

Fig.1 shows the sample graphs of Press Interval sequences of a user. The flat axis represents keystrokes "okamotoken[R]", whose length is 11 including a return key. The character R in [ ] stands for a return key, and is included in the password. The figure below has 25 graphs in it. The average of this user's Press Interval is about 0.1 second.

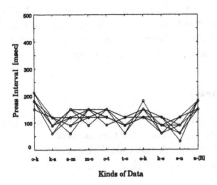

Fig. 1. Press Interval (Characteristic 1)

Release Interval (Characteristic 2) Release Interval is the interval between two key-releases. Release Interval is defined as well as Press Interval. Denoting Release Interval by R_i, R_i is defined as

$$R_i = r_{i+1} - r_i. \tag{2}$$

Fig.2 shows the Release Interval graphs for the same user. These graphs look somewhat different from those in Fig.1.

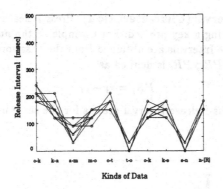

Fig. 2. Release Interval (Characteristic 2)

Release-Press Interval (Characteristic 3) Release-Press Interval is the time from a key-release to the next key-press. For example, if the number of password is L, $(L-1)$ Release-Press Intervals are also obtained from the password. We know that this interval can identify a user the most clearly of the four kinds of characteristics[4]. Denoting Release-Press Interval by RP_i, RP_i is defined as

$$RP_i = p_{i+1} - r_i. \tag{3}$$

Fig.3 shows the Release-Press Interval graphs for the same user. The values of Release-Press Interval may be negative according to the data because a user press a key before releasing the previous key. A user who hits key fast often types like this.

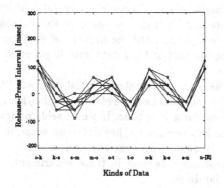

Fig. 3. Release-Press Interval (Characteristic 3)

Press-Release Interval (Characteristic 4) Press-Release Interval is the time interval during keeping a key pressed. For example, if the number of password is L, L Press-Release Intervals are obtained from the password. Denoting Press-Release Interval by PR_i, PR_i is defined as

$$PR_i = r_i - p_i. \tag{4}$$

Fig.4 shows the Press-Release Interval graphs for the same user.

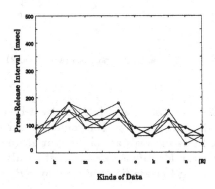

Fig. 4. Press-Release Interval (Characteristic 4)

4.4 Database

The time of personal characteristics is registered into a database in advance. A database can store at most M values, but first we register more than M data in it because the system needs to remove wrong data, which is unusual typing data. The most distant data from the mean of the database is removed one after another as a wrong data until the number of data becomes M. This will decrease the intersection part of legal data and illegal ones. Thus we can make the database more exact.

We think that the user can remake a partial database. A legal user may be seldom able to log in the system because of changing his/her characteristics while he/she uses the system for a long time. If a user seldom logs in the system, the user types a password to remake his/her database according to circumstances. The fresh database is made whenever the user wishes to do. We think that this keeps high acceptance rate. But in these experiments, we do not approve remaking a partial database.

4.5 $WNorm_k$ and $Norm_k$

$Norm_k$ denotes Characteristic k. The value of k stands for the kind of characteristics ($k = 1, 2, 3, 4$). The norm is the summation of the errors between an input data and the average of database given from the password. Though we used the weighted $Norm_k$ only in the previous work[4], we use two kinds of $Norm_k$. One is weighted $Norm_k$ ($WNorm_k$) and the other is non-weighted $Norm_k$ ($Norm_k$), because we found out that $Norm_k$ is better than $WNorm_k$ in Identification Experiment 1 (as shown in section 5).

First, $WNorm_k$ is defined as follows[4]:

$$
\begin{cases}
WNorm_k = \sum_{n=1}^{L-1} \left(\dfrac{1}{\sigma_{kn}} \times |error_{kn}|^2 \right) & (k = 1, 2, 3); \\[4mm]
WNorm_4 = \sum_{n=1}^{L} \left(\dfrac{1}{\sigma_{4n}} \times |error_{4n}|^2 \right) & (k = 4).
\end{cases}
\tag{5}
$$

The value of L stands for the length of the password. The value of $error_{kn}$ stands for the n-th difference between the input data and the mean of database with respect to Characteristic k. The value of σ_{kn} stands for the variance of $error_{kn}$. Hence $WNorm_k$ shows the weighted distance and the less scattered $error_{kn}$ is regarded as important one. The reason why the $|error_{kn}|^2$ is weighted is that the system reflects the tendency of a legal database, and that we aim to increase an acceptance rate for a legal user.

On the other hand, $Norm_k$ is defined as follows:

$$
\begin{cases}
Norm_k = \sum_{n=1}^{L-1} |error_{kn}|^2 & (k = 1, 2, 3); \\[4mm]
Norm_4 = \sum_{n=1}^{L} |error_{4n}|^2 & (k = 4).
\end{cases}
\tag{6}
$$

Since we remove the weight from $|error_{kn}|^2$, each of errors is treated in the same importance. The system can compute $Norm_k$ more efficiently than computes $WNorm_k$.

5 Identification Experiment 1

5.1 Outline

In this section, we compare $WNorm_k$ with $Norm_k$. The characteristics used in this experiment is only Characteristic 3 ($WNorm_3$ or $Norm_3$) because it is the best one according to the previous work[4].

5.2 System

We use the identification system in Workstation (SS5-110MHz, Memory 32Mbyte, OS SunOS4.1.4). This system is implemented by C programming language, and can measure Press-Time and Release-Time used Library of X-Window every 30[msec]. If Press Interval is over 0.8[sec], the user is required to input it again. The time 0.8[sec] seems to be enough because of a precondition that we use a usual password.

5.3 Method

Twenty users have taken part in two experiments, the personal identification experiment and the discrimination experiment from other users. The password used in these experiments is the same for each user and it has ten letters "okamotoken". Since each user uses the same password, we need not think the difference of passwords, and then the use of the password reduces a user's burden. If all of the twenty users use the distinguish passwords, a user need to input not only a own password but also other kinds of nineteen passwords to do two experiments. Furthermore, a user also has to practice to type the passwords till he/she is used to typing. This becomes to impose a burden on a user, and gives bad influence on some experiments. Therefore, each user has only to input one password and then this can reduce a burden.

The experimental flow is following.

1. Practice a password until a user can type smoothly.
2. Type the password more than twenty five times to make a database.
3. Type it every twenty time to do the two experiments, the personal identification experiment and the discrimination one from other users.

5.4 Results

We got the data which are 400 (20×20)-time identification and 7600 ($20 \times 19 \times 20$)-time discrimination. Fig.5 shows the result of using $WNorm_3$, Fig.6 shows the result of using $Norm_3$. As mentioned above, the reason why we consider only Characteristic 3 ($WNorm_3$ and $Norm_3$) is due to [4].

The acceptance rate is 87.5% for $WNorm_3$ and 88.7% for $Norm_3$ at the intersection point of in a point of FAR (False Acceptance Rate) curve and FRR (False Rejection Rate) curve in Fig.5 and 6 We call the threshold at such an intersection point the *optimal threshold*.

Table 1 arranges the acceptance rate according to FRR for two figures above.

Table 1. The Result of Acceptance Rate in Experiment 1

	Acceptance Rate	
	$WNorm_3$	$Norm_3$
FRR $\leq$ 0.1%	15.5%	24.9%
FRR $\leq$ 0.5%	33.2%	42.5%
FRR $\leq$ 1.0%	45.6%	50.9%
Optimal Threshold	87.5%	88.7%

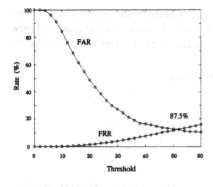
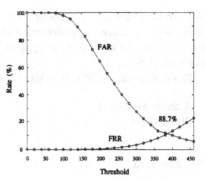

Fig. 5. The Result of $WNorm_k$ **Fig. 6.** The Result of $Norm_k$

5.5 Consideration

Though it is ideal that two curves have no intersection part in these figures, these two curves actually have an intersection in real systems. Hence it is important how the system divides these two curves. This leads high acceptance rate for a legal user and high rejection rate for illegal ones.

We see that the result using $Norm_3$ gives higher acceptance rate than that using $WNorm_3$. Now consider this result more deeply. $WNorm_k$ reflects each user's tendency of database because of the weights. But the database is easy to change because typing keyboard is a changeable characteristic with user's own will. The weight $1/\sigma_{kn}$ is made from a database. If the data has good balance, then the σ_{kn} gets small, and then the weight $1/\sigma_{kn}$ gets large. Hence the data seem to be important. However as well as an identification, the important data is really not the well-balanced data but the data separated from those of other users. Even if the data are well-balanced, if those are similar with others', then that data make little sense. Using $WNorm_k$ may spoil the discrimination data from other users and then the system may have a week point.

All of the twenty testers used the same password in these experiments. However, if they really use a different password from others and are used to typing it, the experimental result will get more superior. If a user is accustomed to typing a password, its data usually have good balance. Furthermore, if we wish to increase the acceptance rate of this system, we prepare a few passwords to get more personal characteristics.

6 Identification Experiment 2

6.1 Outline

We sometimes share a password among the fixed users. Therefore, we experiment with a share password in a fixed member. If the set of users are fixed, the system can learn user's characteristics using his/her database and can derive the optimum norm from four kinds of $Norm_k$. This improves the performance of the system without learning, which uses only $Norm_3$ described in section 5. Hence we investigate the identification system which can learn the user's characteristics from each database.

The characteristic of this system is as following:

- A share password
- Fixed users

6.2 Method

In this section, we apply *the linear discriminant function*[5] to this system. This method had been introduced by R.A.Fisher in 1936 as a strategy for separating two groups. This system learns user's characteristics from each database, and then we put four kinds of the characteristics together. For example, some users have good Characteristic 2, but others have good Characteristic 3. All of the twenty users in this experiment have the same password "okamotoken" .

We have to consider how to put the four kinds of $Norm_k$ (non-weighted) together. If the system does not learn user's characteristics, we cannot decide the best way that put them together, because the way of learning are different for the set of users. For another set of users, the way differs from that one. The system can not learn the user's characteristics for the random users. However, if the user's member is fixed, then we can derive the norm which is more superior to $Norm_3$. Hence that norm can be the best way for identification. With that way, we can find out how to put them together for a fixed member. So we derive $Norm$ from the four kinds of $Norm_k$ for a fixed member using the linear combination as follows:

$$Norm = \sum_{k=1}^{4} (a_k \times Norm_k) \tag{7}$$

The coefficient a_k ($k = 1, 2, 3, 4$) of the linear combination can be are derived by learning the users' databases. Each a_k is the optimum coefficient. For example, some users have the largest coefficient a_4 for Characteristic 4, others have the largest a_2 for Characteristic 2. We use the linear discriminant function that is the best way that finds a_1, a_2, a_3 and a_4 in this system.

6.3 Result

Fig.7 shows the result without learning (using only $Norm_3$) which is the same figure as Fig.6. Fig.8 shows the result with learning using the linear discriminant function. The acceptance rate of optimum threshold in these figures was 88.7% without learning and 92.3% with learning. These say that the effect for learning is recognized.

Table 2 arranges the acceptance rate according to FRR for two figures above.

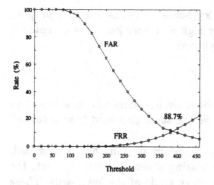

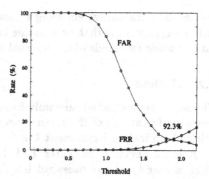

Fig. 7. The Result without Learning   **Fig. 8.** The Result with Learning

Table 2. The Result of Acceptance Rate in Experiment 2

	Acceptance Rate	
	Without Learning	With Learning
FRR ≤ 0.1%	24.9%	24.8%
FRR ≤ 0.5%	42.5%	49.6%
FRR ≤ 1.0%	50.9%	64.5%
Optimal Threshold	88.7%	92.3%

6.4 Consideration

As a whole, the acceptance rate with learning is higher than that without learning from Table 2. We can admit the effect to learn the user's characteristics, when a user use a share password in the fixed member. The system picks out each characteristic of the fixed member and learns a tendency of each user's data. We see that not only $Norm_3$ but also $Norm_1, Norm_2$ and $Norm_4$ are effective characteristics because the system uses only $Norm_3$ without learning, but on the other hand it uses more than two kinds of $Norm_k$ with learning. Some users use all four $Norm_k$, others use only two $Norm_k$, because he/she uses only effective $Norm_k$ (Note that all of the users use $Norm_3$).

Furthermore, the acceptance rate got higher 92.2% from 88.7% in the optimum threshold. We can admit the experimental effect using the linear discriminant function from this result.

7 Identification Experiment Using Unusual Keyboard

7.1 Outline

In this section, we have compared the results using two kinds of the unusual keyboards with the result using user's usual keyboard. We cannot necessarily use an identification system using a usual keyboard at a usual place. A purpose of this experiment is that we examine how high each user keeps the acceptance rate by using two kinds of the unusual keyboards.

7.2 Method

The seven users selected randomly from above twenty users have taken part in this experiment. All of the seven users have the same password "okamotoken" using Identification Experiment 1 or 2.

A usual keyboard indicates TYPE 5c keyboard of Sun that is standard. First, a user types the password ten times using a usual keyboard. Next, the user also types it ten times to every two other kinds of the keyboards. These unusual keyboards which are Unusual Keyboard 1 and 2 is almost same size, but has the different key touch. Unusual Keyboard 1 has the same key touch as a usual keyboard, but on the other hand Unusual Keyboard 2 has a heavier key touch than a usual one. We use only $Norm_3$ in this experiment as well as in Identification Experiment 1 in section 5.

7.3 Result

The decrease of the acceptance rate depends upon the key touch. Some keyboards may decrease little acceptance rate, but some may decrease quite a lot because of their touch difference.

Table 3 arranges the mean of $Nomr_3$, which each user types ten times to every three kinds of the keyboards.

Table 3. $Norm_3$ for Different Keyboard

USER	$Norm_3$ (Characteristic 3)		
	Usual Keyboard	Unusual Keyboard 1	Unusual Keyboard 2
user1	145.0	147.3	232.1
user2	166.3	192.4	290.5
user3	218.6	229.6	353.9
user4	320.2	386.3	374.0
user5	196.5	198.9	394.3
user6	233.8	213.6	423.0
user7	217.2	133.0	282.1
average	214.0	214.5	335.7

7.4 Consideration

We see that Unusual Keyboard 1 gives the similar values to a usual keyboard but Unusual Keyboard 2 gives the higher values than it. We have to think the reason why the difference between Unusual Keyboard 1 and Unusual Keyboard 2 is large in the experimental result. As mentioned above, Unusual Keyboard 1 has the similar key touch as a usual keyboard but Unusual Keyboard 2 has a heavier key touch than it. As the key touch of Unusual Keyboard 2 is different from a usual one, its $Norm_3$ is larger than that of Unusual Keyboard 1. Furthermore, we also think that the key release has an influence on next key press. We cannot handle a motion of releasing a key in a password even if we remember the rhythm of typing it, because it is difficult for us to be conscious of the motion.

8 Conclusion

We have investigated a bad influence of weighting and a new method using the linear discriminant function which are concerned with extraction of personal characteristics and confirmed the efficiency by experimenting. Unlike previous study[3] which collect the reference data by typing "Free-Style" text, we collect one by typing each own password many times. A part of four kinds of interval in a password will be not independence and separate but related to before and behind interval. We intend to regard a password as data mass.

We found the following results.

- We confirmed that $Norm_k$ is better than $WNorm_k$.
- We confirmed that the system can learn user's characteristics by using the linear discrimination function.
- We found that the performance of this system depends upon not a keyboard size but a key touch.

References

1. R.Joyce, and G.Gupta.: Identity Authentication Based on Keystroke Latencies. Communications of the ACM, Vol.33, No.2 (1990) pp.168–176
2. M.Kasukawa, H.Kakuda, and Y.Mori.: An Identity Authentication Method Based on Arpeggio Keystrokes. Transaction of IPSJ, Vol.34, No.5 (1993) pp.1198–1205
3. F.Monrose, A.Rubin.: Authentication via Keystroke Dynamics. 4th ACM Conference on Computer and Communcations Security (1998)
4. K.Omote, E.Okamoto.: User Identification Schemes Using Biometrics of Keystroke. Proceedings of CSS '98 (1998) pp.213–218
5. Alvan R.Feinstein.: Multivariable Analysis, An Introduction. Yale University (1996) pp.325–330,437–439

Boundary Conditions that Influence Decisions about Log File Formats in Muli-application Smart Cards

Constantinos Markantonakis *

Information Security Group,
Royal Holloway, University of London,
Egham, Surrey, TW20 0EX,
United Kingdom
C.Markantonakis@rhbnc.ac.uk

Abstract. In real world smart card applications, smart card log files are mainly used for storing receipts for the successful or otherwise completion of certain events. In traditional single application smart card environments, the decision on which events to be logged was made by the application developer. We believe that in today's multi–application environments the situation is rather more complicated. If more than one application shares the same smart card, a whole range of new events require logging. In this paper we provide suggestions as to the new events to be logged. Furthermore, we propose a standard format for smart card log files in order to make dispute reconciliation procedures easier and faster, and also to efficiently manage the valuable log file space. Finally, we provide some results from an implementation of the proposed standard format in a Java Card.

Keywords: Log Files; Audit Files; Logging Events; Implementation Results; Smart Cards; Java Cards, Multi–Application.

1 Introduction

The importance of audit log files in computer environments needs little description. Log files can be used to:

- record security critical events to enable users to be held responsible for their security critical actions,
- resolve possible disputes about events,
- enable security breaches to be detected after they have happened, and
- recover the system to a safe state after fatal failures.

Good overviews on how logging and auditing are essential computer security mechanisms appear in [2, 3]. Additional guidelines describing good practice for

---

* The author's research is funded by Mondex International Ltd. This work is the opinion of the author and does not necessarily represent the view of the funding sponsor.

the use of auditing in computerised systems are given in [16, 17]. With respect to traditional computer security, log files are also regarded as a mechanism for providing the required evidence to determine whether the system was ever placed in a non–secure state. Thus, when the log file records the before–and–after state of the objects changed, it is possible to recover the system to a safe state. A particular problem arises at this stage: what information/events to log?

In "traditional" multi–application smart card environments the applications are pre–installed in advance. Since the application structures are agreed in advance and since the applications had "trustworthy" origins, it was evident which events were to be logged. The decision as to which events to log was typically made by the application developers which in most cases were the card issuers. Since there was trust in the card issuers, the card holders did not have to worry about logging events.

We believe that the new technology of application code interpretation [8, 9, 13, 14] introduces further complexities which require revised and more dynamic log file handling mechanisms. To justify the need for such mechanisms we simply have to consider some of the new concepts introduced and realise the inapplicability of the existing logging techniques. The need for new techniques arises both because there are new events to be logged (both in the operating system and the application level), and because standardisation of the log file format or even the logged events is probably required.

In [10] we dealt with the complex and realistic issue of securely extracting the log files from the card (when the log files are full) and storing them in some other medium. A similar approach to maintaining log files in untrusted environments is presented in [20]. In [11] we presented the main characteristics and implementation details of a mechanism which dynamically (as the application is executed in the card) identifies and logs a number of security relevant events. In this paper we very briefly present the main characteristics of the different types of log files along with an overview of the events to be logged. Subsequently, we describe a possible standard format for the smart card log files. Finally, we provide some implementation results on the performance of the proposed standard format in one of the most powerful Java Card development kits.

2 Smart Card Log Files Types

In a real world smart card environment, log files will mainly be used in order to allow the Smart Card Operating System (SCOS) to recover from fatal failures and to provide evidence regarding the progress of certain events. The latter implies that the log files might also be used in an auditing procedure in order to resolve certain disputes. In this section we provide a classification of smart card log files.

The uses of log files in a smart card environment may be divided according to the following criteria:

- purpose: audit trail or recovery log,
- architectural level: operating system or application.

The details of each category are presented in figure 1 and analysed in the following paragraphs. However in order to better understand the issues involved consider an example of a multi–application smart card that holds the following applications: an electronic purse application, a loyalty points application used in conjunction with the purse application, a health care application and an application for digitally signing emails or other information. We will refer to this example in the following sections.

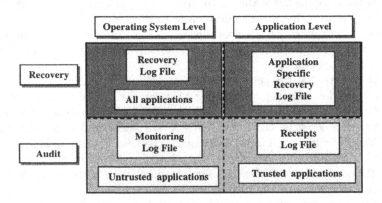

Fig. 1. The different types of Log Files.

In the recovery category we come across two different types of log files, depending on whether the log files are handled by the SCOS or by the smart card application. The benefit from the existence of a recovery mechanism at the SCOS level is that smart card applications will not have to replicate their own recovery code. On the other hand the disadvantage is the complexity of such a mechanism. An example of a recovery log file maintained by a smart card application is the "Pending log" in the Mondex electronic purse [15].

Consider the case that the electronic purse application of our example crashes. Then, a recovery mechanism that will take advantage of the information stored in this log file could help the application to recover the next time that is selected for execution. Clearly these log files will typically not be used as evidence in cases of dispute and the details of the recovery mechanism are outside the scope our research.

A different type of smart card application recovery is proposed by Trane and Lecomte in [22]. The authors suggest that certain recovery information is kept in the smart card terminal and thus recovery is achieved the next time the card interacts with the terminal.

In the audit trail category (i.e. the light gray area in figure 1) we encounter two different types of log files, again depending whether the log files are handled by the SCOS or by the smart card application.

An example of a log file maintained by the SCOS in the audit trail category is the Monitoring log file. The Monitoring log file will hold information for identifying licensing or pay–as–you–use details of the downloaded smart card applications. For example consider the following scenario: the electronic purse application of our example offers certain smart card programming primitives which will be used by other applications (i.e. in our example by the loyalty application). It could be the case that these primitives are offered at a pay–as–you–use basis. Therefore, the loyalty application will be charged for the number of times it used the purse's functionality. When the Monitoring log file is examined it could provide all the required information as to which applications used which primitives and how much they should be charged.

The Receipts log file is an example of an audit log file maintained in the application level. It will hold information generated by smart card applications. In our example this log file will of course hold the payment receipts generated from the electronic purse applications. Similarly, the email signing application should also generate evidence on the digital signatures performed. Generally, the Receipts log file will remain in the smart card application space and it will be subject to auditing. Permission to make entries in this log file will only be granted to authorised and trusted applications (i.e. the applications carrying valid application certificates).

Note that our proposals are relevant to log files in the audit trail category. For example, we propose an architecture that will identify certain events to be placed in the Monitoring and Receipt log file. Additionally, we are interested in what happens when the smart card log files become full. We are not interested in providing smart card application recovery since it would either require knowledge of the application structure or the internals of the SCOS.

In the following subsection we provide an overview of the new events that require logging in a multi–application environment.

2.1 What information should be logged?

The question as to which information should be logged in a smart card environment is significantly more difficult to answer than it would be for a conventional computer system. The main reasons are the hardware and software characteristics of the smart card processors [21, 7] that seriously restrict both the amount and the nature of the events to be logged. In this section we highlight the issues that influence the decision on which events should be logged.

The decision is highly dependent on the following factors:

- the system's security policy,
- the space allocated for the log file,
- how often the log file is downloaded, and
- who decides which events should be logged.

If, for example, a large number of events are characterised as critical and should be monitored then the risk of running out of log file space is increased. On the other hand, if just a few events are monitored, there is always a chance that some important events might be missed. A summary of the type of events that might be logged in a smart card environment is given in Table 1.

Precisely which events and actions are to be logged depend on the system's security policy. The policy should be sufficiently detailed to enable identification of the types of event which require logging.

Another important factor influencing the nature and the amount of information to be logged is the space allocated for the smart card log files. Ideally, the allocated space should permit logging of all significant events without the need for overly frequent download operations, which probably means at least 4–5 kbytes. Although this memory space would allow the logging of sufficient information, more realistically speaking the space offered by the smart card manu-facturers and the service providers would be within the range of 1.0–1.5 kbytes. Another difference compared with computer based logging is that it is not possible to have a data reduction tool in a smart card. Thus, the data to be stored should be reduced to the minimum prior to their storage in the smart card log files.

Table 1. Suggested events to be logged.

Receipts	Application Monitoring
Transaction Type (credit, debit, etc.)	Application Download/Delete
Total transaction value	Use of Cryptographic Primitives Primitives
Transaction status	Freeze Operations
Type of Currency	Certificate Updates/Changes
Transaction time/datestamp	Intrusion Detection Results
Transaction ID, code	Invalid PIN presentations
Digital Signatures	Certain EEPROM Updates
Log File Downloads	File Deletions/Updates
	Reading, Browsing Files
	Flushing Memory Space

In a smart card environment it is impossible to implement the post–selection method described in [16] (i.e. log as many events as possible and decide later which ones should be audited). Obviously, the main restriction is the limited space assigned for the log files. Thus, the events to be logged should be pre–selected. The main advantage of the pre–selection method is that only a number of pre–selected events are logged. As a result, there is more efficient log file space management. The obvious disadvantage is that it is relatively difficult to predict the events which might be of security interest at a future date. Hence, if the table containing the events to be logged is stored in ROM of the card (in order to be more adequately protected) it would be impossible to add new events. If the table is stored in the EEPROM of the card then it becomes easier to securely update the list of logged events but on the other hand it becomes more vulnerable.

The supporting infrastructure for downloading log files is another critical factor. For example, if the log files can be downloaded from the card often, then the amount of data to be logged may increase. The log file downloading procedure could be enhanced by offering the facility in secure public terminals or even over–the–air transmission (by using mobile phones).

Finally, another influencing factor is which entity decides the events that should be logged. Is it the card holder, the card issuer or the service providers? Actually in an environment of so many conflicting interests it is not easy to find a solution acceptable to all interested parties, and the answer to the above question should appear in the system's security policy document. For that reason we do not try to explicitly define which events should be logged. Instead we just provide an indication of the events which are of significance for the above entities.

3 Smart card Log File Format Standardisation

In this section we present a possible standard format for a smart card log file. A similar proposal for a standard audit trail format is presented in [1] but for a traditional computerised system. The two main issues addressed in [1] are extensibility and portability of the audit log files.

3.1 Why standardise a log file format?

A standardised log file format in a smart card environment introduces several benefits.

Firstly, it makes automated log file analysis and reconciliation easier. Suppose that two smart cards from different issuers perform a financial transaction. Subsequently one of the smart card holders denies the transaction and contacts an arbitrator. The arbitrator's task when accessing the log files is simplified if the log files conform to the standardised log file format. As a result the dispute resolution procedure can be speeded up.

Secondly, the extensibility issue is successfully addressed when the data attributes of the log file record are not strictly standardised. This will allow the system to migrate and easily adopt any future changes which might require more or different events to be logged.

Thirdly, it will allow more effective log file storage management. If, for example, an upper limit is set on the amount of data which can be stored in the log file by each application, then the problem of an application over–using valuable log file space is eliminated.

Similarly, addressing the portability issue is of importance since it will allow the log files to be transferred and handled by various systems (PCs, Electronic Wallets, or large Audit Log Storage Servers). For example, a smart card log file should have a format that should not make difficult its smooth transportation between the above entities.

We also believe that if the smart card log files conform to a standardised log file format, the use of log file filter engines [1], which take as input the smart card log files as they are created and translate (post–processing) them into a standard format, is made redundant. Post–processing the smart card log files might be a problem since the whole log file might be encrypted and/or signed before being extracted from the card. Thus, it would be difficult to make the smart card log files conform to a standardised format after they have been extracted from the card. This implies that log file processing has to take place while the log files are still stored in the card.

3.2 The Content of the Log File

We start by specifying the characteristics of the smart card log file. The log file can be a *transparent* file (i.e. a simple linear file) [19]. Its size will depend on the number of applications downloaded in the card and the technical characteristics of the card; realistically speaking a smart card log file will probably have size around 1.0–1.5 kbytes.

We assume that the log file consists of a list of data entries (DEs), which represent a record for each event to be logged, and each data entry is assigned a unique serial number in numerically increasing order. The maximum number of data entries (w) depends on the size of each data entry and the size of the log file. Thus, if we denote the ith data entry in the log file by DE_i, the log file F will have the following form

$$F = (DE_1, DE_2, \cdots, DE_j)$$

for some $j < w$.

Ideally each data entry should correspond to a single (not multiple) auditable "event", and take the form specified below. Note that we assume that the card's LFUM (i.e. the Log File Update Manager as described in [11]) is the only entity responsible for updating the entries in the log file. We further assume that:

$$DE_i = DeBind; f_{K_{AC}}(DE_{i-1} \parallel DeBind) \parallel E$$

$$DeBind = AID \parallel N_C \parallel (LE_1; LE_2; \ldots; LE_e)$$

The separator character between fields is ';' and "$\parallel$" represents concatenation of data items. The *DeBind* variable contains the unique application identifier (AID) field as described in the ISO 7816-5 standard [18], N_C is a monotonically increasing sequence number, and the actual data field of the logged events is $(LE_1; LE_2; \cdots; LE_e)$, where e is the number of logged events. Each LE_i will either hold data from a transaction receipt or the unique event identification information (EID) for the corresponding event to be logged.

The $f_{K_{AC}}(DE_{i-1} \parallel DeBind)$ field represents the result of applying the MAC function f using as input a secret key K_{AC} and as data the previously generated data entry (DE_{i-1}) along with the current *DeBind*. The secret key K_{AC} could be generated by a trusted entity but will be known only internally to the card. The latter could be securely installed in the card either during the personalisation phase or at any later stage during the card life cycle when the card owner decides to activate the logging policy. The $f_{K_{AC}}(DE_{i-1} \parallel DeBind)$ field links the previous data entries with the most recent one. Finally, 'E' denotes the end of this data entry.

In general log files contain a timestamp. A smart card will not possess a real time clock and it will rely on external devices for the value of the timestamp. Hence in this environment a timestamp would be of limited value. Thus, it was decided to make the inclusion of a timestamp optional, and to provide ordering of the events using a sequence number. We assume that the sequence number N_C is generated using a counter held internally to the card, and which cannot be reset or otherwise modified by any external events. In particular, downloading the log file contents to an Audit Log Storage Server (ALSS) [10] will not involve resetting the sequence number counter. The sequence numbers of the data entries (DEs) can thus be used to order downloaded data entries, and gaps in numbers can be used to indicate lost or deleted log file data entries.

As previously stated, the purpose of including a message authentication code (MAC) of the previous generated data entry is to effectively link pairs of successively generated data entries. Eventually, this will create a chain of linking information [6] of all the generated data entries, and along with the sequence number N_C, will act as a protection mechanism against unauthorised insertion or deletion of LEs.

3.3 Size of the Log file Data Entry (DE) Field

The length in bytes of each field of the data entry (DE) is as follows. The end character (E), along with the field separating characters (;) will be one byte each. The AID field could be represented as a 16 byte string, as defined in [18] and N_C could be represented as a 4 byte number (i.e. '0001', the counter for the initial data entry). The MAC of the linking information ($DeLink$) field will have 4 bytes length.

Let us assume that the size of the unique event identification (EID) information of each event is around 5 bytes (e.g. 'ID056', indicates an Update operation). Furthermore, let us assume that applications contain an average number (e) of events to be monitored (e.g. $e = 15$ events). A simple calculation will reveal that the whole logged event (LE) field of each data entry along with the separator characters between the fields will have an expected length in bytes of $LeLength = (5e) + (e - 1)$.

If an average number (e) of events is used then the formula which derives the total space occupied by each data entry is the following:

$$DE \text{ Space} = 6e + 25$$

where e represents the number of events ($LE_1; LE_2; \ldots; LE_e$) to be stored in each data entry, and 25 is the number of bytes for separator characters between the fields, the rest of the actual fields and finally the ending character.

The proposed standard log file format maintains a high level of flexibility towards successfully adapting itself and logging information from a variety of different applications. Therefore, although it is suggested that the LE field of each data entry should contain an average number e of events, nothing excludes that this space is left to the discretion of the application designer. In this case the formula which derives the total space occupied by each data entry is the following:

$$DE\ Space = 26 + q$$

where q represents the total number of bytes in the (LE) entries and 26 is the number of bytes for the rest of the fixed fields (separator and other characters). Following this approach the application developers are left with the task of how effectively they can use these LE entries, to better suit their needs. Additionally, the extensibility issue is more adequately addressed.

3.4 Number of Data Entries in Each Log File

There are two different options when calculating the number of data entries in each log file or when trying to effectively manage the log file space, depending on the number of events logged in each data entry.

The first proposal assumes that the log file contains a fixed number of events $e = 15$ in each data entry. The maximum space (in bytes) of each data entry will be calculated according to the above formula i.e. DE Space = $(6 \times 15 + 25) = 115$. With this approach the number of data entries in a log file (of size LFS) is the following: $w = \lfloor (LFS/DE\ Space) \rfloor$. For example in a log file of 1.5 kbytes we will get $w = \lfloor (1536/115) \rfloor = 13$ entries, which is not very restrictive at all.

The second approach assumes that each data entry might not contain the maximum number of logged events. For example if a data entry logs 5 out of its 15 events and another data entry logs 10 out of 15, and so on, there might be space for extra data entries. The number of the data entries to be claimed will be assigned dynamically by the smart card operating system. On average around 2–3 extra data entries will be added by the existence of such a mechanism.

3.5 Practical Issues of the Log File Standard Format

Our proposals indicate a small number of events which need logging in a truly multi-application smart card environment. We believe that logging such information in the smart card log files, provides extra evidence on the successful or otherwise completion of certain events.

As previously stated, the smart card log file size (LFS) in a real world implementation, will only be around 1–1.5 kbytes. With our proposed standard log file format, the maximum number of data entries will vary within the range 8–16. This implies that the smart card log file will hold a reasonable number of data entries before it fills up and needs to be downloaded or simply over-written.

Furthermore, if the second approach defined in §3.4 is to be followed when allocating the data entries, and all the unused space of each data entry along with the truncated space resulting from formula w is collected, then extra space for a small number of additional data entries might be created.

4 Implementation Details

In this section we examine the results of an experimental implementation of the log file format described in the first part of the paper.

4.1 Implementation Analysis

In a practical implementation the log file data entries (DEs) would be generated either by the SCOS or by applications. As the standard requires that information from either source will have to be transformed into the specified format, both the SCOS and the applications will have to be provided with this functionality.

The required functionality can be implemented in the form of a function (GenerateDataEntry) that will perform the following steps:

- take as input the LEs information,
- generate the format required by the standard (DE), and
- append the new entry to the log file.

Ideally this function should implemented as a SCOS primitive-service for the following reasons:

- it will be more adequately protected if masked in ROM,
- it will run faster since it could be written in machine native code, and
- it will probably eliminate the possibility of having applications passing information outside their own application space.

The last observation is of particular importance since in both Java Card API and particularly in Multos [8], applications are restricted within their own application space. Thus, it would be more efficient to call a SCOS primitive that will securely receive the logging information and perform the above steps, rather than having applications passing information to other applications, outside their own application space. Finally, it will be more secure for the log file (which hold information from multiple applications) to be accessed through an SCOS call, rather directly by each smart card application.

For implementing the standardised log file format functionality we used the GemXpresso development kit [5] from Gemplus. The development kit allows Java Card applications to be implemented and downloaded in the GemXpresso card or in a simulator.

Among the most attractive advantages [12] of the development kit are the following: Firstly, the ability to define "shared application libraries" that their functionality could be shared among various smart card applications. This provides an indirect method of extending the functionality of the smart card operating system. Secondly, it provides the means to write "client" applications that will interface with the smart card applications very easy by using a cut-down version of an RPC (Remote Procedure Call) [4] style protocol called DMI (Direct Method Invocation) [23].

Our implementation consists of three different components: the shared library, the smart card application (i.e. "applet"), and the client application accessing the "applet".

The "shared" library contains the actual class, i.e. `GenerateDataEntry`, that implements the standard log file format functionality. Access to the log file should only be provided to this library. This function accepts the $(LE_1; LE_2; \cdots; LE_e)$ entries and constructs the data entry DE (128bytes) entries as required by the standard. Additionally, prior to generating a data entry it also verifies that the currently running application has not previously called the `GenerateDataEntry` function. This will prevent a single application filling up the space of the log file by repeatedly calling the (`GenerateDataEntry`) command.

However, certain information need to be stored in the card in order to be able to construct the messages in the format specified by the standard. For example, the `GenerateDataEntry` must be provided with the means to get the unique application identifier (AID) of the currently running application. Additionally, a unique sequence counter is required in order to serialise the data entries, as described in §3.2. Furthermore, the standard requires that a hash of the previous data entry DE_{i-1} will always be available in order to link the previous generated messages with the current one. If the log file is overwritten in a cyclic mode, then the previously generated hash will be always available. In case the log file is to be downloaded as described in [10] then a hashed copy of the last data entry $(h(DE_{i-1}))$ should be securely stored in the card.

The log file we used for our implementation is relatively small consisting of 240 bytes when running on the card and 800 bytes when running on the simulator. This design decision is enforced by the restriction on generating large files in the card.

The smart card application simply calls the `GenerateDataEntry` function of the shared library and returns the results to the client. It does not perform any further operations since in that case the timing measurements involved will not be so accurate.

The client application simply interfaces the smart card application with the outside world and it also measures the time spend while the DataEntry (DE) is created, i.e. when the `GenerateDataEntry` is called.

4.2 Results and Performance Evaluation

We have generated a set of results depending on the actual size of the log files and whether the application runs on the card or on the simulator. The implementation and the testing of the client and smart card application used a 400Mhz PC with 128 MB of RAM under Windows NT.

The results when the "GenerateDataEntry" is called in the simulator with a log file of 800 bytes and after ten consecutive executions are presented in Table 2.

The "Connect" figure indicates the time spent by the client application in order to connect with the reader or the simulator. The next value indicates the time spent in the "GenerateDataEntry" procedure. The "Disconnect" figure represents the time spent to close the connection with the reader or the simulator.

When the client communicates with the Java Card, the results do not make much use of the fast PC because the operations were mostly I/O intensive. For this reason the results should not be considered as absolute. For example, the figure for the *CreateDataEntry* function also includes the aforementioned communication overhead when calling the function from the client application and receiving the results back from the card. Obviously when the function will be simply called within the smart card application it will not have to send any information back to the "client" and the execution will be considerably faster. At the time of implementation we did not had access to the tools that will accurately measure the time spent when a function is executed in the card without been invoked within a client application.

Table 2. GemXpresso simulator results using log file of 800 bytes

Part name	Time (ms)	Std. Deviation
Connect	11266	46.81
CreateDataEntry	200	3.34
Disconnect	400	5.84
Total	11866	55.99

When the "GenerateDataEntry" is called in the card with a log file of 282 bytes we get the results given in Table 3.

Table 3. GemXpresso card results using log file of 282 bytes

Part name	Time (ms)	Std. Deviation
Connect	2704	39.96
CreateDataEntry	2293	62.81
Disconnect	120.5	4.86
Total	5117.5	107.63

As we can see from Tables 2 and 3 the "Disconnect" figure presents smaller variations compared to the connect figure. The results from the simulator are provided in order to demonstrate that our implementation actually works with large files as proposed in the standard. For a log file of 800 bytes we could successfully store 6 data entries. Since we could not create a relatively large file in the card, we had to modify the application code in order to run the tests for only two data entries.

The sizes in bytes of each of the above three entities when implemented in Java and downloaded in the Java Card are as follows: The library is 1287 bytes and the application calling the "GenerateDataEntry" function is 253 bytes. The

client application is 5500 kbytes, it is also implemented in Java and runs on the
PC. The client is using the proxy interface (2691 kbytes) as generated by the
GemXpresso development kit, the proxy interface automatically transforms the
client and card requests in Application Data Protocol Data Units packets.

5 Conclusions

In this paper we have outlined the nature of the events which might require
logging in a truly multi–application smart card environment. We also explained
the advantages of a standardised log file format. The format described in this
paper successfully meets the requirements of today's multi application smart
card log files since it effectively utilises the relatively small amount of smart
card memory. Subsequently, from our implementation results we conclude that
a possible real implementation of the standard log file format is totally feasible.
In order to get a better understanding of the issues involved, a more realistic
implementation, which will provide the functionality as part of the operating
system, is required. Currently we are experimenting with the MULTOS smart
card operating system and we hope that by the end of 1999 we will be able to
provide results analogous to those described here for the MULTOS operating
system. Furthermore, we are currently trying to estimate the actual overheads
involved in the communication link between the "client" application and the
smart card application. This will enable us to provide more accurate figures on
the performance of the proposed functionality.

6 Acknowledgments

The author would like to thank Dieter Gollmann and Chris Mitchell for their
helpful comments.

References

1. Matt Bishop. A Standard Audit Log Format. In *Proceedings of the 19th National
   Information Systems Security Conference*, pages 136–145, 1995. Also available in
   http://seclab.cs.ucdavis.edu/~bishop/scriv/index.html.
2. Matt Bishop, Christofer Wee, and Jeremy Frank. Goal Oriented Auditing and Logging. Technical report, Department of Computer Science, Univeristy of California at Davis, 1996. Also available in
   http://seclab.cs.ucdavis.edu/~bishop/scriv/index.html.
3. David Bonyun. The Role of a Well-Defined Auditing Process in the Enforcement
   of Privacy Policy and Data Security. In *Proceedings of the 1981 IEEE Symposium
   on Security and Privacy*, pages 19–26, 1981.
4. George Coulouris, Jean Dollimore, and Tim Kindberg. *Distributed Systems: Concepts and Design*. Addison-Welsey Publishing Company Inc., 1994.
5. Gemplus. *GemXpresso Reference Manual*. Gemplus, 1998.
6. Stuart Haber and W.Scott Stornetta. How to time-stamp a digital document.
   Journal of Cryptology, 3(2):99–111, 1996.

7. Pieter Hartel and E. de Jong Frz. Smart cards and card operating systems. In *Conference Proceedings*, pages 725–730. Int. Conf. UNIFORUM'96, San-Francisco, California, February 1996.

8. MAOSCO. *MULTOS Reference Manual Ver 1.2*. MAOSCO, 1998.

9. MAOSCO. Multos, the smart card gets smarter. `www.multos.com`, July 1998.

10. Constantinos Markantonakis. Secure Log File Download Mechanisms for Smart Cards. In *Lecture Notes in Computer Science*, volume to be published. Third Smart Card Research and Advanced Application Conference Cardis'98, September 1998.

11. Constantinos Markantonakis. An architecture of Audit Logging in a Multi-application Smart card Environment. In *EICAR' 99 Conference Proceedings, ISBN: 87-98727-0-9*. EICAR'99 E-Commerce and New Media Managing Safety and Malware Challenges Effectively, Aalborg, Denmark, March 1999.

12. Constantinos Markantonakis and Simeon Xenitellis. Implementation of a Secure Log File Download Manager for the Java Card. In *Kluwer Academic Publishers, to be published*. CMS'99 Communications and Multimedia Security, Katholieke Universiteit Leuven, Belgium, September 1999.

13. Sun Microsystems. The Java Card API Ver 2.0 specification. `http://www.javasoft.com/products/javacard/`, 1998.

14. Sun Microsystems. *The Java Card API Ver 2.1 Specification*. JavaSoft, 1999.

15. Mondex. Brief description of the mondex log file. `http://www.mondex.com/mondex/cgi-bin/printpage.plenglish+global-technology-security`, 1996.

16. National Computer Security Center (NCSC). A Guide to Understanding Audit in Trusted Systems. Technical report, Department of Defense (DoD), NCSC-TG-001, Library no. S-228-470, July 1987.

17. U.S. Department of Defence. Trusted Computer System Evaluation Criteria. Technical report, U.S Department of Defence, Computer Security Center, December 1985.

18. International Standard Organisation. *ISO/IEC 7816-5, Information technology - Identification cards - Integrated Circuit(s) Cards with Contacts, Part 5, Numbering System and Registration Procedure for Application Identifiers*. International Organization for Standardization, 1994.

19. International Standard Organisation. *ISO/IEC 7816-4, Information technology - Identification cards - Integrated circuits(s) cards with contacts - Inderindustry Commands for Interchange*. International Organization for Standardization, 1995.

20. Bruce Schneier and John Kelsey. Cryptographic Support for Secure Logs on Untrusted Machines. In *Usenix Press*, pages 53–62. The Seventh USENIX Security Symposium Proceedings, January 1998.

21. Gustavus J Simmons. *Contemporary Cryptology; The Science of Information Integrity*. Institute of Electrical and Electronic Engineer, December 1991. Chapter 12.

22. Patrick Trane and Sylvain Lecomte. Failure Recovery Using Action Logs for Smart Cards Transactions Based Systems. In *Third IEEE International On-Line Testing Workshop*, July 1997.

23. Jean-Jacques Vandewalle and Eric Vetillard. Developing Smart Card Based Applications Using Java Card. In *Springer Verlag*. Third Smart Card Research and Advanced Application Conference - CARDIS'98, September 1998. to be published.

Send Message into a Definite Future

Wenbo Mao

Hewlett-Packard Laboratories,
Filton Road, Stoke Gifford, Bristol BS34 8QZ, United Kingdom
wm@hplb.hpl.hp.com

Abstract. Rivest et al proposed a time-lock puzzle scheme for encrypting messages which can only be decrypted in the future. Such a puzzle specifies an algorithm for decrypting the message locked in and the specified algorithm has a well understood time complexity. However, that time-lock puzzle scheme does not provide a means for one to examine whether a puzzle has been formed in good order. Consequently, one may foolishly waste a lengthy time on trying to solve an intractable problem. This weakness prohibits that scheme from applications that involve mutually untrusted parties. We propose a new time-lock puzzle scheme which includes an efficient protocol that allows examination of the time needed for decrypting the message locked in.

1 Introduction

Rivest et al [5] proposed a timed-release crypto puzzle scheme (time-lock puzzle). The goal is to encrypt a message so that it cannot be decrypted until a pre-determined period of time has passed.

Several applications have been observed in [5]:

- A bidder in an auction wants to seal his bid so that it can only be opened after the bidding period is closed.
- A homeowner wants to give his mortgage holder a series of encrypted mortgage payments. These might be encrypted digital cash with different decryption dates, so that one payment becomes decryptable (and thus usable by the bank) at the beginning of each successive month.
- A key-escrow scheme can be based on timed-release crypto, so that the government can get the message keys, but only after a fixed, pre-determined period.
- An individual wants to encrypt his diaries so that they are only decryptable after fifty years (when the individual may have forgot the decryption key).

The time-lock puzzle scheme proposed in [5] has its security based on the computational problem of finding an element in a multiplicative group modulo a composite integer. The scheme provides a prescribed instruction leading to the target element for solving the puzzle (the element will reveal a key for decrypting the locked message). If a puzzle solver follows the instruction provided, the problem has a well-understood time complexity. Otherwise, the problem is believed to be intractable if the composite modulus used is sufficiently large.

In this paper we shall point out a major weakness of that time-lock puzzle scheme: the time required to solve a puzzle is not pre-determinable from a solver's point of view. In fact, apart from the puzzle maker, no one can be sure if a puzzle is indeed solvable until it has actually been solved. A time-lock puzzle in usual applications typically requires a lengthy time to solve. So to start solving a puzzle without knowing if it has been created in good order can mean to foolishly risk wasting a lengthy period of time, unless the puzzle maker is trusted to have created the puzzle as bona-fide. However, in most applications where a timed-release crypto puzzle finds a role to play, a puzzle maker should not be trusted by a solver (this is the case in the first three applications listed above). Therefore, a time-lock puzzle scheme which does not allow a solver to pre-determine solveability within the time length declared has a serious limitation in applications.

We will propose a new time-lock puzzle scheme which can allows a solver to examine with confidence the time length that is needed to solve a puzzle.

2   The previous time-lock puzzle

We first introduce the time-lock puzzle scheme proposed in [5], and then analyse a weakness with it.

Suppose Alice has a message M that she wants to encrypt with a time-lock puzzle for T seconds. She generates a composite modulus $n = pq$ with p and q being two large primes. She sets t such that $t = TS$ with S being the number of squarings modulo n per second that can be performed by the solver. Alice then generates a random key K for a conventional cryptosystem. She encrypts M with the key K to form ciphertext C_M. She now picks a random element a modulo n, and encrypts K as

$$C_K = K + a^e \pmod{n},$$

where

$$e = 2^t \pmod{\phi(n)}.$$

She finally outputs the time-lock puzzle (n, a, t, C_K, C_M) and erases any other variables created during this computation. With the knowledge of $\phi(n)$, Alice can construct the puzzle efficiently.

Assume that finding M from C_M without the correct key is computationally infeasible, so is computing $\phi(n)$ from n. Then computing $a^{2^t} \pmod{n}$ as recommended by the scheme will be the only known way to solve the puzzle. Once $a^{2^t} \pmod{n}$ is reached, K will be revealed which will further lead to decryption of C_M. This computation requires t steps of squaring modulo n. Obviously, t must be a tractable quantity.[1]

---

[1] The idea of repeated squaring an element in a group modulo a composite integer has also been used in a key escrow scheme with a time control feature [1], which utilises

An advantage of this puzzle scheme is that the process of repeated squaring is "intrinsically sequential". There seems no obvious way to parallelise it to any large degree. So the time needed to solve a puzzle can be well controlled by Alice.

We should note that in this scheme there is no control whatsoever over Alice in terms of enforcing her to form a puzzle to be bona fide. She can introduce errors into a puzzle which is undetectable to a solver (let Bob be the solver) before he realises that he has been spending a lengthy period of time with fruitless result. For instance, Alice can set t in the puzzle to be infeasibly large. Note that it is trivially easy for Alice to set the order of a modulo n and that of 2 modulo $\phi(n)$ to be sufficiently large; this will guarantee that t can also be sufficiently large without making e and a^e to show a sign of cycle. Thus to reach $a^e \pmod{n}$ by following Alice's instruction will take infeasibly many steps.

In usual applications of timed-release cryptography (e.g., key escrow with time-delayed key recovery feature), the puzzle maker's successful cheating means a total defeat of the system security. The problem here is not that Alice is untrustworthy; it is the absence of a means for Alice to show her honesty. Thus, in applications where the puzzle maker and solver do not trust each other (this is likely, for instance, these two parties should not trust each other in the first three applications listed in Section 1), [5] suggests that a commonly trusted third party be used to check that Alice has formed a puzzle correctly. The use of a commonly trusted third party greatly limits the usefulness of the time-lock puzzle scheme.

3 A new time-lock puzzle scheme

The new time-lock puzzle scheme proposed here is equipped with a protocol that allows Alice to efficiently prove her correctness in creation of a puzzle. Solution of a puzzle now means successful factoring of a large composite integer that she has generated for the puzzle. So the new scheme is in fact a timed-release factorisation of a large integer. We relate the factorisation problem to that of computing a "small" discrete logarithm which has a well understood and controllable time complexity.

To create a puzzle, Alice shall setup $n = pq$ with p and q being large primes such that -1 has the positive Jacobi symbol modulo n. (The Jacobi symbol of an element modulo n can be efficiently evaluated.) Let $\left(\frac{x}{n}\right)$ denote the Jacobi symbol of x modulo n. Then we require

$$\left(\frac{-1}{n}\right) = 1. \tag{1}$$

Alice shall disclose an element of a hidden order modulo n. (Almost all elements in the multiplicative group modulo n have hidden orders.) Let e be this element and t be the hidden order. This means we require

$$e^t \equiv 1 \pmod{n}. \tag{2}$$

---

the difficulty of computing a square-root of a groups element as a means to prevent unauthorised wire-tapping of communications taken place prior to a warranted date.

In addition, we require that the element e be chosen to satisfy

$$\left(\frac{e}{n}\right) = -1. \tag{3}$$

Since Jacobi symbol is multiplicative, we have

$$1 = \left(\frac{1}{n}\right) = \left(\frac{e^t (\bmod n)}{n}\right) = \left(\frac{e}{n}\right)^t = (-1)^t.$$

Therefore the secret order t must be even.

The following method can be used to setup n, e, and t to satisfy the requirements in (1-3).

First, Alice chooses two primes u and v. She should then test the primality of the following two quantities

$$p = 2p'u + 1, \quad q = 2q'v + 1, \tag{4}$$

for some odd numbers p' and q'. The procedure completes once both p and q are found to be primes. Let $n = pq$ and $t = 2uv$. Notice that u, v, p' and q' are odd; we have

$$\left(\frac{-1}{n}\right) = \left(\frac{-1}{p}\right)\left(\frac{-1}{q}\right) = (-1)^{(p-1)/2}(-1)^{(q-1)/2} = (-1)^{p'u}(-1)^{q'v} = 1.$$

So requirement (1) is met.

Alice then finds an element a modulo p of order $p-1 = 2p'u$, and an element b modulo q of order $(q-1)/2 = q'v$. It is easy for her to find such a and b. Let

$$c = a^{p'} \bmod p,$$

$$d = b^{2q'} \bmod q.$$

We know

$$c^{2u} \equiv 1 \pmod{p},$$

$$d^v \equiv 1 \pmod{q}.$$

Applying the Chinese remainder theorem, Alice can compute $e < n$ satisfying

$$e \equiv c \pmod{p} \quad \text{and} \quad e \equiv d \pmod{q}.$$

Obviously,

$$e^t \equiv e^{2uv} \equiv 1 \pmod{n}.$$

This meets the requirement (2).

The fact that a generates a group of $p-1$ elements with half of them being quadratic non-residues renders itself to be a quadratic non-residue (modulo p). Therefore

$$\left(\frac{a}{p}\right) = -1.$$

Further, since p' is odd, we know

$$\left(\frac{c}{p}\right) = \left(\frac{(a^{p'} \bmod p)}{p}\right) = \left(\frac{a}{p}\right)^{p'} = (-1)^{p'} = -1.$$

But d is a square number (modulo p), therefore

$$\left(\frac{d}{q}\right) = 1.$$

These yield

$$\left(\frac{e}{n}\right) = \left(\frac{e \bmod p}{p}\right)\left(\frac{e \bmod q}{q}\right) = \left(\frac{c}{p}\right)\left(\frac{d}{q}\right) = (-1)(1) = -1.$$

This meets the requirement (3).

Using e, n and t, Alice can prove her knowledge of t and at the same time showing the size of t without disclosing t to a verifier (Bob). This is via an interactive knowledge proof protocol (which is specified in A, and the protocol is due to Damgård [2]). The protocol is very efficient. There are also efficient protocols for Alice to show that n is the product of two primes (e.g., [6]).

Proof of the size of t and the two-prime-product structure of n are all what Alice has to do in order to show her honesty in construction of her time-lock puzzle. When t is relatively "small" ($\leq 2^{130}$), such a proof indicates the time needed for finding this secret order of e. The time complexity can be measured by $\sqrt{t}$ based on using the best known algorithms: Shank's baby-step-giant-step algorithm (e.g., page 105 [3]), Pollard's catching-kangaroo algorithm [4]. These algorithms make use of the fact that the discrete logarithms to be extracted are much smaller than the size of the (main) group in question.

After Bob has verified and accepted Alice's proofs, the time-lock puzzle will be accepted as the pair (e, n). Below we show that any message encrypted using n (e.g., using the RSA cryptosystem) can be decrypted after the puzzle is solved. The instruction to solve the puzzle is to extract t as the discrete logarithm of 1 to the base e modulo n.

Suppose that t has been revealed with $\sqrt{t}$ steps of computation. Then Bob can compute

$$f = e^{t/2} \bmod n.$$

Here $t/2 = uv$ is odd. We know from (2)

$$f^2 \equiv e^t \equiv 1 \,(\bmod\, n).$$

We also know from (3)

$$\left(\frac{f}{n}\right) = \left(\frac{e^{uv} (\bmod\, n)}{n}\right) = \left(\frac{e}{n}\right)^{uv} = (-1)^{uv} = -1,$$

while from (1)

$$\left(\frac{1}{n}\right) = \left(\frac{-1}{n}\right) = 1.$$

Therefore
$$f \not\equiv \pm 1 \,(\mathrm{mod}\, n).$$

From this we can conclude that f must be a non-trivial square-root of 1. Since $f^2 \equiv 1 \,(\mathrm{mod}\, n)$, we can write

$$(f+1)(f-1) = kn, \tag{5}$$

for some k. With $f \neq \pm 1$ and $0 < f < n$, we know

$$0 < f - 1 < f + 1 < n.$$

Thus (5) indicates that either $f - 1$ or $f + 1$ must contains a non-trivial factor of n.

So n is factored and the time-lock puzzle solved! The solving time is set under the control of an evidence which is thoroughly examinable by the solver before setting out to solve the puzzle.

4 Discussion

We discuss a number of issues which are related to the proposed scheme.

i) The time complexity measurement for extracting discrete logarithm (i.e., extracting t needs $T = \sqrt{t}$ steps of multiplications modulo n) reaches the lowest known to date on exploitation of extracting "small" discrete logarithms. Any algorithm using fewer operations will provide a breakthrough improvement on solving the "small" discrete logarithm problem.

ii) Since squaring modulo n takes exactly the same time measurement as multiplication modulo n, the time control T in our scheme (which is the number of multiplication) will be exactly equal to that in [5] which is the number of squaring modulo n.

iii) Space complexity (measured in number of bits need to be stored): Pollard's catching-kangaroo algorithm [4] is $O(\log t \log n)$ while Shank's baby-step-giant-step algorithm (e.g., page 105 [3]) is $O(t^{1/2} \log n)$. So for a large t (e.g., $t \approx 2^{100}$) the space requirement of the catching-kangaroo algorithm is trivial while that of the baby-step-giant-step algorithm becomes prohibitive for practical use of the algorithm. For a small t, the baby-step-giant-step algorithm is preferred because it is deterministic.

iv) Neither of the two algorithms that we suggest to use can be parallelised very well. Running m processes independently gives a speedup of only $\sqrt{m}$. So for instance, in order to achieve a speedup of one thousand folds by parallelisation, one million processes are needed. Van Oorschot and Wiener [7] suggested a parallelised catching-kangaroo algorithm which uses m kangaroos, each digs a hole after many jumps (rather than after each jump as in Pollard's original algorithm). That algorithm achieves a linear speedup; however, its space complexity becomes $O(t^{1/2} \log n)$ (number of bits to be stored in a central storage shared by all processors). When t is large (i.e., when parallelisation is most effective), this space requirement is prohibitive for a practical use of the parallelised algorithm.

v) Pollard's rho algorithm for extracting discrete logarithms (e.g., page 106 [3]) is not usable here since it requires to know the order of the group (the discrete logarithm of 1 to the base e) in first place.

vi) For $t \leq 2^{130}$, $T \leq 2^{65}$ which allows the scheme to be usable in applications which fall in a wide range of time frame. However, for a very long term time control, one should consider the effect of Moore's Law. This consideration should be universally observed for any time-lock puzzle which is based on a computational complexity problem, as long as Moore's Law remains to be true.

vii) Alice can set the size of n to be arbitrarily large by setting large p' and q' in (4). They can also be selected as primes, which makes $p - 1$ and $q - 1$ to be non-smooth, a standard means for countering a number of known factorisation algorithms based on exploiting the smoothness of $\phi(n)$. Under such a setting, to factor n without using the prescribed instruction is well understood to be more costly.

5 Conclusion

We have constructed an integer factorisation-based time-lock puzzle scheme. The construction allows the puzzle maker to prove to a solver the time needed for solving a puzzle.

Acknowledgments

I would like to thank my colleague Nigel Smart for discussions which led to the study of the parallelised catching-kangaroo algorithm.

References

1. Burmester, M., Desmedt, Y. and Seberry, J. Equitable key escrow with limited time span (or, how to enforce time expiration cryptographically). Extended abstract. Advances in Cryptology — Proceedings of ASIACRYPT 98 (K. Ohta and D. Pei eds.), Lecture Notes in Computer Science, Springer-Verlag 1514 (1998) pages 380–391.

2. Damgård, I.B. Practical and provably secure release of a secret and exchange of signatures. Advances in Cryptology: Proceedings of EUROCRYPT 93 (T. Helleseth, ed.), Lecture Notes in Computer Science, Springer-Verlag, 765 (1994) pages 201–217.

3. Menezes, A.J., van Oorschot, P.C. and Vanstone, S.A. *Handbook of Applied Cryptography.* CRC Press. 1997.

4. Pollard, J.M. Monte Carlo method for index computation (mod p), *Mth. Comp.*, Vol.32, No.143 (1978), pages 918–924.

5. Rivest, R.L., Shamir, A. and Wagner, D.A. Time-lock puzzles and timed-release crypto. Manuscript. Available at (http://theory.lcs.mit.edu/~rivest/RivestShamirWagner-timelock.ps).

6. van de Graaf, J. and Peralta, R. A simple and secure way to show the validity of your public key. Advances in Cryptology — Proceedings of CRYPTO 87 (E. Pomerance, ed.), Lecture Notes in Computer Science, Springer-Verlag 293 (1988) pages 128–134.
7. van Oorschot, P.C. and M.J.Wiener M.J. Parallel collision search with cryptanalytic applications. *J. of Cryptology*, Vol.12, No.1 (1999), pages 1–28. (http://theory.lcs.mit.edu/ rivest/RivestShamirWagner-timelock.ps).

A Proof of the size of a secret order

The basic technique is due to Damgård [2]. We specify a simplified variation to suit our case of showing the order of an element. In our variation, the number in question is protected in computational (statistical) zero-knowledge.

Let t be in the interval $[a, b] = \{x | a \leq x \leq b\}$ where $c = b - a$ and $c \leq a$. In protocol Size specified below, Alice can convince Bob that t, the discrete logarithm of 1 modulo n to the base e, is in the interval $[a - c, b + c]$.

Protocol Size(e, n)

Execute the following k times:

1. Alice picks $0 < s_1 < c$ at uniformly random, and sets $s_2 = s_1 - c$; she sends to Bob the pair

$$E_1 = e^{s_1} \bmod n, \quad E_2 = e^{s_2} \bmod n;$$

2. Bob selects $d = 0$ or $d = 1$ at uniformly random, and sends d to Alice;
3. Alice sets

$$u_1 = s_1, \quad u_2 = s_2 \qquad \text{for } d = 0;$$
$$u_1 = s_1 + t, u_2 = s_2 + t \text{ for } d = 1;$$

and sends u_1, u_2 to Bob;
4. Bob verifies

$$E_1 \equiv e^{u_1} (\bmod\, n), \quad E_2 \equiv e^{u_2} (\bmod\, n),$$

and

$$0 < u_1 < c, \qquad -c < u_2 < 0 \quad \text{for } d = 0;$$
$$c \leq u_1 \leq b + c, 0 \leq u_2 \leq b + c \text{ for } d = 1;$$

In Bob's evaluation of $e^x \bmod n$, the case of $x < 0$ should be evaluated by performing $(1/e)^{|x|} \bmod n$.

Set, for instance, $[a, b] = [2^{\ell-1}, 2^\ell]$. Then $c = 2^{\ell-1}$. Size will prove that the order of the element e has a size not exceeding $\ell + 1$ binary bits. The probability for this to hold is at least $1 - 1/2^k$.

Efficient Accumulators Without Trapdoor
Extended Abstract

Tomas Sander *

International Computer Science Institute
1947 Center Street, Berkeley, CA 94704, USA
sander@icsi.berkeley.edu

Abstract. In 1994 Benaloh and de Mare introduced the notion of one way accumulators that allow to construct efficient protocols for proving membership in a list and related problems like time stamping and authentication. As required by Benaloh et al. unlike in signature based protocols no central trusted authority is (should be) needed. Accumulator based protocols do further improve on hash tree based protocols for proving membership in a list as verification and storage requirements are independent of the number of items in the list. Benaloh's et al. accumulator construction was based on exponentiation modulo a RSA modulus $N = PQ$.

As already noted by Benaloh et al. the party (or parties) who generated the needed RSA modulus N during system set up knows a factorization of N. This knowledge allows this party to completely bypass the security of accumulator based protocols. For example a time stamping agency could forge time stamps for arbitrary documents.

Thus these parties need to be trusted in (at least) two ways. First that they do not abuse their knowledge of the trapdoor and secondly to have had adequate security in place during system set up, which prevented outside attackers from getting hold of P and Q.

In this paper we describe a way to construct (generalized) RSA moduli of factorization unknown to anybody. This yields (theoretically) efficient accumulators such that "nobody knows a trapdoor" and the two above mentioned trust requirements in the parties who set up the system can be removed.

Keywords: one way accumulator, RSA modulus.

1 Introduction

The basic idea of the accumulator based authentication scheme suggested by Benaloh and de Mare in [2] is as follows: Let N be a RSA modulus and let $x \in (\mathbb{Z}/N\mathbb{Z})^{\times}$. Further let $\mathcal{L} = \{y_1, \dots y_m\} \subset (\mathbb{Z}/N\mathbb{Z})^{\times}$ be a list of items for which membership to $\mathcal{L}$ should later be proved. The "accumulated hash value" z of the list $\mathcal{L}$ is defined to be the value $z = x^{y_1 \cdot y_2 \cdots y_m} \mod N$. Assume now that Victor has obtained z over an authenticated channel. To prove membership

---

* Current address: InterTrust, Santa Clara, CA, USA.

of an element y to $\mathcal{L}$ Alice presents to Victor a y'th root r of z. Victor accepts if $r^y = z$.

Assume y in fact belongs to the list, i.e. let $y = y_i$. Then a y'th root of z can be computed by raising x to the power $a = \prod_{y \in \mathcal{L} \setminus \{y_i\}} y$. Thus if Alice has obtained a or computed it herself using her knowledge of the other items in $\mathcal{L}$ she can present x^a to Victor to prove that y contributed to the accumulated hash z, i.e. that $y \in \mathcal{L}$.

Barić and Pfitzmann [1] showed how this basic scheme can be turned into a provably secure scheme: assuming (variants of) the RSA assumption they describe an accumulator based protocol such that forging membership proofs is unfeasible. We postpone the discussion of the security of accumulator based protocols to Section 3.

A main advantage of accumulator based protocols for proving membership in a list over protocols using Merkle's authentication trees [7] is improved efficiency: in hash tree based protocols the elements of the list $\mathcal{L}$ to be authenticated are inserted as leaves of a hash tree. Assume Victor has obtained the root R of the hash tree over an authenticated channel. To prove membership of y to $\mathcal{L}$ Alice presents to Victor a hash path leading from the leaf y to the root R and Victor verifies the correctness of this path. Thus Alice needs to store $\log \sharp \mathcal{L}$ many values and Victor's work during checking the correctness of the path also depends logarithmically on the number of items $\mathcal{L}$. In contrast to this the complexity of accumulator based protocols is independent of the number of items hashed together: the length of a hash chain that Alice needs to present to prove membership to $\mathcal{L}$ collapses to one.

Although accumulator based schemes are more efficient they also have a serious disadvantage over hash tree based authentication schemes. Hash tree based schemes are secure under the sole cryptographic assumption that the hash function is strongly collision resistant: *no party* is able to forge membership proofs. This is no longer the case in accumulator based protocols. All efficient accumulator constructions known today are based on RSA type functions. The problem is that the RSA modulus N needs to be *constructed* during system setup. RSA moduli are typically found by choosing two random large primes P and Q and multiplying them together. Thus the party who provides N during system setup, knows also the "trapdoor" P and Q. [1] Unfortunately a party who knows P and Q can completely bypass the system's security. It can forge membership proofs for arbitrary values $y \in (\mathbb{Z}/N\mathbb{Z})^\times$. To do this the party computes a t s.t. $ty \equiv 1 \bmod \varphi(N)$. Now z^t is a y'th root of z that "proves" to Victor that $y \in \mathcal{L}$.

This is an undesirable property in many applications and already Benaloh et al. asked in their original paper [2] if accumulators "without trapdoor" could be

---

[1] Even if secure multiparty computation protocols are employed during system setup still a colluding curious coalition of this group could recover P and Q. Thus the general problem that someone "knows the trapdoor" remains also in the distributed setting.

constructed. The main result of this paper is that this question has an affirmative answer.

Let us first describe two examples to illustrate weaknesses of accumulators that have a (known) trapdoor. In [2] an efficient round based time stamping protocol using accumulators was suggested. In each round the submitted documents are hashed together using the accumulator. Now the time stamping agency (or the parties that provided the modulus N) can forge arbitrary time stamps for any document of their choice. Thus the time stamping agency needs to be fully trusted to not abuse its knowledge of the trapdoor or to have it destroyed after system set up.

A different example is provided by an electronic cash scheme that was recently suggested by Sander and Ta-Shma in [9]. This system offers payers unconditional anonymity but unlike previous (blind signature based) schemes it is not vulnerable to the blackmailing [11] and the bank robbery attack. The serious bank robbery attack was described by Jakobsson and Yung in [6]. In this attack the secret key which the bank uses to sign coins is compromised, e.g. by an insider attack. In the scheme described in [9] users submit during withdrawal hash values of electronic coins to the bank which inserts them as leaves into a hash tree. Its root R is distributed to merchants by the bank. During payment a user demonstrates to the merchant the validity of a coin C by proving with a zero knowledge argument that there is a hash path from C to R.

This system is resistant to the bank robbery attack as there are no critical bank secrets that allow to forge money. To speed up the system it is asked in [9] whether accumulators without trapdoor exist that could substitute the hash trees. As remarked in [9] parties that know the trapdoor of an accumulator can forge arbitrary amounts of electronic money. Thus the resistance of the scheme against the bank robbery attack would at least partially be lost if existing accumulators were used. At least during system set up the scheme would again be vulnerable to the bank robbery attack. These examples motivate why it is useful to have accumulators without trapdoor.

1.1 Accumulators without trapdoor

A construction of an accumulator that does not have a trapdoor was given by Nyberg in [8]. This accumulator is provably secure. However the accumulator is not space efficient (and thereby also not time efficient): the accumulated hash of N items has length $N \log(N)$. Thus the potential performance advantages offered by accumulator based protocols are not preserved.

All efficient accumulators that have been found so far are based on exponentiation modulo a RSA modulus. Note that the RSA type accumulator described by Benaloh et al. would not have an (exploitable) trapdoor if we had an algorithm that outputs RSA moduli $N = PQ$ such the parties executing the algorithm had no way to extract P and Q. To the best of our knowledge no such algorithm is known today. Unfortunately we can not present such an algorithm here either.

We avoid the problem of constructing such RSA moduli and suggest to use "generalized RSA moduli of unknown complete factorization " instead for the construction of accumulators. By abuse of notation we call these numbers RSA-UFOs. We call a number N a RSA-UFO if N has at least two large prime factors P and Q such that it is infeasible for *any* coalition of players including those that generated N to find a splitting of N into factors N_1, N_2, such that $P \mid N_1$ and $Q \mid N_2$, i.e. it is infeasible to construct a splitting of N into factors that "separate" P and Q.

In this paper we suggest a (theoretically) efficient algorithm that constructs a RSA-UFO with very high probability: given a security parameter k and an element ϵ there is an algorithm with running time polynomial in k and $\log(1/\epsilon)$ that outputs an element that is a RSA-UFO w.r.t. two primes P and Q of bit length at least k with probability $1 - \epsilon$. We call such a number a (k, ϵ)-RSA-UFO. By choosing ϵ sufficiently small any practically desirable degree of certainty can be achieved. The algorithm that outputs (k, ϵ)-RSA-UFOs is described in Section 2.

Using the techniques of Barić and Pfitzmann [1] we construct in Section 3 an accumulator using (k, ϵ)-RSA-UFOs such that "nobody knows a trapdoor". We show that this accumulator is provably secure under a natural RSA type assumption.

2 Constructing Generalized RSA Moduli Of Unknown Factorization

Our general strategy for the construction of (k, ϵ)-RSA-UFOs is as follows: We first prove that a constant fraction of integers with bit length (roughly) $\leq 3k$ has at least two distinct prime factors of bit length $\geq k$. Now one picks uniformly and independently $r = O(\log \frac{1}{\epsilon})$ integers $a_1, a_2, \ldots, a_r$ of bit length $\leq 3k$. The probability that at least one of these numbers has two distinct prime factors of length at least k is $\geq 1 - \epsilon$. Thus the product $A := a_1 \cdot a_2 \cdot \ldots \cdot a_r$ has now at least one factor a_i which has two large distinct prime divisors P and Q w.v.h.p.. To avoid cheating the random choices should have been made by choosing the bits in the expansion of the numbers according to a public or publicly verifiable random source (see Paragraph 2.1 for a brief discussion on ways how this could be achieved).

We argue now that it is reasonable to believe that A is a (k, ϵ)-RSA-UFO. W. v. h. p. one of the factors, let's say w.l.o.g. a_1, has at least two distinct k bit prime divisors P and Q. Assume now that a coalition of players were able to produce a splitting of A into factors N_1, N_2 s.t. $P \mid N_1$ and $Q \mid N_2$. The probability that two (and thus polynomially many) randomly chosen numbers have a common large prime factor is negligible. Thus $\gcd(N_1, a_1)$ and $\gcd(N_2, a_1)$ yield a splitting of a large number a_1 into two factors, such that each of them contains a different large prime divisor of a_1.

Now a player could simulate the choices of the other random numbers $a_2, \ldots, a_r$ himself. He thereby had an algorithm that - with a non-negligible probability of success - splits a random number B, which has two large prime factors P and Q

into factors which separate P and Q. This seems to be a completely infeasible task for all available factoring algorithms.

Theorem 1. *Let $\xi \in (\frac{1}{3}, \frac{5}{12})$. Then the number of integers $\leq x$ that have two distinct prime factors $\geq x^{\xi}$ is $x(\frac{1}{2}\ln^2 \frac{1}{2\xi} + O(\frac{1}{\ln x}))$.*

Proof. Let $A_t := \{y \leq x \mid t \mid x\}$ be the set of all multiples of t that are smaller or equal then x.

Let $G(x,\xi) = \bigcup_{p,q \geq x^{\xi}, p \neq q} \#A_{pq}$ be the number of all integers $\leq x$ that are divisible by 2 distinct primes $p, q \geq x^{\xi}$.

Furthermore let $\pi(x)$ denote the number of primes $\leq x$, ln the logarithm to the base e, and $\lfloor \cdot \rfloor$ the greatest integer function.

Now let $\xi \in (\frac{1}{3}, \frac{5}{12})$. Then distinct sets A_{pq}, A_{rs}, for $p \neq q, r \neq s$ are disjoint as $A_{pq} \cap A_{rs} = A_{\text{lcm}(p,q,r,s)}$ and $\text{lcm}(p,q,r,s)$ has at least three prime divisors $> \sqrt[3]{x}$ in this case.

Hence

$$G(x,\xi) = \sum_{x^{\xi} \leq p < q \leq \sqrt{x}} \#A_{pq} = \frac{1}{2}\Big(\sum_{x^{\xi} \leq p,q \leq \sqrt{x}} \#A_{pq} - \sum_{x^{\xi} \leq p \leq \sqrt{x}} \#A_{p^2} \Big).$$

As $\#A_t = \lfloor \frac{x}{t} \rfloor$ we get

$$G(x,\xi) = \frac{1}{2}\Big(\sum_{x^{\xi} \leq p,q \leq \sqrt{x}} \lfloor \frac{x}{pq} \rfloor - \sum_{x^{\xi} \leq p \leq \sqrt{x}} \lfloor \frac{x}{p^2} \rfloor \Big)$$

$$= \frac{1}{2}\Big(\sum_{x^{\xi} \leq p,q \leq \sqrt{x}} \frac{x}{pq} + O(\pi^2(\sqrt{x})) - \sum_{x^{\xi} \leq p \leq \sqrt{x}} \frac{x}{p^2} + O(\pi(x)) \Big)$$

$$= \frac{1}{2}x\Big(\sum_{x^{\xi} \leq p,q \leq \sqrt{x}} \frac{1}{pq} - \sum_{x^{\xi} \leq p \leq \sqrt{x}} \frac{1}{p^2} + O(\frac{1}{\ln x}) \Big).$$

It is $\sum_{p \leq \sqrt{x}} \frac{1}{p^2} = O(1)$, as the series converges.

To estimate the other sum we use the well known number theoretic fact that there is a constant B s.t. $\sum_{p \leq y} \frac{1}{p} = \ln \ln y + B + O(\frac{1}{\ln y})$ (cf. e.g. [5]). From this identity one obtains that $\sum_{y^{\xi} \leq p \leq \sqrt{y}} \frac{1}{p} = \ln \ln y^{\frac{1}{2}} - \ln \ln y^{\xi} + O(\frac{1}{\ln x}) = \ln \frac{1}{2\xi} + O(\frac{1}{\ln x})$.

Now

$$\sum_{x^{\xi} \leq p,q \leq \sqrt{x}} \frac{1}{pq} = \Big(\sum_{x^{\xi} \leq p \leq \sqrt{x}} \frac{1}{p} \Big)^2 = \Big(\ln \frac{1}{2\xi} + O(\frac{1}{\ln x})\Big)^2 = \ln^2 \frac{1}{2\xi} + O(\frac{1}{\ln x}).$$

All together we obtain

$$G(x,\xi) = \frac{1}{2}x\Big(\ln^2 \frac{1}{2\xi} + O(\frac{1}{\ln x}) \Big)$$

$$= x\Big(\frac{1}{2}\ln^2 \frac{1}{2\xi} + O(\frac{1}{\ln x}) \Big),$$

and the claim is proved.

Now we can describe an algorithm CONSTRUCT RSA-UFO that on input $(k, \frac{1}{2^l})$ outputs a $(k, \frac{1}{2^l})$-RSA-UFO modulus of bit length $O(kl)$. First a ξ that is slightly larger then $\frac{1}{3}$ is picked. Let k be sufficiently large. According to Theorem 1 there is a constant p depending only on the previous choice of ξ such that a randomly chosen number of bit length $\leq (\lfloor \frac{1}{\xi} \rfloor + 1)k$ is a k-RSA-UFO with probability $\geq p$. Thus one chooses a list $\mathcal{M}$ of $(\lfloor \frac{1}{p} \rfloor + 1)l$ random numbers of bit length $(\lfloor \frac{1}{\xi} \rfloor + 1)k$. This can be done by choosing each bit of their bit representation according to a public random source. Then the probability that none of the elements in $\mathcal{M}$ is a k-RSA-UFO is $\leq \frac{1}{2^l}$. Hence $N = \prod_{m \in \mathcal{M}} m$ is a $(k, \frac{1}{2^l})$ RSA-UFO modulus of bit length $O(kl)$ and the algorithm CONSTRUCT RSA-UFO outputs N.

A more careful determination of the constants will be contained in the full version of the paper.

Certainly there are some improvements to this general algorithm possible in order to obtain RSA-UFOs of smaller bit length. One may test the sampled numbers in $\mathcal{M}$ for primality and throw those found prime away. Furthermore one can divide out small prime factors from the elements in $\mathcal{M}$ to decrease the bit length of the output.

We think it's a theoretically interesting question if there is an algorithm that produces $(k, \frac{1}{2^l})$ RSA-UFO moduli of bit length smaller then $O(kl)$.

For practical applications it would be interesting to find RSA-UFOs of bit length as small as possible for parameters that give adequate security for practical applications. For example it would be interesting to have RSA-UFOs offering security comparable to usual 1024 bit RSA moduli. We ask: what is the smallest bit length of $(512, \frac{1}{2^{80}})$-RSA-UFOs? Our basic algorithm sketched above will yield an output of bit length (much) greater then 40.000 bits for these parameters, and may thus be too large for practical applications.

2.1 On the generation of public random strings

Our construction depends upon the availability of publicly verifiable random strings.

We describe an example to give an idea how such strings may be found in practice. Assume in November U.S. Congress publicly decides that such a string should be generated. Now Congress may fix a future date, e.g. noon of December 15 and decide on using the stock quotations at NYSE and Nasdaq at this future time for the generation of a random string. Also the procedure how to generate from the stock market data the random string is fixed by Congress in November. At noon of Dec. 15 a snapshot of the stock market is taken and using a (predetermined) hash function hashed to a shorter string. (If more random bits are needed one may also repeat this procedure on several (predetermined) dates and concatenate the obtained strings.)

The U.S. Congress decision is public and everybody knows when this snapshot will be taken. The stock market seems to be quite unpredictable and in particular it does not seem possible that any reasonable coalition of players is able to manipulate the stock market in such a way that the output of the hash functions can be substantially influenced (e.g., forced to belong to a small set of values). The hash function should further extract the (assumed) randomness in the stock market data and produce bits that are close to uniform. Important for us is that later any party can verify that the procedure to construct the random string was correctly executed.

This is certainly only a heuristic argument, but it makes plausible that practically useful ways to generate a certified public random string seem to very well exist.

3 Accumulators From Generalized RSA Moduli

In the last section we described an algorithm that outputs RSA-UFOs. In this section we construct from this provably secure accumulators using the techniques developed in [1].

Intuitively an accumulator is secure if given a list $\mathcal{L}$ that hashed to an accumulated value z it is infeasible to produce a membership proof for an element x not belonging to $\mathcal{L}$. Assume we are given a (regular) RSA modulus N and a basis $x \in (\mathbb{Z}/N\mathbb{Z})^\times$ for the exponentiation. The original protocol to authenticate elements in a list $\mathcal{L} = \{y_1, \ldots, y_m\}$ via accumulators is not secure in the sense described above. Recall that to authenticate a value y Alice needs to present a y'th root of z to Victor. If Alice knows $\mathcal{L}$ she can not only authenticate the elements $y_1, y_2, \ldots, y_m$, which do belong to $\mathcal{L}$, but also, e.g., the element $y := y_1 \cdot y_2$, which may not belong to $\mathcal{L}$. A y'th root of z is given by

$$x^{\prod_{y \in \mathcal{L} \setminus \{y_1, y_2\}} y}.$$

More general: for every subset $S \subset \mathcal{L}$ the value $y_S := \prod_{y \in S} y$ can be authenticated by the value

$$x^{\prod_{y \in \mathcal{L} \setminus S} y}.$$

Although the list $\mathcal{L}$ contains only m items thus later potentially 2^m items can be authenticated later. This is clearly unsatisfactory.

Barić and Pfitzmann solve this (and more) problems by restricting the inputs to the accumulator function to prime numbers $e < N$. When Alice authenticates a value y she has as before to present a y'th root a of z. But in the verification step Victor not only checks that $a^y = z$ but also that $y < N$ and that y is prime.

From now on we will restrict ourselves to accumulators where the inputs are restricted to prime numbers and Victor checks primality during verification.

Barić and Pfitzmann introduce an even stronger notion of security for accumulators by requiring that it should be unfeasible to construct any list $\mathcal{L}$ such that a membership proof for an element not in $\mathcal{L}$ can be forged (thus here an adversary is allowed to choose the list $\mathcal{L}$ himself). In [1] an accumulator that achieves this requirement is called collision free.

This strong notion of security is appropriate for the applications sketched before. In the time stamping application it should not be possible to produce time stamps on documents that have not been submitted at the time when their accumulated hash was computed, no matter how the submitted documents were chosen. In the electronic cash application users submit (hash values of) coins to the bank which inserts them as leaves into a hash tree. To assure that users are not able to forge money they should not be able to submit any set of coins to the bank, such that they can later prove that a new, forged coin belongs to the tree. To substitute trees by accumulators it is necessary to preserve this property also for accumulator based schemes.

Barić's and Pfitzmann's definition of collision freeness considers only adversaries, that do not know the factorization of N. Their accumulator is not collision free for parties that do know the trapdoor P and Q. We consider here a stronger definition of security by requiring that *no coalition* of polynomial time players should be able to forge membership proofs. Thus also bank insiders can not forge electronic coins and the time stamping agency can not forge time stamps. We call accumulators achieving this stronger security property *universally collision free*.

Definition 1. *A family of accumulators* $(x, N, \mathcal{P})$ *is called universally collision free if for all* $m \geq 1$, *it is infeasible for any coalition of probabilistic time bounded players to find a list* $\mathcal{L} = \{y_1, \ldots y_m\} \subset \mathcal{P}$, *a* $y \in \mathcal{P}, y \notin \mathcal{L}$, *and an* $a \in (\mathbb{Z}/N\mathbb{Z})^\times$ *such that a authenticates y, i.e.* $a^y \equiv x^{y_1 \cdots y_m} \bmod N$, *with non-negligible probability of success.*

In [1] a strong RSA assumption is formulated under which collision freeness of the accumulator against adversaries not knowing the trapdoor P, Q is proved. We formulate here a strong RSA-UFO assumption under which we then prove the security of the accumulator against any coalition of polynomial time players.

Strong RSA Problem: Given $x \in (\mathbb{Z}/N\mathbb{Z})^\times$ find a prime $e < N$ and an element s s.t. $s^e = x$.

Strong RSA-UFO Assumption: Let CONSTRUCT RSA-UFO be the probabilistic algorithm described in Section 2 that outputs a composite number N. Then for all probabilistic polynomial time algorithms $\mathcal{A}$, all polynomials P and all sufficiently large k:

$$Pr[a^e = x \bmod N \wedge e \text{ prime } \wedge \log e < 2k :$$

$$N = \text{CONSTRUCT RSA-UFO}(k, \frac{1}{2^k}); x \in_R \mathbb{Z}/N\mathbb{Z}; (a, e) \leftarrow \mathcal{A}(N, z)] < \frac{1}{P(k)}.$$

Unlike in the usual RSA assumption an adversary is also allowed to pick the exponent e.

We suggest here an RSA type accumulator, as described before, where the modulus N is an RSA-UFO, i.e. an output of the algorithm CONSTRUCT RSA-UFO with security parameters k and $\frac{1}{2^k}$, x is a random number modulo $\tilde{N}$ that has also been constructed according to a public random source, and the inputs to the accumulator are restricted to the set $\mathcal{P}$ of primes e with $\log e < 2k$.

Theorem 2. *Under the strong RSA-UFO assumption this accumulator is universally collision free.*

Proof. The proof proceeds as in [10, 2, 1]. Assume that given a RSA-UFO N and x an adversary can find a collision of the accumulator with the parameters (N, x), i.e. he can find a list of primes $\mathcal{L} = \{y_1, \ldots, y_m\} \subset \mathcal{P}$, a prime $y \notin \mathcal{L}, y \in \mathcal{P}$ and an element a s.t. $a^y = x^{y_1 \cdots y_m}$. As y is prime there are integers u, v s.t. $u(y_1 \cdots y_m) + vy = 1$, thus $(a^u x^v)^y = x$. Thus the adversary can extract an e'th root of x, for the prime $e = y$, breaking the strong RSA-UFO assumption.

3.1 On the strong RSA-UFO assumption

The strong RSA-UFO assumption is obtained from the strong RSA assumption formulated in [1] by considering RSA-UFOs $\tilde{N}$ instead of "usual" RSA moduli N. Note that variants of the (regular) strong RSA assumption have recently also been used in [4] and [3]. The strong RSA assumption has not been tested for too long although in [1] several arguments that could support the soundness of this assumption have been presented. We want to give here at least a relative argument: if one believes the strong RSA assumption it is also reasonable to believe the strong RSA-UFO assumption.

RSA moduli $N = PQ$ where P and Q are primes of the same bit length k are commonly used in cryptographic applications as they constitute the hardest known instances for factoring algorithms, where the running time is measured in the bit length of the input. We argue now that it is reasonable to believe that computing e'th roots of (random) elements $x \in (\mathbb{Z}/\tilde{N}\mathbb{Z})^\times$ for a $(k, \frac{1}{2^k})$- RSA-UFO $\tilde{N}$, where $\tilde{N}$ has been obtained as an output of CONSTRUCT RSA-UFO and $\log e < 2k$ is not easier then extracting e'th roots in $\mathbb{Z}/N\mathbb{Z}$ for usual RSA moduli $N = PQ$ for k-bit primes P and Q and $e < N$.

Assume an adversary were able to extract an e'th root of a random element x modulo $\tilde{N}$, where $\tilde{N}$ is an output of the algorithm CONSTRUCT RSA-UFO. Recall that $\tilde{N}$ was obtained as a product of randomly chosen large numbers $a_1, \ldots, a_r$. W.v.h.p. there is now one factor, say a_1, that has two distinct prime divisors P and Q of length at least k. Thus if an adversary can compute an e'th root y of $x \mod \tilde{N}$ then, given a_1, he can in particular compute an e'th root $y \mod a_1$ of $x \mod a_1$. Now if x is a random number modulo $\tilde{N}$ then $x \mod a_1$ is random number modulo a_1.

Thus even if an oracle would give an adversary a complete factorization of $\tilde{N}$ for free except the factorization of the divisor $P \cdot Q$ of a_1, the adversary would still be left with the problem to compute an e'th root of $x \mod P \cdot Q$, where P and Q are primes of bit length at least k and $e < PQ$. This problem doesn't look any easier then computing an e'th root modulo a "true" RSA modulus which is the product of two primes of length exactly k.

3.2 Universally collision free accumulators under the random oracle assumption

Motivated by the fact that the strong RSA assumption still needs to be tested in [1], assuming the random oracle hypothesis, also a collision free accumulator under the *normal* RSA assumption is constructed. By replacing RSA moduli by RSA-UFOs the same construction (with the same proofs) yields easily an accumulator that can be proved to be universally collision free under the normal RSA assumption formulated for RSA-UFO moduli and the random oracle hypothesis. As this is completely analogous to the construction in [1] we omit here further details.

We further refer the reader directly to [1] for a discussion of a conversion algorithm that allows to convert (many) inputs into prime numbers, i.e. suitable inputs for the accumulator.

4 Conclusion

We showed how to make a provably secure accumulator without a (known) trapdoor. Our solution is based on an algorithm that outputs integers with large prime divisors such that "nobody knows a complete factorization". We think it is a theoretically and practically interesting question if there is an algorithm that produces such integers of a smaller bit length then the algorithm described in this paper.

5 Acknowledgments

It's a pleasure to thank Amnon Ta-Shma for many helpful and stimulating discussions that lead to this paper. I would further like to thank Stuart Haber and Moti Yung for interesting conversations on several aspects of this work.

References

1. Baric and Pfitzmann. Collision-free accumulators and fail-stop signature schemes without trees. In *EUROCRYPT: Advances in Cryptology: Proceedings of EURO-CRYPT*, volume 1233, 1997.
2. Josh Benaloh and Michael de Mare. One-way accumulators: A decentralized alternative to digital signatures (extended abstract). In Tor Helleseth, editor, *Advances in Cryptology—EUROCRYPT 93*, volume 765 of *Lecture Notes in Computer Science*, pages 274–285. Springer-Verlag, 1994, 23–27 May 1993.
3. J. Camenisch and M. Michels. A group signature scheme with improved efficiency. *Lecture Notes in Computer Science*, 1514:160–??, 1998.
4. Eiichiro Fujisaki and Tatsuaki Okamoto. Statistical zero knowledge protocols to prove modular polynomial relations. In Burton S. Kaliski Jr., editor, *Advances in Cryptology—CRYPTO '97*, volume 1294 of *Lecture Notes in Computer Science*, pages 16–30. Springer-Verlag, 17–21 August 1997.

5. G. Hardy and E. Wright. An introduction to the theory of numbers. Oxford University Press, 5th edition, 1985.

6. M. Jakobsson and M. Yung. Revokable and versatile electronic mony. In Clifford Neuman, editor, *3rd ACM Conference on Computer and Communications Security*, pages 76–87, New Delhi, India, March 1996. ACM Press.

7. R. Merkle. Protocols for public key cryptosystems. In IEEE, editor, *Proceedings of the 1980 Symposium on Security and Privacy, April 14–16, 1980 Oakland, California*, 1109 Spring Street, Suite 300, Silver Spring, MD 20910, USA, 1980. IEEE Computer Society Press.

8. K. Nyberg. Fast accumulated hashing. In Dieter Grollman, editor, *Fast Software Encryption: Third International Workshop*, volume 1039 of *Lecture Notes in Computer Science*, pages 83–87, Cambridge, UK, 21–23 February 1996. Springer-Verlag.

9. T. Sander and A. Ta-Shma. Auditable, anonymous electronic cash. In *To appear in the proceedings of Crypto '99. Forthcoming volume in Lecture Notes in Computer Science*, 1999.

10. Adi Shamir. On the generation of cryptographically strong pseudorandom sequences. *ACM Transactions on Computer Systems*, 1(1):38–44, February 1983.

11. S. von Solms and D. Naccache. On blind signatures and perfect crimes. *Computers and Security*, 11(6):581–583, October 1992.

Evolutionary Heuristics for Finding Cryptographically Strong S-Boxes

W. Millan, L. Burnett, G. Carter, A. Clark and E. Dawson

Information Security Research Centre,
Queensland University of Technology,
GPO Box 2434, Brisbane 4001
Queensland, Australia
FAX: +61-7-3221 2384
Email: {millan,burnett}@isrc.qut.edu.au

Abstract. Recent advances are reported in the use of heuristic optimisation for the design of cryptographic mappings. The genetic algorithm (GA) is adapted for the design of regular substitution boxes (s-boxes) with relatively high nonlinearity and low autocorrelation. We discuss the selection of suitable GA parameters, and in particular we introduce an effective technique for breeding s-boxes. This *assimilation* operation, produces a new s-box which is a simple and natural compromise between the properties of two dissimilar parent s-boxes. Our results demonstrate that assimilation provides rapid convergence to good solutions. We present an analysis comparing the relative effectiveness of including a local optimisation procedure at various stages of the GA. Our results show that these algorithms find cryptographically strong s-boxes faster than exhaustive search.

1 Introduction

Confidence in the security of modern and future electronic communications rests on the belief that the cryptographic algorithms employed are able to resist cryptanalytic attacks. Since the introduction of the original Data Encryption Standard (DES) by NIST in the 1970's, there has been an increasing research effort devoted to discovering structures and components that can be utilised in the design of ciphers to achieve this goal of security.

Substitution boxes (s-boxes) are the most widely used method to provide nonlinearity in block ciphers. An s-box is a mapping from n input bits to m output bits. When $n = m$, the mapping is reversible and is said to be bijective. However, in many block ciphers the s-boxes are not bijective, the Data Encryption Standard (DES) [8] being a well-known example in which the number of input bits (six) is greater than the number of output bits (four). In this paper, we investigate regular s-boxes in which the number of input bits is greater than or equal to the number of output bits, and in particular we focus on the case of eight bits in and four bits out.

In order to resist modern cryptanalytic attacks based on linear approximation [4] and differential characteristics [1], s-box mappings should satisfy several cryptographic properties. In this paper we concentrate on *nonlinearity* and *auto-correlation*, which are described in detail in Section 2. It is known that randomly generated s-boxes usually lack serious cryptographic weaknesses, however random s-boxes rarely possess exceptionally good properties, so they may not be considered appropriate for use. Specific properties can be optimised by dedicated construction methods, however other security properties such as algebraic complexity can be overlooked.

In this paper we present viable alternative methods for the construction of secure s-boxes, using two heuristic search algorithms: the genetic algorithm and hill climbing. These algorithms are described and some options for their use are discussed in Section 3. Our results, which are presented in Section 4, clearly show that the GA is a very effective method for s-box design. At the heart of the GA is a breeding process for regular s-boxes, which we call *assimilation*. The effect of hill-climbing at various stages of the algorithm is also investigated. Finally in Section 5 we make some concluding remarks.

2 Boolean Function and S-Box Theory

Block ciphers manipulate binary vectors to achieve cryptographic security. The main technique used to introduce nonlinear complexity is the use of substitutions, in which one vector is replaced with another. A list of substitution vectors is called a Substitution Box, or s-box. A general s-box is typically described by the number of input bits, n and the number of output bits, m. An s-box is then a look-up-table of 2^n m-bit vectors.

A description of s-boxes in terms of their component Boolean functions ($f(x)$) allows the use of the fast transform techniques in the calculation of cryptographic properties. From this view an s-box is a set of m single output Boolean functions, each taking n inputs. The m columns of the look-up-table are the truth tables of these component Boolean functions. The properties of these output functions and their non-zero linear combinations can be calculated using the well-known Walsh-Hadamard Transform (WHT): $\hat{F}(\omega) = \sum_x (-1)^{f(x)} \cdot (-1)^{L_\omega(x)}$, where $L_\omega(x) = \omega_1 x_1 \oplus \omega_2 x_2 \oplus \cdots \oplus \omega_n x_n$ is a linear function defined by ω. The WHT can be calculated in $n2^n$ operations, and shows the Hamming distance between the function $f(x)$ and all linear functions and their complements (together called affine functions).

Some other definitions are necessary. The Hamming weight of a Boolean function is the number of 1's in the truth table: $hwt(f) = \sum_x f(x)$. A function is balanced if it has equal numbers of 1's and 0's in the truth table, or equivalently if $hwt = 2^{n-1}$.

The Hamming distance between two Boolean functions is a count of the number of truth table positions in which they differ: $d(f(x), g(x)) = \#\{x : f(x) \neq g(x)\}$, where $\#\{\}$ indicates the cardinality of the set. The statistical correlation between two functions is directly related to the Hamming distance by $c(f, g) = \frac{d(f,g)}{2^n} \in [-1, 1]$. Hence the Walsh Hadamard Transform shows the correlation to all affine functions: $c(L_\omega(x), f(x)) = 2^{-n}\hat{F}(\omega)$.

The *nonlinearity* of a single output Boolean function $f(x)$ is the maximum Hamming distance to any affine function, and it is given directly by $nonlin = \frac{1}{2}(2^n - WH_{max})$, where WH_{max} is the largest magnitude value in the Walsh Hadamard transform $\hat{F}(\omega)$. High nonlinearity (small WH_{max}) is required for the security of cryptographic mappings. The nonlinearity of an s-box is defined to be the minimum nonlinearity of all non-zero linear combinations of the output [9].

The autocorrelation function provides a measure of self-similarity for a Boolean function. It is defined by $r(s) = \sum_x (-1)^{f(x)} \cdot (-1)^{f(x+s)}$, but it may be calculated efficiently by $r(s) = 2^{-n} \sum_\omega \hat{F}(\omega)(-1)^{L_\omega(s)}$ [2]. The maximum absolute value of the autocorrelation function, AC_{max}, is a measure of the resistance the function has against differential attacks. A small value of AC_{max} is required for cryptographic security. To study the overall autocorrelation of an s-box, we use the largest AC_{max} of all non-zero linear combinations of the output.

Nonlinearity and AC_{max} are the most important cryptographic properties and in this paper we investigate them for all linear combinations of the outputs of the s-boxes we construct. We confine our attention to s-boxes with $n > m$ that have the property which generalises balance to multiple output functions: that all possible m-bit vectors occur as output an equal number of times, namely 2^{n-m}. S-boxes with this property are called *regular* [10]. It is known that an s-box is regular if and only if all nonzero linear combinations of the output are balanced Boolean functions. Hence in our investigations only balanced Boolean functions are considered.

Determining the upper bound of nonlinearity (and lower bound of AC_{max}) is an important open problem for single-output balanced Boolean functions, and also for regular s-boxes. The construction methods we present in this paper provide examples of extreme values that we now know can be attained, so they may be considered as our current conjectures for the true bounds.

3 Genetic Algorithm and the GA with Hill Climbing

3.1 The Genetic Algorithm

As its name suggests, the genetic algorithm (GA) is modelled on our understanding of the biological evolutionary process. During a GA, a set (or pool) of solutions (in our case regular s-boxes) is manipulated using breeding schemes, mutation operators and selection processes, in order to improve the characteristics (cryptographic properties) of the elements in the set. Genetic Algorithms are well known and are extensively used to solve difficult optimisation problems in science and engineering. The application of GAs to cryptology is only recent.

An extremely important component of any evolutionary process is the "fitness function", which is used to quantitatively evaluate proposed solutions. Not only must a fitness function give an accurate indication of the suitability of a given solution, but also it must allow the accurate comparison of two solutions, to enable rational selection. An example of a problem that does not appear to have a suitable fitness function is the subset-sum problem [3], and that problem prevented the GA in that case from being effective. For the GA experiments reported here, the fitness function we use is simply the nonlinearity or autocorrelation value of the s-box.

The first step in implementing an evolutionary process is to decide on a solution representation. The representation used here is the truth table of the output Boolean functions where $m \times 2^n$ represents the size of a single s-box. This representation is natural and enables the use of fast Walsh-Hadamard techniques for the fitness calculation.

The generation of an initial pool for a GA requires a method to generate regular s-boxes uniformly at random. In our experiments, this randomisation is achieved by randomly swapping the elements of an array containing 2^{n-m} copies of each of the 2^m possible output vectors.

Another essential ingredient of any evolutionary process is a suitable breeding scheme and mutation operator. Here, these two processes are combined in a single operation. Breeding occurs between two "parent" s-boxes to create a single "offspring" s-box. This *assimilation* process tends to produce offspring which retain characteristics of the parents. The breeding algorithm we have developed is described in Section 3.3.

In brief, the genetic algorithm proceeds as follows. Firstly an initial pool of solutions is generated. In general, this may involve random generation, or the use of some simple heuristic capable of generating "better-than-random" solutions which are still sub-optimal. Once a pool has been established, a loop ("generation") consisting of parental selection, breeding strategy, mutation and the selection of survivors for the next generation is iterated until a satisfactory solution is obtained.

3.2 Genetic Algorithms with Hill Climbing

The hill climbing approach to Boolean function design was introduced in [7] as a means of improving the nonlinearity of a given Boolean function. It was subsequently adapted for s-boxes, and investigations involving bijective s-boxes have been outlined in [5]. Essentially, hill climbing bijective or regular s-boxes involves making a well chosen swap of two output vectors, and retaining the swap only if the modification improves a cryptographic property. The process is iterated until no possible swap can improve the s-box. At that point we say that a *locally maximum* s-box has been obtained.

Genetic algorithms which included hill climbing have been previously applied in [6] to random generations of single output balanced Boolean functions. We improve that work by applying the approach to the investigation of random regular s-boxes with $n = 8$ input bits and $m \in \{2, \cdots, 8\}$ output bits.

We now discuss the two specific genetic algorithms in this paper which utilise hill climbing.

Combined Method For this variation of the genetic algorithm, the initial pool of P s-boxes is generated randomly and the fitness of each s-box is calculated. In each iteration all possible pairings of the s-boxes in the solution pool are used as parents for breeding. Hence, there are $\frac{P(P-1)}{2}$ offspring produced each generation. In the combined method, each child is hill climbed after birth to a local maximum, and the fitness of all parents and offspring are compared. The best P from the combined pool of parents and children are retained and the process is repeated. Where an s-box from the parent pool and offspring pool have the same fitness, preference is given to the s-box from the offspring pool, since this should encourage diversity in the breeding pool from generation to generation.

As the pool converges, s-boxes in the pool may become very similar, reducing the chance for further improvement. For this reason, it is necessary to ensure that no duplicate s-boxes are inserted in the new breeding pool during the selection stage.

Hill Climbing Final Pool This variation of the Genetic algorithm follows the same procedure as that outlined above, with the exception that, rather than hill climbing each child produced from the breeding process, the hill climbing is performed only on the final pool when the genetic algorithm has been allowed to run until some specific stopping criteria has been reached.

It is often found that evolutionary processes, such as the two described above, converge to a pool of very similar solutions. Our genetic algorithm programs have been written to monitor convergence by calculating the total Hamming distance, for all m component output functions, between all pairs of parents. When this average "s-box distance" between parents falls below a suitable threshold value (for example $4 * m$), the pool has almost completely converged and the algorithm stops.

3.3 The Assimilation Operation

For a breeding operation we seek a method to obtain a regular s-box that shares some of the features of two regular, n bit in, m bit out, parent s-boxes. In particular, we desire a pair of good parents to have a reasonable expectation of producing a good child. The following process, called *assimilation*, meets these criteria by aiming to assign each child vector output equal to that of one of the parents.

Let the parent s-boxes be denoted $b_1(x), b_2(x)$ and let the resulting child be $b_c(x)$. For given input vector x_j, the following breeding procedure applies. Note that the number of occurrences of $b_i(x_j)$ $i = 1, 2$ in $b_c(x_j)$ will be denoted by $\#(b_i(x_j))$.

Case 1: $\quad b_1(x_j) = b_2(x_j)$

- If $\#(b_1(x_j)) < 2^{n-m}$ then $b_c(x_j) = b_1(x_j)$
$\#(b_1(x_j))$ is incremented by one.

- Else $b_c(x_j)$ is chosen to be $b_1(x_k)$ $k \neq j$ such that $\#\{b_1(x_k)\}$ is least. $\#(b_1(x_k))$ is incremented by one.

Case 2: $b_1(x_j) \neq b_2(x_j)$

- If $\#(b_1(x_j)) = 2^{n-m}$ and $\#(b_2(x_j)) = 2^{n-m}$ then $b_c(x_j)$ is chosen to be $b(x_k)$ $k \neq j$ such that $\#\{b(x_k)\}$ is least. $\#(b(x_k))$ is incremented by one.
- else if $\#(b_1(x_j)) < \#(b_2(x_j))$ then $b_c(x_j) = b_1(x_j)$ and $\#(b_1(x_j))$ is incremented by one.
- else if $\#(b_1(x_j)) > \#(b_2(x_j))$ then $b_c(x_j) = b_2(x_j)$ and $\#(b_2(x_j))$ is incremented by one.
- else if $\#(b_1(x_j)) = \#(b_2(x_j))$ and $\#(b_i(x_j))$ i $= 1,2 < 2^{n-m}$ then $b_c(x_j)$ is randomly chosen from $b_1(x_j)$ and $b_2(x_j)$, and the appropriate $\#(b_i(x_j))$ is incremented by one.

In general, $b_c(x_j) = b_1(x_j)$ or $b_2(x_j)$. However, on some occasions, this will not be the case, and $b_c(x_j)$ will be a mutation. This algorithm ensures that all output vectors occur equally often, so that the child will be regular.

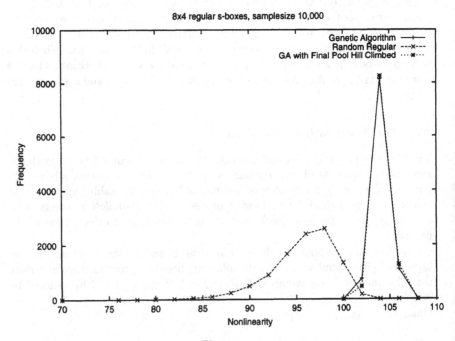

Figure 1

4 Experimental Results

All GA experiments we report here were conducted with a pool of 10 s-boxes. The experiments conducted to produce Figures 1 to 4, and Figure 6 have been conducted on a sample size of 10,000 regular s-boxes. That is, 1,000 separate GAs were conducted, each with a pool of 10. That means 10,000 random s-boxes are initially generated and 10,000 final pool s-boxes were examined. During these experiments, we are comparing the performance and efficiency of three variations: the plain GA, a GA with final pool hill climbing, and a GA with all children hill climbed. The standard method we compare these techniques to is random generation.

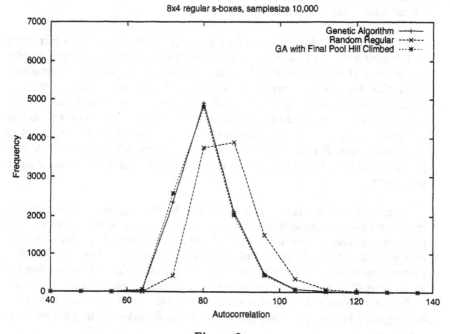

Figure 2

The combined version in which all children are hill climbed requires much more time to run and produces little or no improvement in the properties of the final pool, compared to the plain GA. This behaviour is consistent over many experiments. For this reason the combined method has not been included in the results we now present. Instead, we concentrate on the plain GA, with and without hill climbing the final pool.

The bulk of experiments have been performed on 8 * 4 s-boxes. In addition, to assess the effect that additional outputs have on achievable property values, we have examined the change in performance as m is varied from 2 to 8. We now discuss these results in more detail.

Figures 1 and 2 show frequency distributions for the properties of s-box nonlinearity and overall autocorrelation, respectively. Recall that we are seeking higher nonlinearity and lower AC_{max}. It is clear that a plain GA is easily able to generate regular s-boxes with higher nonlinearity or lower autocorrelation than randomly generated s-boxes. The effect of hill climbing applied to the final pool seems slight. More precisely, there were few GAs with final pools whose best value could be improved by hill climbing. This suggests that the plain GA is able to obtain a locally maximum(minimum) s-box with high probability. The application of hill climbing is able to improve those s-boxes which are not optimum, however, no improvement is possible for s-boxes which are already at a local maximum(minimum).

Figures 3 and 4 show how the best property values change with increasing m, for nonlinearity and autocorrelation respectively. Adding one output doubles the number of distinct output functions, considering all linear combinations. Given that the s-box properties are defined to be the worst over all of these linear combinations, we would expect the attainable property value to decrease as m increases. Our experiments confirm this and show a fairly linear reduction in values. For all m, the GA was able to obtain better s-boxes than random generation alone. For those m for which random generation and the GA have produced the same best autocorrelation, the GA produced that best value much more often.

Figure 5 illustrates the dynamics of a GA by showing how the best nonlinearity and the average distance between parents vary as a GA proceeds. The graph shows sudden rises in the value of the best nonlinearity achieved, as well as a gradual decrease in the distances between members of the pool. The last property value improvement occurs before the final collapse into convergence around the 40th iteration. This behaviour is typical of evolutionary processes. In Figure 6 we show a frequency distribution of the number of iterations until convergence. This graph shows that the GA usually takes 10-40 generations to converge.

Some more detailed results on the effectiveness of hill climbing applied to a GA appear in Table 1 which shows the value and frequency of the best nonlinearity achieved for various m. The table shows the highest nonlinearity obtained in 1,000 GAs with a pool of 10, and the number of s-boxes (out of 10,000 in the final pools) that attained that value. The third column shows the percentage of final GA s-boxes that were able to be improved by hill climbing. The table data suggests that over 75% of GAs were able to achieve a local maximum, without resorting to hill climbing.

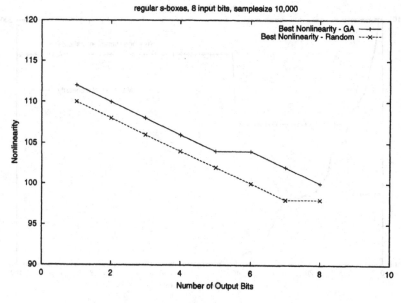

Figure 3

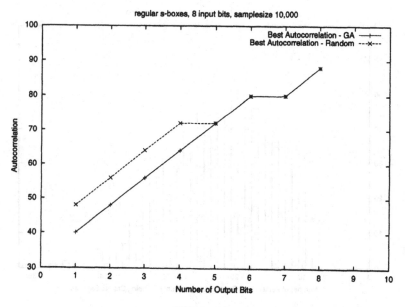

Figure 4

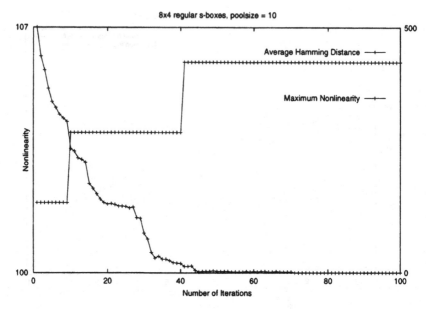

Figure 5

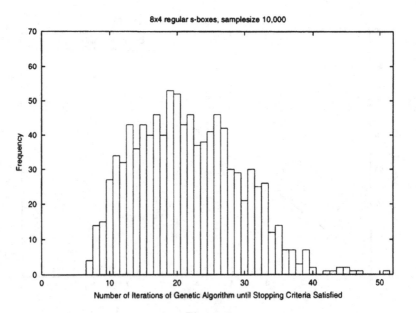

Figure 6

S-Box size	GA Best Non-Linearity	GA Frequency	GA with HC % Improvement
8 x 2	110	342	25%
8 x 3	108	454	21%
8 x 4	106	1108	15%
8 x 5	104	2013	19%
8 x 6	104	9	0%
8 x 7	102	17	0%

Table 1: Value and Frequency of best nonlinearity for GAs, and rate of improvement with Hill Climbing at end.

5 Conclusions

In this paper we have introduced the use of evolutionary heuristics for the generation of regular s-boxes with high nonlinearity and low autocorrelation. These properties are a necessary condition for the security of ciphers which employ them.

The GA we describe uses a new technique for breeding regular s-boxes. The regular property is ensured for the child, which shares the properties of two parents. This assimilation process assists convergence to good s-boxes.

Our experiments demonstrate that a simple GA is able to generate more secure s-boxes than a random search. When combined with hill climbing, the genetic algorithm can sometimes be more effective, although the benefit seems to be much less that has been observed for the combination applied to single output Boolean functions.

References

1. E. BIHAM and A. SHAMIR Differential Cryptanalysis of DES-like Cryptosystems *Advances in Cryptology - Crypto'90*, LNCS vol. 537, pages 2-21, Springer-Verlag, 1991.

2. C. CARLET Partially-Bent Functions. *Advances in Cryptology - Crypto'92, Proceedings*, LNCS vol. 740, pages 280-291, Springer-Verlag, 1993.

3. A. CLARK, E. DAWSON and H. BERGEN Combinatorial Optimisation and the Knapsack Cipher *Cryptologia*, Vol. 20, No. 1, pages 85-93, 1996.

4. M. MATSUI Linear Cryptanalysis Method of DES Cipher *Advances in Cryptology - Eurocrypt'93*, LNCS vol. 765, pages 386-397, Springer-Verlag, 1993.

5. W. MILLAN How to Improve the Nonlinearity of Bijective S-boxes *Australasian Conference on Information Security and Privacy ACISP '98*, LNCS vol. 1438, pages 181-192, Springer-Verlag, 1998.

6. W. MILLAN, A. CLARK and E. DAWSON Heuristic Design of Cryptographically Strong Balanced Boolean Functions *Advances in Cryptology - Eurocrypt'98* LNCS vol. 1403, pages 489-499, Springer-Verlag, 1998.

7. W. MILLAN, A. CLARK and E. DAWSON Smart Hill Climbing Finds Better Boolean Functions *Workshop on Selected Areas of Cryptology, SAC '97, Proceedings*, pages 50-63, 1997.

8. NATIONAL INSTITUTE OF STANDARDS AND TECHNOLOGY (NIST) Data Encryption Standard *U.S. Department of Commerce FIPS* Publication 46, January 1977.

9. K. NYBERG On the Construction of Highly Nonlinear Permutations. *Advances in Cryptology - Eurocrypt'92, Proceedings* LNCS vol. 658, pages 92-98, Springer-Verlag, 1993.

10. X.-M. ZHANG, Y. ZHENG Difference Distribution Table of a Regular Substitution Box. *Workshop on Selected Areas in Cryptology, SAC'96, Proceedings*, pages 57-60, 1996.

Incremental Authentication
of Tree-Structured Documents*

Patrik Ekdahl[1] and Ben Smeets[2]

[1] Department of Information Technology, Lund University,
PO Box 118, S-221 00 Lund, Sweden.
Email: Patrik.Ekdahl@it.lth.se
[2] Ericsson Mobile Communications AB,
Scheelevägen 7, S-221 83 Lund, Sweden.
Email: Ben.Smeets@ecs.ericsson.se

Abstract. A new message authentication code (MAC) is described that exploits the tree structure present in many modern document formats, e.g. SGML and XML. The new code supports incremental updating of the cryptographic checksum in the process of making incremental changes to the document. Theoretical bounds on the probability of a successful substitution attack are derived. Through experimental results we demonstrate that for randomly chosen messages the success probability of such an attack will be smaller and is easily identified.

1 Introduction

In the past years much research has been devoted to efficient message authentication schemes that allow the receiver to check the integrity of a received message [6]. The classical paradigm for most schemes is that the actual data carrying the information is considered to be unstructured and modeled as a sequence of symbols. However most information stored in and manipulated by practical information systems is represented by structured data. For example this is the case for information stored in databases, web browsable pages, and or SGML/XML documents. Conventional authentication schemes would typically compute a cryptographic checksum over the raw data that represents the information. By doing so the structure hidden in this data is not used in the protection scheme. This conventional approach is appropriate for many documents but for several emerging ways to store documents this approach has disadvantages. The latter is most prominent for information laid out via web pages and for large work documents where only parts are updated in the process of producing the final document to be released. For such cases it would be advantageous to have authentication mechanisms that exploit the structure of the data to allow easy updating of the overall cryptographic checksum in the process of having incremental changes of the document. In this paper an authentication mechanism

---

* This work was supported in part by the Swedish National Board for Industrial and Technical Development, grant #97-9630.

is described that achieves this in the setting of tree structured documents. We combine several known authentication techniques with a novel technique to authenticate the tree structure of the document. The resulting mechanism could be useful for, for example, SGML structured documents. We provide the technical foundation behind the mechanisms and provide confirming experimental results that demonstrate the practicality.

1.1 Message Authentication Codes

Message Authentication Codes (MACs) form the cryptographic tool for the authentication of messages. These messages are generated through some process which we refer to as the source. A MAC is a hash function h that takes as inputs a secret key k and a source message s. The source message can be of varying length. This, usually large, input is reduced by the hash function to a smaller cryptographic checksum value which often is called the tag, t, i.e. we can write, $t = h_k(s)$. MACs have been studied in great detail and many solutions for unstructured messages are known [6]. Incremental MACs, that is, MACs that support incremental authentication of messages, was first proposed by Bellare, Goldreich and Goldwasser [1], and have been studied in a few papers, e.g. [2],[3]. Here we combine several known techniques with a novel technique to authenticate the structure of the document.

Consider a set of hash functions $\mathcal{H}$ in which every element has an index, such that hash function $h_k \in \mathcal{H}$ is the k:th function in the set. The secret shared between the sender and the receiver is indexing which one function will be actually used in the authentication. Thus the index k is the secret key. Here, as usual, we assume that the key is chosen uniformly among all the possible values.

The source value s and the computed tag $t = h_k(s)$ are combined to form the channel message m. The channel message is the final message sent. Here we follow the convention that the channel message is just the concatenation of s and t, i.e.,

$$m = (s, t) = (s, f_k(s)).$$

1.2 MAC Security Parameters

Before proceeding to the specifics of tree-structures we should precise what kind of attacks we can expect and how well a MAC provides protection. The two basic attacks considered here are the *impersonation* attack and the *substitution* attack. Consider the following scenario:

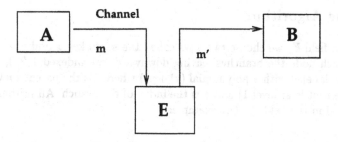

Fig. 1. The sender A and receiver B are intercepted by the adversary E

Here A is the sender, B is the receiver and E is the adversary. In the impersonation attack E sends a message to B, hoping for it to be accepted as valid. Since E does not know the key, she can not tell whether a certain message m is valid or not. She can only make a guess based on the probability that m is valid. Recall that the key is chosen uniformly. We define the probability of success of the impersonation attack as

$$P_I = \max_m P(m \text{ is valid}). \tag{1}$$

In the substitution attack E first intercepts a valid message m from A then she substitute m with m' which she sends to B. The probability of success here is defined as

$$P_S = \max_{m',m\ m\neq m'} P(m' \text{ is valid } |m \text{ is observed}) \tag{2}$$

For MACs of the type considered here it has been shown that $P_S \geq P_I$. (For a proof see e.g.[4].) Thus we mainly focus on the value of P_S. Note that (2) define P_S as a maximum over the message m. So P_S is defined as the worst-case substitution. We might consider P_S as an average over m, as the adversary has no control over which message A will send.

1.3 Problem Formulation

Consider a tree with data only in the leaves. The set of all leaves in the tree is denoted by $\mathcal{L}$. Further we assume that the data in the leaves can be represented by a number m_l from some finite prime field, so $m_l \in \mathbb{F}_p$ where p is a prime and $l \in \mathcal{L}$. The numbers m_l are called leaf messages.

We want to authenticate not only the messages but also the structure of the tree. This means that an adversary should not be able to exchange any leaf message m_l for another, say m'_l, and she (the adversary) should not be able to move a branch in the tree so that the tree obtains a different structure. At least not without a very high probability of disclosure. Furthermore we want the authentication mechanisms to be incremental. That means that if we change a small subset of the tree, the calculations needed to compute the new tag for this tree is proportional to the amount of changes made.

2   The Algorithm

From the field $\mathbb{F}_p$ we choose two keys called the *depth* key x and the *width* key y. For each node the branches leaving downwards are indexed $1, 2, 3, \ldots$. Every branch is labeled with a polynomial $(x^i + y^j)$ where i is the parent's level in the tree (the root is at level 1) and j is the index of the branch. All arithmetics are performed in the field $\mathbb{F}_p$. For example:

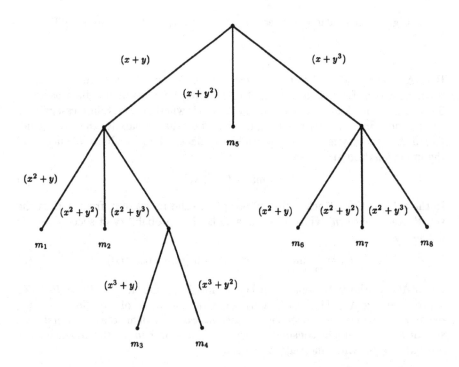

For each leaf l we construct the leaf polynomial $Q_l(x, y)$ as the product of the branch labels from the root down to that specific leaf. For instance, the polynomial of leaf 1 in the example above is $Q_1(x, y) = (x + y)(x^2 + y)$. In this way every leaf polynomial will be unique. If we denote the source tree (including both structure and messages) by s the tag is calculated by the hash function $h_{x,y}(s)$ as

$$t = h_{x,y}(s) = \sum_{l \in \mathcal{L}_s} Q_l(x, y) * m_l \tag{3}$$

where m_l is the message in leaf l and $\mathcal{L}_s$ is the set of all leaves in s.

When we choose keys x and y it is possible that a branch label will evaluate to zero. This will severely damage the authentication capability. Therefore we define that if a branch label evaluates to zero we replace it with 1. By construction (3) we get a polynomial in two indeterminants (the keys) with a degree depending on the depth and width of the tree. If the sender transmits almost the same tree several times, only changing a few leaf messages each time, the opponent can reduce the complexity and derive an equation system that is easy to solve. To prevent this we must introduce an extra key, called the bias key k_b, that is added to $h_{x,y}(s)$. This method to use an extra key in combination with a MAC is due to Wegman and Carter [8]. The final tag is thus calculated as

$$t = h_{x,y}(s) + k_b \qquad (4)$$

Since k_b acts as a one-time-pad it must be changed every time the tree is changed. This can be solved with a pseudo-random key schedule known to the sender and receiver. A technical way for solving this problem has been devised and studied by Rogaway [7].

2.1 The Incremental Property

Suppose we have a source tree s with a tag t. If we want to exchange a leaf message m_l for m_l' we first calculate m_l's contribution to the tag as $t_{m_l} = Q_l(x,y)m_l$ and subtract this from the total tag t. Then we calculate m_l''s contribution as $t_{m_l'} = Q_l(x,y)m_l'$ and add this to t to form the final tag for the new tree. By this property, the amount of computation needed is proportional to the amount of change (and depth of the changed leaves). The incremental property is very useful if for instance the tree leaves are distributed over a physically wide area, for example the Internet where access times can be a problem.

3 Theoretical Aspects

We will now look at the probabilities of the impersonation and substitution attack. First, for a uniformly distributed discrete random variable, the probability of any event can be calculated as

$$P = \frac{\text{number of favorable outcomes}}{\text{total number of outcomes}}$$

Let $h_{k_1,k_2}(s)$ be the evaluated polynomial with keys k_1 and k_2 for source tree s. The tag is calculated as $t = h_{k_1,k_2}(s) + k_b$. Considering the impersonation attack, we see that for every pair (k_1, k_2) there exists a bias key k_b such that the opponents tag t' is correctly verified by the receiver. Thus if the opponent sends (s', t') and there is p^2 pairs (k_1, k_2), we have

$$P_I = \frac{|\{k_1, k_2, k_b; t' = h_{k_1,k_2}(s') + k_b\}|}{|\{k_1, k_2, k_b\}|} = \frac{p^2}{p^3} = \frac{1}{p} \qquad (5)$$

This confirms the intuitive notion that if we apply a one-time-pad, the adversary has no better chance in succeeding than guessing from p elements. For the substitution attack we assume that the adversary observes (s, t). This information is used to narrow down the possible key space, and we get

$$P_S = \frac{|\{k_1, k_2, k_b; t' = h_{k_1,k_2}(s') + k_b \text{ and } t = h_{k_1,k_2}(s) + k_b\}|}{|\{k_1, k_2, k_b; t = h_{k_1,k_2}(s) + k_b\}|} \tag{6}$$

If we subtract t' from t we get $t'' = h_{k_1,k_2}(s'')$ as the restriction in the nominator. Here s'' is known by the opponent and assume $t'' = 0$ without loss of generality. We see that (6) can be calculated by counting the number of roots of a polynomial in two variables (k_1, k_2) over a finite field $\mathbb{F}_p^2$. The denominator equals p^2 by the same argument as for the nominator in the impersonation attack. Let $N(h_{x,y}(s))$ denote the number of roots to the equation $h_{x,y}(s) = 0$ and we can write

$$P_S = \frac{N(h_{x,y}(s))}{p^2} \tag{7}$$

The following theorem from the theory of finite fields is useful in getting a bound on the value of P_S.

Theorem 1 (see proof in [5]). *Let $f \in \mathbb{F}_p[x_1, ..., x_n]$ with $\deg(f) = d \geq 0$. Then the equation $f(x_1, ..., x_n) = 0$ has at most dp^{n-1} solutions in $\mathbb{F}_p^n$.*

Let $D(s)$ denote the maximum depth of source tree s and let $B(s)$ be a number such that no node in s has more than $B(s)$ children. We can upper bound the degree of $h_{x,y}(s)$ by

$$\deg(h_{x,y}(s)) \leq \max(\sum_{i=1}^{D(s)} i, D(s)B(s)) \tag{8}$$

The first term in the maximum is the highest power the depth key can have and the second is the highest power the width key can have. Theorem (1) now gives us an upper bound on P_S.

$$P_S = \frac{N(h_{x,y}(s))}{p^2} \leq \frac{\deg(h_{x,y}(s))p}{p^2} = \frac{\deg(h_{x,y}(s))}{p} \leq \frac{\max(\sum_{i=1}^{D(s)} i, D(s)B(s))}{p} \tag{9}$$

So the probability of a substitution attack is upper bounded by properties of the (source) tree. Although this upper bound is tight in general, the average success probability for a substitution attack will be essentially smaller.

We can get a more reasonable estimate of P_S by investigating the average number of solutions to $h_{x,y}(s) = 0$. This question is addressed by the following result.

Theorem 2 (see proof in [5]). *For a positive integer d, let Ω_d be the set of all $f \in \mathbb{F}_p[x_1, ..., x_n]$ with $\deg(f) \leq d$. Let $\omega(d)$ be the number of n-tuples $(i_1, ..., i_n)$*

of nonnegative integers with $i_1 + \cdots + i_n \leq d$. Then the cardinality $|\Omega_d|$ of Ω_d is $p^{\omega(d)}$. For $f \in \Omega_d$ let $N(f)$ be the number of solution of $f(x_1, ... x_n) = 0$ in $\mathbb{F}_p^n$. Then we have

$$\frac{1}{|\Omega_d|} \sum_{f \in \Omega_d} N(f) = p^{n-1} \tag{10}$$

In our case $n = 2$ and the expected value of $N(h_{x,y}(s))$ is p. From (7) we see that a reasonable estimate of P_S would be $1/p$. Next we consider the variance of $N(h_{x,y}(s))$. The proof of the following theorem can be found in [5].

Theorem 3. *With the notation used in theorem (2), we have*

$$\frac{1}{|\Omega_d|} \sum_{f \in \Omega_d} (N(f) - p^{n-1})^2 = p^{n-1} - p^{n-2} \tag{11}$$

As stated, for the proposed algorithm $n = 2$ and the variance of the number of roots of $h_{x,y}$ equals $p - 1$. To conclude, the result tells us that the expected value of $N(h_{x,y}(s))$ equals p and that the standard deviation is approximately $\sqrt{p}$. The estimate $P_S \approx 1/p$ becomes apparent with increasing p.

4 Simulation Results

In this section we present some simulation results to show that the estimate of P_S indeed is a useful security parameter. We do this by experimentally sampling the distribution of the number of roots, $N(Q(x,y))$ to polynomials in two indeterminates over a finite field.

Figure 2 shows the relative frequency of $N(Q(x,y))$ for five hundred randomly chosen polynomials $Q(x,y)$ with maximum degree 3, 4 and 5, generated with coefficients in $\mathbb{F}_p$ where $p = 503$. Note that the probability mass is well centered around p as theorems (2) and (3) predict. The bound for P_S in (9) is a theoretically correct bound but for practical applications it is indeed rather pessimistic. In Fig. 3 we have increased the field to $p = 1009$. The same well centered probability mass but it is relatively more narrow since the standard deviation only growths as $\sqrt{p}$.

The last Fig. 4, shows the distribution of key collisions for 160 randomly generated trees. For each tree we have chosen a key pair (x_0, y_0) and calculated the tag $t_0 = h_{x_0,y_0}(s))$. The histogram shows the relative frequency of the number of key pairs (x, y) that satisfies $t_0 = h_{x,y}(s)$. In this simulation we used $p = 4001$. The degree of $h_{x,y}(s)$ was in the range $4 \ldots 40$ approximately, and number of leaves was in the range $5 \ldots 2000$ (also approximately). All leaf messages was randomly chosen in $\mathbb{F}_p$.

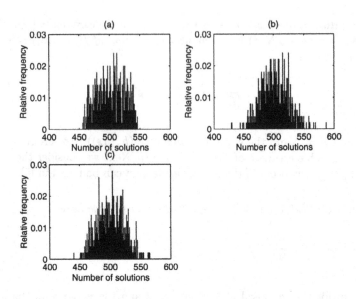

Fig. 2. Relative frequency of the number of roots to 500 randomly chosen polynomials. $p = 503$. (a) Max degree 3. (b) Max degree 4. (c) Max degree 5.

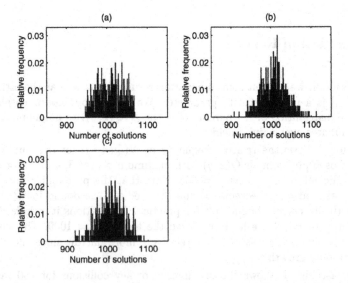

Fig. 3. Relative frequency of the number of roots to 500 randomly chosen polynomials. $p = 1009$. (a) Max degree 3. (b) Max degree 4. (c) Max degree 5.

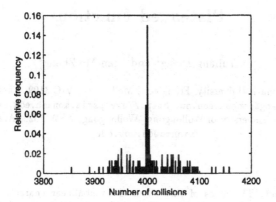

Fig. 4. Relative frequency of number of key collisions for 160 randomly generated trees with p=4001.

5 Conclusions

This paper describes an incremental message authentication code that exploits the tree structure of the data representing a document. We derived theoretical upperbounds on the probability that an adversary may succeed in a substitution attack. In particular we show by experiments that an estimate of the success probability is a useful security parameter in addition to the maximal success probability.

References

1. M. Bellare, O. Goldreich, S. Goldwasser. *Incremental cryptography: the case of hashing and signing.* Lecture Notes in Computer Science 839. (CRYPTO '94). Springer-Verlag 1994.
2. M. Bellare, O. Goldreich, S. Goldwasser. *Incremental cryptography with application to virus protection.* Proceedings of the 27th Annual Symposium on Theory of Computing, ACM, 1995.
3. M. Bellare, D. Micciancia. *A New Paradigm for Collision-Free Hashing: Incremental at Reduced Cost.* Lecture Notes in Computer Science 1233. (EUROCRYPT '97). Springer-Verlag 1997.
4. T. Johansson. *Contributions to Unconditionally Secure Authentication.* Ph.D. Thesis. Dept of Information Theory. Lund University. 1994.
5. R. Lidl, H. Niederreiter. *Finite Fields* Vol. 20 in Encyclopedia of Mathematics and its Applications. Addison-Wesley. 1983.
6. A.J. Menezes, P. van Oorshot, S. Vanstone. *Handbook of applied cryptography.* CRC Press series on discrete mathematics and its applications. 1996.
7. P. Rogaway. *Bucket Hashing and its Application to Fast Message Authentication.* Lecture Notes in Computer Science 963. (CRYPTO '95). Springer-Verlag 1995.
8. M.N. Wegman, J.L. Carter. *New Hash Functions and Their Use in Authentication and Set Equality.* Journal of Computer and System Science, 22 (pp 265-279). 1981.

Plateaued Functions

Yuliang Zheng[1] and Xian-Mo Zhang[2]

[1] Monash University, Frankston, Melbourne, VIC 3199, Australia
yuliang.zheng@monash.edu.au, http://www.pscit.monash.edu.au/~yuliang/
[2] The University of Wollongong, Wollongong, NSW 2522, Australia
xianmo@cs.uow.edu.au

Abstract. The focus of this paper is on nonlinear characteristics of cryptographic Boolean functions. First, we introduce the notion of plateaued functions that have many cryptographically desirable properties. Second, we establish a sequence of strengthened inequalities on some of the most important nonlinearity criteria, including nonlinearity, propagation and correlation immunity, and prove that critical cases of the inequalities coincide with characterizations of plateaued functions. We then proceed to prove that plateaued functions include as a proper subset all partially-bent functions that were introduced earlier by Carlet. This settles an open question that arises from previously known results on partially-bent functions. In addition, we construct plateaued, but not partially-bent, functions that have many properties useful in cryptography.

Key Words

Bent Functions, Cryptography, Nonlinear Characteristics, Partially-Bent Functions, Plateaued Functions.

1 Motivations

In the design of cryptographic functions, one often faces the problem of fulfilling the requirements of a multiple number of nonlinearity criteria. Some of the requirements contradict others. The most notable example is perhaps bent functions — while these functions achieve the highest possible nonlinearity and satisfy the propagation criterion with respect to every non-zero vector, they are not balanced, not correlation immune and exist only when the number of variables is even.

Another example that clearly demonstrates how some nonlinear characteristics may impede others is partially-bent functions introduced in [2]. These functions include bent functions as a proper subset. Partially-bent functions are interesting in that they can be balanced and also highly nonlinear. However, except those that are bent, all partially-bent functions have non-zero linear structures, which are considered to be cryptographically undesirable.

The primary aim of this paper is to introduce a new class of functions to facilitate the design of cryptographically good functions. It turns out that these cryptographically good functions maintain all the desirable properties of partially-bent functions while possess no non-zero liner structures. This class of functions are called *plateaued functions*. To study the properties of plateaued functions, we establish a sequence of inequalities concerning nonlinear characteristics of functions. We show that plateaued functions can be characterized by the critical cases of these inequalities. In particular, we demonstrate that plateaued functions reach the upper bound on nonlinearity given by the inequalities.

We also examine relationships between plateaued functions and partially-bent functions. We show that partially-bent functions must be plateaued while that the converse is not true. This disproves a conjecture and motivates us to construct plateaued functions without non-zero linear structures. Other useful properties of plateaued functions include that they exist for both even and odd numbers of variables, can be balanced and correlation immune.

2 Boolean Functions

Definition 1. *We consider functions from V_n to $GF(2)$ (or simply functions on V_n), V_n is the vector space of n tuples of elements from $GF(2)$. Usually we write a function f on V_n as $f(x)$, where $x = (x_1, \ldots, x_n)$ is the variable vector in V_n. The* truth table *of a function f on V_n is a $(0,1)$-sequence defined by $(f(\alpha_0), f(\alpha_1), \ldots, f(\alpha_{2^n-1}))$, and the* sequence *of f is a $(1,-1)$-sequence defined by $((-1)^{f(\alpha_0)}, (-1)^{f(\alpha_1)}, \ldots, (-1)^{f(\alpha_{2^n-1})})$, where $\alpha_0 = (0, \ldots, 0, 0)$, $\alpha_1 = (0, \ldots, 0, 1)$, \ldots, $\alpha_{2^{n-1}-1} = (1, \ldots, 1, 1)$. The* matrix *of f is a $(1,-1)$-matrix of order 2^n defined by $M = ((-1)^{f(\alpha_i \oplus \alpha_j)})$ where $\oplus$ denotes the addition in $GF(2)$. f is said to be* balanced *if its truth table contains an equal number of ones and zeros.*

Given two sequences $\tilde{a} = (a_1, \cdots, a_m)$ and $\tilde{b} = (b_1, \cdots, b_m)$, their *component-wise product* is defined by $\tilde{a} * \tilde{b} = (a_1 b_1, \cdots, a_m b_m)$ and the *scalar product* of $\tilde{a}$ and $\tilde{b}$, denoted by $\langle \tilde{a}, \tilde{b} \rangle$, is defined as the sum of the component-wise multiplications, where the operations are defined in the underlying field. In particular, if $m = 2^n$ and $\tilde{a}$, $\tilde{b}$ are the sequences of functions f and g on V_n respectively, then $\tilde{a} * \tilde{b}$ is the sequence of $f \oplus g$ where $\oplus$ denotes the addition in $GF(2)$.

An *affine* function f on V_n is a function that takes the form of $f(x_1, \ldots, x_n) = a_1 x_1 \oplus \cdots \oplus a_n x_n \oplus c$, where $\oplus$ denotes the addition in $GF(2)$ and $a_j, c \in GF(2)$, $j = 1, 2, \ldots, n$. Furthermore f is called a *linear* function if $c = 0$.

A $(1,-1)$-matrix A of order m is called a *Hadamard* matrix if $AA^T = mI_m$, where A^T is the transpose of A and I_m is the identity matrix of order m. A Sylvester-Hadamard matrix of order 2^n, denoted by H_n, is generated by the following recursive relation

$$H_0 = 1, \quad H_n = \begin{bmatrix} H_{n-1} & H_{n-1} \\ H_{n-1} & -H_{n-1} \end{bmatrix}, \quad n = 1, 2, \ldots.$$

Let ℓ_i, $0 \leq i \leq 2^n - 1$, be the i row of H_n. Then ℓ_i is the sequence of a linear function $\varphi_i(x)$ defined by the scalar product $\varphi_i(x) = \langle \alpha_i, x \rangle$, where $\alpha_i \in V_n$ is the binary representation of integer i, $i = 0, 1, \ldots, 2^n - 1$.

The *Hamming weight* of a $(0,1)$-sequence ξ, denoted by $W(\xi)$, is the number of ones in the sequence. Given two functions f and g on V_n, the *Hamming distance* $d(f, g)$ between them is defined as the Hamming weight of the truth table of $f(x) \oplus g(x)$, where $x = (x_1, \ldots, x_n)$.

Definition 2. *The* nonlinearity *of a function f on V_n, denoted by N_f, is the minimal Hamming distance between f and all affine functions on V_n, i.e., $N_f = \min_{i=1,2,\ldots,2^{n+1}} d(f, \varphi_i)$ where $\varphi_1, \varphi_2, \ldots, \varphi_{2^{n+1}}$ are all the affine functions on V_n.*

The following characterizations of nonlinearity will be useful (for a proof see for instance [6]).

Lemma 1. *The nonlinearity of f can be expressed by*

$$N_f = 2^{n-1} - \frac{1}{2} \max\{|\langle \xi, \ell_i \rangle|, 0 \leq i \leq 2^n - 1\}$$

where ξ is the sequence of f and ℓ_i is the ith row of H_n, $i = 0, 1, \ldots, 2^n - 1$.

Definition 3. *Let f be a function on V_n. For a vector $\alpha \in V_n$, denote by $\xi(\alpha)$ the sequence of $f(x \oplus \alpha)$. Thus $\xi(0)$ is the sequence of f itself and $\xi(0) * \xi(\alpha)$ is the sequence of $f(x) \oplus f(x \oplus \alpha)$. Set $\Delta(\alpha) = \langle \xi(0), \xi(\alpha) \rangle$, the scalar product of $\xi(0)$ and $\xi(\alpha)$. $\Delta(\alpha)$ is also called the* auto-correlation *of f with a shift α.*

Definition 4. *Let f be a function on V_n. We say that f satisfies the* propagation criterion *with respect to α if $f(x) \oplus f(x \oplus \alpha)$ is a balanced function, where $x = (x_1, \ldots, x_n)$ and α is a vector in V_n. Furthermore f is said to satisfy the propagation criterion of degree k if it satisfies the propagation criterion with respect to every non-zero vector α whose Hamming weight is not larger than k (see [7]).*

The *strict avalanche criterion (SAC)* [11] is the same as the propagation criterion of degree one.

Obviously, $\Delta(\alpha) = 0$ if and only if $f(x) \oplus f(x \oplus \alpha)$ is balanced, i.e., f satisfies the propagation criterion with respect to α.

Definition 5. *Let f be a function on V_n. $\alpha \in V_n$ is called a* linear structure *of f if $|\Delta(\alpha)| = 2^n$.*

For any function f, $\Delta(\alpha_0) = 2^n$, where $\alpha_0 = 0$, the zero vector on V_n. Hence the zero vector is a linear structure of every function on V_n. It is easy to verify that the set of all linear structures of a function f form a subspace of V_n, whose dimension is called the *linearity* of f. It is also well-known that if f has non-zero linear structures, then there exists a nonsingular $n \times n$ matrix B over $GF(2)$ such that $f(xB) = g(y) \oplus h(z)$, where $x = (y, z)$, $x \in V_n$, $y \in V_p$, $z \in V_q$, $p + q = n$, g

is a function on V_p and g has no non-zero linear structures, h is a linear function on V_q. Hence q is equal to the linearity of f.

There exist a number of equivalent definitions of correlation immune functions [1,4]. It is easy to verify that the following definition is equivalent to Definition 2.1 of [1]:

Definition 6. *Let f be a function on V_n and let ξ be its sequence. Then f is called a kth-order correlation immune function if and only if $\langle \xi, \ell \rangle = 0$ for every ℓ, the sequence of a linear function $\varphi(x) = \langle \alpha, x \rangle$ on V_n constrained by $1 \leq W(\alpha) \leq k$.*

The following lemma is the re-statement of a relation proved in Section 2 of [2].

Lemma 2. *For every function f on V_n, we have*

$$(\Delta(\alpha_0), \Delta(\alpha_1), \ldots, \Delta(\alpha_{2^n-1}))H_n = (\langle \xi, \ell_0 \rangle^2, \langle \xi, \ell_1 \rangle^2, \ldots, \langle \xi, \ell_{2^n-1} \rangle^2).$$

where ℓ_i is the ith row of H_n, $j = 0, 1, \ldots, 2^n - 1$.

3 Bent Functions and Partially-bent Functions

Notation 1 *Let f be a function on V_n, ξ the sequence of f and ℓ_i denote the ith row of H_n, $i = 0, 1, \ldots, 2^n - 1$. Set $\Im = \{i \mid 0 \leq i \leq 2^n - 1, \langle \xi, \ell_i \rangle \neq 0\}$, $\Re = \{\alpha \mid \Delta(\alpha) \neq 0, \alpha \in V_n\}$ and $\Delta_M = \max\{|\Delta(\alpha)||\alpha \in V_n, \alpha \neq 0\}$*

It is easy to verify that $\#\Im$, $\#\Re$ and Δ_M are invariant under any nonsingular linear transformation on the variables, where $\#$ denotes the cardinal number of a set.

Since $\sum_{j=0}^{2^n-1}\langle \xi, \ell_j \rangle^2 = 2^{2n}$ (Parseval's equation, Page 416, [5]) and $\Delta(\alpha_0) = 2^n$, neither $\Im$ nor $\Re$ is an empty set. $\Im$ reflects the correlation immunity property of f, while $\Re$ reflects its propagation characteristics and Δ_M forecasts the avalanche property of the function. Therefore information on $\#\Im$, $\#\Re$ and Δ_M is useful in determining important cryptographic characteristics of f.

Definition 7. *A function f on V_n is called a bent function [8] if $\langle \xi, \ell_i \rangle^2 = 2^n$ for every $i = 0, 1, \ldots, 2^n - 1$, where ℓ_i is the ith row of H_n, $i = 0, 1, \ldots, 2^n - 1$.*

A bent function on V_n exists only when n is even, and it achieves the maximum nonlinearity $2^{n-1} - 2^{\frac{1}{2}n-1}$. From [8] and Parseval's equation, we have the following:

Theorem 1. *Let f be a function on V_n and ξ denote the sequence of f. Then the following statements are equivalent: (i) f is bent, (ii) for each i, $0 \leq i \leq 2^n - 1$, $\langle \xi, \ell_i \rangle^2 = 2^n$ where ℓ_i is the ith row of H_n, $i = 0, 1, \ldots, 2^n - 1$, (iii) $\#\Re = 1$, (iv) $\Delta_M = 0$, (v) the nonlinearity of f, N_f, satisfies $2^{n-1} - 2^{\frac{1}{2}n-1}$, (vi) the matrix of f is an Hadamard matrix.*

An interesting theorem of [2] explores a relationship between $\#\Im$ and $\#\Re$. This result can be expressed as follows.

Theorem 2. *For any function f on V_n, we have $(\#\Im)(\#\Re) \geq 2^n$, where the equality holds if and only if there exists a nonsingular $n \times n$ matrix B over $GF(2)$ and a vector $\beta \in V_n$ such that $f(xB \oplus \beta) = g(y) \oplus h(z)$, where $x = (y, z)$, $x \in V_n$, $y \in V_p$, $z \in V_q$, $p + q = n$, g is a bent on V_p and h is a linear function on V_q.*

Based on the above theorem, the concept of *partially-bent* functions was also introduced in the same paper [2].

Definition 8. *A function on V_n is called a partially-bent function if $(\#\Im)(\#\Re) = 2^n$.*

One can see that partially-bent functions include both bent functions and affine functions. Applying Theorem 2 together with properties of linear structures, or using Theorem 2 of [10] directly, we have

Proposition 1. *A function f on V_n is a partially-bent function if and only if each $|\Delta(\alpha)|$ takes the value of 2^n or 0 only. Equivalently, f is a partially-bent function if and only if $\Re$ is composed of linear structures.*

Some partially-bent functions have a high nonlinearity and satisfy the SAC or the propagation criterion of a high degree. Furthermore, some partially-bent functions are balanced. All these properties are useful in cryptography.

4 Plateaued Functions

Now we introduce a new class of functions called plateaued functions. Here is the definition.

Definition 9. *Let f be a function on V_n and ξ denote the sequence of f. If there exists an even number r, $0 \leq r \leq n$, such that $\#\Im = 2^r$ and each $\langle \xi, \ell_j \rangle^2$ takes the value of 2^{2n-r} or 0 only, where ℓ_j denotes the jth row of H_n, $j = 0, 1, \ldots, 2^n - 1$, then f is called a rth-order plateaued function on V_n. f is also called a plateaued function on V_n if we ignore the particular order r.*

Due to Parseval's equation, the condition $\#\Im = 2^r$ can be obtained from the condition "each $\langle \xi, \ell_j \rangle^2$ takes the value of 2^{2n-r} or 0 only, where ℓ_j denotes the jth row of H_n, $j = 0, 1, \ldots, 2^n - 1$". For the sake of convenience, however, we mentioned both conditions in Definition 9.

The following result can be obtained immediately from Definition 9.

Proposition 2. *Let f be a function on V_n. Then we have (i) if f is a rth-order plateaued function then r must be even, (ii) f is an nth-order plateaued function if and only if f is bent, (iii) f is a 0th-order plateaued function if and only if f is affine.*

Th following is a consequence of Theorem 3 of [10].

Proposition 3. *Every partially-bent function is a plateaued function.*

In the coming sections we characterize plateaued functions and disprove the converse of Proposition 3.

5   Characterizations of Plateaued Functions

First we introduce Hölder's Inequality [3]. It states that for real numbers $a_j \geq 0$, $b_j \geq 0$, $j = 1, \ldots, k$, p and q with $p > 1$ and $\frac{1}{p} + \frac{1}{q} = 1$, the following is true: $(\sum_{j=1}^{k} a_j^p)^{1/p}(\sum_{j=1}^{k} b_j^q)^{1/q} \geq \sum_{j=1}^{k} a_j b_j$ where the quality holds if and only if there exists a constant $\nu \geq 0$ such that $a_j = \nu b_j$ for each $j = 1, \ldots, k$.

In particular, set $p = q = 2$ in Hölder's Inequality. We conclude

$$\sum_{j=1}^{k} a_j b_j \leq \sqrt{(\sum_{j=1}^{k} a_j^2)(\sum_{j=1}^{k} b_j^2)} \tag{1}$$

where the quality holds if and only if there exists a constant $\nu \geq 0$ such that $a_j = \nu b_j$ for each $j = 1, \ldots, k$.

Notation 2 *Let f be a function on V_n and ξ denote the sequence of f. Let χ denote the real valued $(0,1)$-sequence defined as $\chi = (c_0, c_1, \ldots, c_{2^n-1})$ where $c_j = \begin{cases} 1 \ if \, \alpha_j \in \Im \\ 0 \ otherwise \end{cases}$ and $\alpha_j \in V_n$, that is the binary representation of integer j. Write*

$$\chi H_n = (s_0, s_1, \ldots, s_{2^n-1}) \tag{2}$$

where each s_j is an integer.

We note that $\chi \begin{bmatrix} \langle \xi, \ell_0 \rangle^2 \\ \langle \xi, \ell_1 \rangle^2 \\ \vdots \\ \langle \xi, \ell_{2^n-1} \rangle^2 \end{bmatrix} = \sum_{j=0}^{2^n-1} \langle \xi, \ell_j \rangle^2 = 2^{2n}$ where the second equal-

ity holds thanks to Parseval's equation. By using Lemma 2, we have $\chi H_n \begin{bmatrix} \Delta(\alpha_0) \\ \Delta(\alpha_1) \\ \vdots \\ \Delta(\alpha_{2^n-1}) \end{bmatrix} =$

2^{2n}. Noticing $\Delta(\alpha_0) = 2^n$, we obtain $s_0 2^n + \sum_{j=1}^{2^n-1} s_j \Delta(\alpha_j) = 2^{2n}$. Since

$$\Delta(\alpha_j) = 0 \ if \ \alpha_j \notin \Re \tag{3}$$

$s_0 2^n + \sum_{\alpha_j \in \Re, j > 0} s_j \Delta(\alpha_j) = 2^{2n}$. As $s_0 = \#\Im$, where $\#$ denotes the cardinal number of a set, we have $\sum_{\alpha_j \in \Re, j > 0} s_j \Delta(\alpha_j) = 2^n(2^n - \#\Im)$. Note that

$$2^n(2^n - \#\Im) = \sum_{\alpha_j \in \Re, j > 0} s_j \Delta(\alpha_j) \leq \sum_{\alpha_j \in \Re, j > 0} |s_j \Delta(\alpha_j)| \leq s_M \Delta_M (\#\Re - 1) \tag{4}$$

Hence the following inequality holds.

$$s_M \Delta_M(\#\Re - 1) \geq 2^n(2^n - \#\Im) \tag{5}$$

From (2),

$$\#\Im \cdot 2^n = \sum_{j=0}^{2^n-1} s_j^2 \text{ or } \#\Im(2^n - \#\Im) = \sum_{j=1}^{2^n-1} s_j^2 \tag{6}$$

Theorem 3. *Let f be a function on V_n and ξ denote the sequence of f. Then*

$$\sum_{j=0}^{2^n-1} \Delta^2(\alpha_j) \geq \frac{2^{3n}}{\#\Im}$$

where the equality holds if and only if f is a plateaued function.

Proof. By using (4), (1) and (6), we obtain

$$2^{2n} \leq \sum_{\alpha_j \in \Re} s_j \Delta(\alpha_j) \leq \sum_{\alpha_j \in \Re} |s_j \Delta(\alpha_j)| \leq \sqrt{(\sum_{\alpha_j \in \Re} s_j^2)(\sum_{\alpha_j \in \Re} \Delta^2(\alpha_j))}$$

$$\leq \sqrt{(\sum_{j=0}^{2^n-1} s_j^2)(\sum_{j=0}^{2^n-1} \Delta^2(\alpha_j))} \leq \sqrt{\#\Im 2^n \sum_{j=0}^{2^n-1} \Delta^2(\alpha_j)} \tag{7}$$

Hence $\frac{2^{3n}}{\#\Im} \leq \sum_{j=0}^{2^n-1} \Delta^2(\alpha_j)$. We have proved the inequality in the theorem.

Assume that the equality in the theorem holds i.e., $\sum_{j=0}^{2^n-1} \Delta^2(\alpha_j) = \frac{2^{3n}}{\#\Im}$. This implies that all the qualities in (7) hold. Hence

$$2^{2n} = \sum_{\alpha_j \in \Re} s_j \Delta(\alpha_j) = \sum_{\alpha_j \in \Re} |s_j \Delta(\alpha_j)| = \sqrt{(\sum_{\alpha_j \in \Re} s_j^2)(\sum_{\alpha_j \in \Re} \Delta^2(\alpha_j))}$$

$$= \sqrt{(\sum_{j=0}^{2^n-1} s_j^2)(\sum_{j=0}^{2^n-1} \Delta^2(\alpha_j))} = \sqrt{\#\Im 2^n \sum_{j=0}^{2^n-1} \Delta^2(\alpha_j)} \tag{8}$$

Applying the property of Hölder's Inequality to (8), we conclude that

$$|\Delta(\alpha_j)| = \nu|s_j|, \, \alpha_j \in \Re \tag{9}$$

where $\nu > 0$ is a constant. Applying (9) and (6) to (8), we have

$$2^{2n} = \sum_{\alpha_j \in \Re} |s_j \Delta(\alpha_j)| = \sqrt{\#\Im 2^n \nu^2 \sum_{j=0}^{2^n-1} s_j^2} = \nu \#\Im 2^n \tag{10}$$

From (10), we have $\sum_{\alpha_j \in \Re} s_j \Delta(\alpha_j) = \sum_{\alpha_j \in \Re} |s_j \Delta(\alpha_j)|$. Hence (9) can be expressed more accurately as follows

$$\Delta(\alpha_j) = \nu s_j, \ \alpha_j \in \Re \tag{11}$$

where $\nu > 0$ is a constant. From (8), it is easy to see that $\sum_{\alpha_j \in \Re} s_j^2 = \sum_{j=0}^{2^n-1} s_j^2$. Hence

$$s_j = 0 \text{ if } \alpha_j \notin \Re \tag{12}$$

Combining (11), (12) and (3), we have

$$\nu(s_0, s_1, \ldots, s_{2^n-1}) = (\Delta(\alpha_0), \Delta(\alpha_1), \ldots, \Delta(\alpha_{2^n-1})) \tag{13}$$

Comparing (13) and (2), we obtain

$$\nu \chi H_n = (\Delta(\alpha_0), \Delta(\alpha_1), \ldots, \Delta(\alpha_{2^n-1})) \tag{14}$$

Further comparing (14) and the equation in Lemma 2, we obtain

$$2^n \nu \chi = (\langle \xi, \ell_0 \rangle^2, \ldots, \langle \xi, \ell_{2^n-1} \rangle^2) \tag{15}$$

Noticing that χ is a real valued $(0,1)$-sequence, containing $\#\Im$ ones and by using Parseval's equation, we obtain $2^n \nu(\#\Im) = 2^{2n}$. Hence $\nu(\#\Im) = 2^n$, and there exists an integer r with $0 \le r \le n$ such that $\#\Im = 2^r$ and $\nu = 2^{n-r}$. From (15) it is easy to see that $\langle \xi, \ell_j \rangle^2 = 2^{2n-r}$ or 0. Hence r must be even. This proves that f is a plateaued function.

Conversely assume that f is a plateaued function. Then there exists an even number r, $0 \le r \le n$, such that $\#\Im = 2^r$ and $\langle \xi, \ell_j \rangle^2 = 2^{2n-r}$ or 0, Due to Lemma 2, we have $\sum_{j=0}^{2^n-1} \Delta^2(\alpha_j) = 2^{-n} \sum_{j=0}^{2^n-1} \langle \xi, \ell_j \rangle^4 = 2^{-n} \cdot 2^r \cdot 2^{4n-2r} = 2^{3n-r}$. Hence we have proved $\sum_{j=0}^{2^n-1} \Delta^2(\alpha_j) = \frac{2^{3n}}{\#\Im}$.

Lemma 3. *Let f be a function on V_n and ξ denote the sequence of f. Then the nonlinearity N_f of f satisfies $N_f \le 2^{n-1} - \frac{2^{n-1}}{\sqrt{\#\Im}}$, where the equality holds if and only if f is a plateaued function.*

Proof. Set $p_M = \max\{|\langle \xi, \ell_j \rangle| \mid j = 0, 1, \ldots, 2^n - 1\}$, where ℓ_j is the jth row of H_n, $0 \le j \le 2^n - 1$. Using Parseval's equation, we obtain $p_M^2 \#\Im \ge 2^{2n}$. Due to Lemma 1, $N_f \le 2^{n-1} - \frac{2^{n-1}}{\sqrt{\#\Im}}$.

Assume that f is a plateaued function. Then there exists an even number r, $0 \le r \le n$, such that $\#\Im = 2^r$ and each $\langle \xi, \ell_j \rangle^2$ takes either the value of 2^{2n-r} or 0 only, where ℓ_j denotes the jth row of H_n, $j = 0, 1, \ldots, 2^n - 1$. Hence $p_M = 2^{n-\frac{1}{2}r}$. By using Lemma 1, we have $N_f = 2^{n-1} - 2^{n-\frac{1}{2}r-1} = 2^{n-1} - \frac{2^{n-1}}{\sqrt{\#\Im}}$.

Conversely assume that $N_f = 2^{n-1} - \frac{2^{n-1}}{\sqrt{\#\Im}}$. From Lemma 1, we have also $N_f = 2^{n-1} - \frac{1}{2} p_M$. Hence $p_M \sqrt{\#\Im} = 2^n$. Since both p_M and $\sqrt{\#\Im}$ are integers and powers of two, we can let $\#\Im = 2^r$, where r is an integer with $0 \le r \le n$. Hence $p_M = 2^{n-\frac{r}{2}}$. Obviously r is even. From Parseval's equation, $\sum_{j \in \Im} \langle \xi, \ell_j \rangle^2 = 2^{2n}$, and the fact that $p_M^2 \#\Im = 2^{2n}$, we conclude that $\langle \xi, \ell_j \rangle^2 = 2^{2n-r}$ for all $j \in \Im$. This proves that f is a plateaued function.

From the proof of Lemma 3, we can see that Lemma 3 can be stated in a different way as follows.

Lemma 4. *Let f be a function f on V_n and ξ denote the sequence of f. Set $p_M = \max\{|\langle \xi, \ell_j \rangle| \mid j = 0, 1, \ldots, 2^n - 1\}$, where ℓ_j is the jth row of H_n, $0 \le j \le 2^n - 1$. Then $p_M \sqrt{\#\Im} \ge 2^n$ where the equality holds if and only if f is a plateaued function.*

Summarizing Theorem 3, Lemmas 3 and 4, we conclude

Theorem 4. *Let f be a function on V_n and ξ denote the sequence of f. Set $p_M = \max\{|\langle \xi, \ell_j \rangle| \mid j = 0, 1, \ldots, 2^n - 1\}$, where ℓ_j is the jth row of H_n, $0 \le j \le 2^n - 1$. Then the following statements are equivalent: (i) f is a plateaued function on V_n, (ii) there exists an even number r, $0 \le r \le 2^n$, such that $\#\Im = 2^r$ and each $\langle \xi, \ell_j \rangle^2$ takes the value of 2^{2n-r} or 0 only, where ℓ_j denotes the jth row of H_n, $j = 0, 1, \ldots, 2^n - 1$, (iii) $\sum_{j=0}^{2^n-1} \Delta^2(\alpha_j) = \frac{2^{3n}}{\#\Im}$, (iv) the nonlinearity of f, N_f, satisfies $N_f = 2^{n-1} - \frac{2^{n-1}}{\sqrt{\#\Im}}$, (v) $p_M \sqrt{\#\Im} = 2^n$, (vi)*
$$N_f = 2^{n-1} - 2^{-\frac{n}{2}-1}\sqrt{\sum_{j=0}^{2^n-1} \Delta^2(\alpha_j)}.$$

Proof. Due to Definition 9, Theorem 3, Lemmas 3 and 4, (i), (ii), (iii), (iv) and (v) hold. (vi) follows from (iii) and (iv).

Theorem 5. *Let f be a function on V_n and ξ denote the sequence of f. Then the nonlinearity N_f of f satisfies*

$$N_f \le 2^{n-1} - 2^{-\frac{n}{2}-1}\sqrt{\sum_{j=0}^{2^n-1} \Delta^2(\alpha_j)}$$

where the equality holds if and only if f is a plateaued function on V_n.

Proof. Set $p_M = \max\{|\langle \xi, \ell_j \rangle| \mid j = 0, 1, \ldots, 2^n - 1\}$. Multiplying the equality in Lemma 2 by itself, we have $2^n \sum_{j=0}^{2^n-1} \Delta^2(\alpha_j) = \sum_{j=0}^{2^n-1} \langle \xi, \ell_j \rangle^4 \le p_M^2 \sum_{j=0}^{2^n-1} \langle \xi, \ell_j \rangle^2$. Applying Parseval's equation to the above equality, we have $\sum_{j=0}^{2^n-1} \Delta^2(\alpha_j) \le 2^n p_M^2$. Hence $p_M \ge 2^{-\frac{n}{2}} \sqrt{\sum_{j=0}^{2^n-1} \Delta^2(\alpha_j)}$. By using Lemma 1, we have proved the inequality $N_f \le 2^{n-1} - 2^{-\frac{n}{2}-1} \sqrt{\sum_{j=0}^{2^n-1} \Delta^2(\alpha_j)}$. The rest part of the theorem can be proved by using Theorem 4.

Theorem 3, Lemmas 3 and 4 and Theorem 4 represent characterizations of plateaued functions.

To close this section, let us note that since $\Delta(\alpha_0) = 2^n$ and $\#\Im \le 2^n$, we have $2^{n-1} - 2^{-\frac{n}{2}-1}\sqrt{\sum_{j=0}^{2^n-1} \Delta^2(\alpha_j)} \le 2^{n-1} - 2^{\frac{n}{2}-1}$ and $2^{n-1} - \frac{2^{n-1}}{\sqrt{\#\Im}} \le 2^{n-1} - 2^{\frac{n}{2}-1}$. Hence both inequalities $N_f \le 2^{n-1} - 2^{-\frac{n}{2}-1}\sqrt{\sum_{j=0}^{2^n-1} \Delta^2(\alpha_j)}$ and $N_f \le 2^{n-1} - \frac{2^{n-1}}{\sqrt{\#\Im}}$ are improvements on a more commonly used inequality $N_f \le 2^{n-1} - 2^{\frac{n}{2}-1}$.

6 Other Cryptographic Properties of Plateaued Functions

By using Lemma 1, we conclude

Proposition 4. *Let f be a rth-order plateaued function on V_n. Then the nonlinearity N_f of f satisfies $N_f = 2^{n-1} - 2^{n-\frac{r}{2}-1}$.*

The following result is the same as Theorem 18 of [12].

Lemma 5. *Let f be a function on V_n $(n \geq 2)$, ξ be the sequence of f, and p is an integer, $2 \leq p \leq n$. If $\langle \xi, \ell_j \rangle \equiv 0 \pmod{2^{n-p+2}}$, where ℓ_j is the jth row of H_n, $j = 0, 1, \ldots, 2^n - 1$, then the degree of f is at most $p - 1$.*

Using Lemma 5, we obtain

Proposition 5. *Let f be a rth-order plateaued function on V_n. Then the algebraic degree of f, denoted by $\deg(f)$, satisfies $\deg(f) \leq \frac{r}{2} + 1$.*

We note that the upper bound on degree in Proposition 5 is tight for $r < n$. For the case of $r = n$, the function, mentioned in Proposition 5, is a bent function on V_n. [8] gives a better upper bound on degree of bent function on V_n. That bound is $\frac{n}{2}$.

The following property of plateaued functions can be verified by noting their definition.

Proposition 6. *Let f be a rth-order plateaued function on V_n, B be any nonsingular $n \times n$ matrix over $GF(2)$ and α be any vector in V_n. Then $f(xB \oplus \alpha)$ is also a rth-order plateaued function on V_n.*

Theorem 6. *Let f be a rth-order plateaued function on V_n. Then the linearity of f, q, satisfies $q \leq n - r$, where the equality holds if and only if f is partially-bent.*

Proof. There exists a nonsingular $n \times n$ matrix B over $GF(2)$ such that $f(xB) = g(y) \oplus h(z)$, where $x = (y, z)$, $y \in V_p$, $z \in V_q$, $p + q = n$, g is a function on V_p and g has no non-zero linear structures, h is a linear function on V_q. Hence q is equal to the linearity of f. Set $f^*(x) = f(xB)$.

Let ξ, η and ζ denote the sequences of f^*, g and h respectively. It is easy to verify $\xi = \eta \times \zeta$, where $\times$ denotes the Kronecker product [13]. From the structure of H_n, each row of H_n, L, can be expressed as $L = \ell \times e$, where ℓ is a row of H_p and e is a row of H_q. It is easy to verify

$$\langle \xi, L \rangle = \langle \eta, \ell \rangle \langle \zeta, e \rangle \tag{16}$$

Since h is linear, ζ is a row of H_q. Replace e by ζ in (16), we have

$$\langle \xi, L' \rangle = \langle \eta, \ell \rangle \langle \zeta, \zeta \rangle = 2^q \langle \eta, \ell \rangle \tag{17}$$

where $L' = \ell \times \zeta$ is still a row of H_n.

Note that f^* is also a rth-order plateaued function on V_n. Hence $\langle \xi, L \rangle$ takes the value of $\pm 2^{n-\frac{1}{2}r}$ or zero only. Due to (17), $\langle \eta, \ell \rangle$ takes the value of $\pm 2^{n-\frac{1}{2}r-q} = \pm 2^{p-\frac{1}{2}r}$ or zero only. This proves that g is a rth-order plateaued function on V_p. Hence $r \leq p$ and $r \leq n-q$, i.e., $q \leq n-r$.

Assume that $q = n - r$. Then $p = r$. From (17), each $\langle \eta, \ell \rangle$ takes the value of $\pm 2^{\frac{p}{2}} = \pm 2^{\frac{r}{2}}$ or zero only, where ℓ is any row of H_p. Hence applying Parseval's equation to g, we can conclude that for each row ℓ of H_p, $\langle \eta, \ell \rangle$ cannot take the value of zero. In other words, for each row ℓ of H_p, $\langle \eta, \ell \rangle$ takes the value of $\pm 2^{\frac{p}{2}}$ only. Hence we have proved that g is a bent function on V_p. Due to Theorem 2, f is partially-bent. Conversely, assume that f is partially-bent. Due to Theorem 2, g is a bent function on V_p. Hence each $\langle \eta, \ell \rangle$ takes the value of $\pm 2^{\frac{p}{2}}$ only, where ℓ is any row of H_p. Note that both ζ and e are rows of H_q hence $\langle \zeta, e \rangle$ takes the value 2^q or zero only. From (16), we conclude that $\langle \xi, L \rangle$ takes the value $\pm 2^{q+\frac{p}{2}}$ or zero only. Recall f is a rth-order plateaued function on V_n. Hence $q + \frac{p}{2} = n - \frac{r}{2}$. This implies that $r = p$, i.e., $q = n - r$.

7 Relationships between Partially-bent Functions and Plateaued Functions

To examine more profound relationships between partially-bent functions and plateaued functions, we introduce one more characterization of partially-bent functions as follows.

Theorem 7. *For every function f on V_n, we have*

$$\frac{2^n - \#\Im}{\#\Im} \leq \frac{\Delta_M}{2^n}(\#\Re - 1)$$

where the equality holds if and only if f is partially-bent.

Proof. From Notation 2, we have $s_M \leq s_0 = \#\Im$. As a consequence of (5), we obtain the inequality in the theorem. Next we consider the equality in the theorem. Assume that the equality holds, i.e.,

$$\Delta_M(\#\Re - 1)\#\Im = 2^n(2^n - \#\Im) \tag{18}$$

From (4),

$$2^n(2^n - \#\Im) \leq \sum_{\alpha_j \in \Re, j > 0} |s_j \Delta(\alpha_j)|$$

$$\leq \Delta_M \sum_{\alpha_j \in \Re, j > 0} |s_j| \leq \Delta_M(\#\Re - 1)\#\Im \tag{19}$$

From (18), we can see that all the equalities in (19) hold. Hence

$$\Delta_M(\#\Re - 1)\#\Im = \sum_{\alpha_j \in \Re, j > 0} |s_j \Delta(\alpha_j)| \tag{20}$$

Note that $|s_j| \leq \#\Im$ and $|\Delta(\alpha_j)| \leq \Delta_M$, for $j > 0$. Hence from (20), we obtain

$$|s_j| = \#\Im \text{ whenever } \alpha_j \in \Re \text{ and } j > 0 \tag{21}$$

and $|\Delta(\alpha_j)| = \Delta_M$ for all $\alpha_j \in \Re$ with $j > 0$.

Applying (21) to (6), and noticing $s_0 = \#\Im$, we obtain $\#\Im \cdot 2^n = \sum_{j=0}^{2^n-1} s_j^2 \geq \sum_{\alpha_j \in \Re} s_j^2 = (\#\Re)(\#\Im)^2$. This results in $2^n \geq (\#\Re)(\#\Im)$. Together with the inequality in Theorem 2, it proves that $(\#\Re)(\#\Im) = 2^n$, i.e., f is a partially-bent function.

Conversely assume that f is a partially-bent function, i.e., $(\#\Im)(\#\Re) = 2^n$. Then the inequality in the theorem is specialized as

$$\Delta_M(2^n - \#\Im) \geq 2^n(2^n - \#\Im) \tag{22}$$

We need to examine two cases. Case 1: $\#\Im = 2^n$. Obviously the equality in (22) holds. Case 2: $\#\Im \neq 2^n$. From (22), we have $\Delta_M \geq 2^n$. Thus $\Delta_M = 2^n$. This completes the proof.

Next we consider a non-bent function f. With such a function we have $\Delta_M \neq 0$. Thus from Theorem 7, we have the following result.

Corollary 1. *For every non-bent function f on V_n, we have*

$$(\#\Im)(\#\Re) \geq \frac{2^n(2^n - \#\Im)}{\Delta_M} + \#\Im$$

where the equality holds if and only if f is partially-bent (but not bent).

Proposition 7. *For every non-bent function f, we have*

$$\frac{2^n(2^n - \#\Im)}{\Delta_M} + \#\Im \geq 2^n$$

where the equality holds if and only if $\#\Im = 2^n$ or f has a non-zero linear structure.

Proof. Since $\Delta_M \leq 2^n$, the inequality is obvious. On the other hand, it is easy to see that the equality holds if and only if $(2^n - \Delta_M)(2^n - \#\Im) = 0$.

From Proposition 7, one observes that for any non-bent function f, Corollary 1 implies Theorem 2.

Theorem 8. *Let f be a rth-order plateaued function. Then the following statements are equivalent: (i) f is a partially-bent function, (ii) $\#\Re = 2^{n-r}$, (iii) $\Delta_M(\#\Re - 1) = 2^{2n-r} - 2^n$, (iv) the linearity q of f satisfies $q = n - r$.*

Proof. (i) $\Longrightarrow$ (ii). Since f is a partially-bent function, we have $(\#\Im)(\#\Re) = 2^n$. As f is a rth-order plateaued function, $\#\Im = 2^r$ and hence $\#\Re = 2^{n-r}$.

(ii) $\implies$ (iii). It is obviously true when $r = n$. Now consider the case of $r < n$. Using Theorem 7, we have $\frac{2^n - \#\mathfrak{S}}{\#\mathfrak{S}} \leq \frac{\Delta_M}{2^n}(\#\mathfrak{R} - 1)$ which is specialized as

$$2^{n-r} - 1 \leq \frac{\Delta_M}{2^n}(2^{n-r} - 1) \tag{23}$$

From (23) and the fact that $\Delta_M \leq 2^n$, we obtain $2^{n-r} - 1 \leq \frac{\Delta_M}{2^n}(2^{n-r} - 1) \leq 2^{n-r} - 1$. Hence $\Delta_M = 2^n$ or $r = n$. (iii) obviously holds when $\Delta_M = 2^n$. When $r = n$, we have $\#\mathfrak{R} = 1$ and hence (iii) also holds.

(iii) $\implies$ (i). Note that (iii) implies $\frac{2^n - \#\mathfrak{S}}{\#\mathfrak{S}} = \frac{\Delta_M}{2^n}(\#\mathfrak{R} - 1)$ where $\#\mathfrak{S} = 2^r$. By Theorem 7, f is partially-bent.

Due to Theorem 6, (iv) $\iff$ (i).

8 Construction of Plateaued Functions and Disproof of The Converse of Proposition 3

Lemma 6. *For any positive integers t and k with $k < 2^t < 2^k$, there exist 2^t non-zero vectors in V_k, say $\beta_0, \beta_1, \ldots, \beta_{2^t-1}$, such that for any non-zero vector $\beta \in V_k$, the 2^t-set $\{\varphi_{\beta_0}(\beta), \varphi_{\beta_1}(\beta), \ldots, \varphi_{\beta_{2^t-1}}(\beta)\}$, contains both zero and one, where φ_β is the linear function on V_k defined by $\varphi_\beta(x) = \langle \beta, x \rangle$.*

Proof. We choose k linearly independent vectors in V_k, say $\beta_1, \ldots, \beta_k$. From linear algebra, $(\langle \beta_1, \beta \rangle, \ldots, \langle \beta_k, \beta \rangle)$ goes through all the non-zero vectors in V_k exactly once while β goes through all the non-zero vectors in V_k.

Hence there exists a unique β^* satisfying $(\langle \beta_1, \beta^* \rangle, \ldots, \langle \beta_k, \beta^* \rangle) = (1, \ldots, 1)$. Furthermore, for any non-zero vector $\beta \in V_k$ with $\beta \neq \beta^*$, $\{\langle \beta_1, \beta \rangle, \ldots, \langle \beta_k, \beta \rangle\}$ contains both one and zero.

Let β_0 be a non-zero vector in V_k, such that $\langle \beta_0, \beta^* \rangle = 0$. Obviously $\beta_0 \notin \{\beta_1, \ldots, \beta_k\}$. If $2^t > k + 1$, choose other $2^t - k - 1$ non-zero vectors in V_k, $\beta_{k+1}, \ldots, \beta_{2^t-1}$, such that $\beta_0, \beta_1, \ldots, \beta_k, \beta_{k+1}, \ldots, \beta_{2^t-1}$ are mutually distinct. It is easy to see that for any non-zero vector $\beta \in V_k$, $\{\beta_0, \beta_1, \ldots, \beta_{2^t-1}\}$ contains both one and zero. This proves the lemma.

The following example proves the existence of rth-order plateaued functions on V_n, where $0 < r < n$, and disproves the converse of Proposition 3. We note that in this section, we will not discuss nth-order and 0th-order plateaued function on V_n as they are bent and affine functions respectively.

Example 1. Let t and k be positive integers with $k < 2^t < 2^k$. Let $\beta_0, \beta_1, \ldots, \beta_{2^t-1}$ be the 2^t non-zero vectors in V_k defined in Lemma 6. Let ξ_j denote the sequence of φ_{β_j}, $j = 0, 1, \ldots, 2^t - 1$. Set $\xi = \xi_0, \xi_1, \ldots, \xi_{2^t-1}$. Let $n = k + t$ and f be the function on V_n whose sequence is ξ.

By using the properties of H_n, it is easy to verify that each $\langle \xi, \ell_i \rangle$ takes the value of $\pm 2^k$ or 0 only, where ℓ_i is the ith row of H_n, $i = 0, 1, \ldots, 2^n - 1$. Using Parseval's equation, we obtain $\#\mathfrak{S} = 2^{2n-2k}$. Let $r = 2n - 2k = 2t$. Then f is a rth-order plateaued function on V_n. Due to $n = k + t$, $r = 2n - 2k = 2t$ and $t < k$, $0 < r < n$ holds.

We now consider $\Delta(\alpha)$ with the function f. Let $\alpha = (\gamma, \beta)$ where $\gamma \in V_t$, $\beta \in V_k$. Note that

$$\Delta(\alpha) = \begin{cases} \sum_{\gamma_j \oplus \gamma_i = \gamma} \langle \xi_j, \xi_i(\beta) \rangle, & \text{if } \gamma \neq 0 \\ \sum_{j=0}^{2^t - 1} \langle \xi_j, \xi_j(\beta) \rangle, & \text{if } \gamma = 0 \text{ but } \beta \neq 0 \end{cases} \tag{24}$$

where $\gamma_j \in V_t$ is the binary representation of an integer j, $j = 0, 1, \ldots, 2^n - 1$.

Since $\varphi_{\beta_j} \neq \varphi_{\beta_i}$ if $j \neq i$, where $\varphi_\beta(x) = \langle \beta, x \rangle$, $\varphi_{\beta_j}(x) \oplus \varphi_{\beta_i}(x \oplus \beta)$ is a non-zero linear function and hence balanced. We have now proved $\langle \xi_j, \xi_i(\beta) \rangle = 0$ for $j \neq i$. Hence $\Delta(\alpha) = 0$ when $\gamma \neq 0$.

On the other hand, for any linear function φ on V_k, we have $\varphi(x) \oplus \varphi(x \oplus \beta) = \varphi(\beta)$. Hence $\langle \xi_j, \xi_j(\beta) \rangle = 2^k$ if and only if $\varphi_{\beta_j}(\beta) = 0$. In addition, $\langle \xi_j, \xi_j(\beta) \rangle = -2^k$ if and only if $\varphi_{\beta_j}(\beta) = 1$. By using Lemma 6, we have $\Delta(\alpha) = \sum_{j=0}^{2^t - 1} \langle \xi_j, \xi_j(\beta) \rangle \neq \pm 2^t \cdot 2^k = \pm 2^n$ for $\beta \neq 0$.

In summary, we have

$$\Delta(\alpha) \begin{cases} = 0 & \text{if } \gamma \neq 0 \\ \neq \pm 2^n & \text{if } \gamma = 0 \text{ and } \beta \neq 0 \\ = 2^n & \text{if } \alpha = 0 \end{cases} \tag{25}$$

Since f is a rth-order plateaued function on V_n and $r < n$, f is not bent, on the other hand, (25) shows that f has non-zero linear structures. Hence we conclude that f is not partially-bent. Hence we have proved that f is plateaued but not partially-bent. This disproves the converse of Proposition 3.

f has some other interesting properties. In particular, due to Proposition 4, the nonlinearity N_f of f satisfies $N_f = 2^{n-1} - 2^{n - \frac{r}{2} - 1}$. Note that the sequence of any non-zero linear function is $(1, -1)$-balanced. Hence each ξ_j and $\xi = \xi_0, \xi_1, \ldots, \xi_{2^t - 1}$ are $(1, -1)$-balanced. This implies that f is $(0, 1)$-balanced. Since the function f is not partially-bent, by using Theorem 2, we have $(\#\Im)(\#\Re) > 2^n$. This proves that $\#\Re > 2^{n-r}$. On the other hand, from (25), we have $\#\Re \leq 2^k = 2^{n - \frac{1}{2}r}$. Thus we can conclude that $2^{n-r} < \#\Re \leq 2^{n - \frac{1}{2}r}$.

We end this example by noting that such functions as f exist on V_n both for n even and odd.

Now we summarize the relationships among bent, partially-bent and plateaued functions. Let $\mathbf{B_n}$ denote the set of bent functions on V_n, $\mathbf{P_n}$ denote the set of partially-bent functions on V_n and $\mathbf{F_n}$ denote the set of plateaued functions on V_n. Then the above results imply that $\mathbf{B_n} \subset \mathbf{P_n} \subset \mathbf{F_n}$, where $\subset$ denotes the relationship of proper subset. We further let $\mathbf{G_n}$ denote the set of plateaued functions on V_n that do *not* have non-zero linear structures and are not bent functions. The relationships among these classes of functions are shown in Figure 1. Example 1 proves that $\mathbf{G_n}$ is nonempty.

Next we consider how to improve the function in Example 1 so as to obtain a rth-order plateaued function on V_n satisfying the SAC and all the properties mentioned in Example 1.

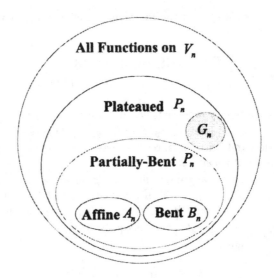

Fig. 1. Relationship among bent, partially bent, and plateaued functions

Example 2. Note that if $r > 2$, i.e., $t > 1$, then from Example 1, $\#\Re \leq 2^{n-\frac{1}{2}r} < 2^{n-1}$. In other words, $\#\Re^c > 2^{n-1}$ where $\Re^c$ denotes the complementary set of $\Re$. Hence there exist n linearly independent vectors in $\Re^c$. In other words, there exist n linearly independent vectors with respect to which f satisfies the propagation criterion. Hence we can choose a nonsingular $n \times n$ matrix A over $GF(2)$ such that $g(x) = f(xA)$ satisfies the SAC (see [9]). The nonsingular linear transformation A does not alter any of the properties of f in Example 1

We can further improve the function in Example 2 so as to obtain a rth-order plateaued functions on V_n having the highest degree and satisfying all the properties in Example 1.

Example 3. Given any vector $\delta = (i_1, \ldots, i_s) \in V_t$, we define a function on V_t by $D_\delta(y) = (y_1 \oplus \bar{i_1}) \cdots (y_t \oplus \bar{i_s})$ where $y = (y_1, \ldots, y_t)$ and $\bar{i} = 1 \oplus i$ indicates the binary complement of i.

Let $\xi_{i_1 \cdots i_p}$, $(i_1, \ldots, i_p) \in V_p$, be the sequence of a function $f_{i_1 \cdots i_p}(x_1, \ldots, x_q)$ on V_q. Let ξ be the concatenation of $\xi_{0 \cdots 00}, \xi_{0 \cdots 01}, \ldots, \xi_{1 \cdots 11}$, namely, $\xi = (\xi_{0 \cdots 00}, \xi_{0 \cdots 01}, \ldots, \xi_{1 \cdots 11})$. It is easy to verify that ξ is the sequence of a function on V_{q+p} given by

$$f(y_1, \ldots, y_p, z_1, \ldots, z_q) = \bigoplus_{(i_1 \cdots i_p) \in V_p} D_{i_1 \cdots i_p}(y_1, \ldots, y_p) f_{i_1 \cdots i_p}(z_1, \ldots, z_q). \quad (26)$$

Let $\delta_j \in V_t$ is the binary representation of integer j, $j = 0, 1, \ldots, 2^t - 1$. Write $\psi_{\delta_0} = \varphi_{\beta_0}, \psi_{\delta_1} = \varphi_{\beta_1}, \ldots, \psi_{\delta_{2^t-1}} = \varphi_{\beta_{2^t-1}}$, where $\beta_0, \beta_1, \ldots, \beta_{2^t-1}$ are the same with those in Example 1 and $\varphi_\beta = \langle \beta, x \rangle$.

Due to (26), the function f on V_{t+k}, mentioned in Example 1 can be expressed as $f(y, z) = \bigoplus_{\delta \in V_t} D_\delta(y)\psi_\delta(z)$.

Case 1: $\bigoplus_{\delta \in V_t} \psi_\delta \neq 0$. Write $\bigoplus_{\delta \in V_t} \psi_\delta = \psi$, where ψ must be a non-zero linear function on V_k. Note that each $D_\delta(y)$ contains $y_1 \cdots y_t$. Hence the term $y_1 \cdots y_t \psi(z)$ survives in the final algebraic normal form representation of $f(y, z)$ and hence the degree of f is $t + 1 = \frac{r}{2} + 1$.

Case 2: $\bigoplus_{\delta \in V_t} \psi_\delta = 0$, i.e., $\bigoplus_{j=0}^{2^t - 1} \varphi_{\beta_j} = 0$. Note that there exist $2^k - 1$ non-zero vectors in V_k and $2^k - 1 > 2^t$. Hence we can replace $\varphi_{\beta_{2^t-1}}$ by any non-zero linear function φ on V_k, that differs from $\varphi_{\beta_0}, \varphi_{\beta_1}, \ldots \varphi_{\beta_{2^t-1}}$. This reduces Case 2 to Case 1.

We have now constructed a rth-order plateaued function with degree $\frac{r}{2} + 1$. Applying the discussions in Examples 1 and 2, we can obtain a rth-order plateaued function on V_n having degree $\frac{r}{2} + 1$ and satisfying all the properties of the function constructed in Example 1.

It should be noted that the function in this example achieves the highest possible algebraic degree given in Proposition 4. Thus the upper bound on the algebraic degree of plateaued functions, mentioned in Proposition 5, is tight.

9 Conclusions

We have introduced and characterized a new class of functions called plateaued functions. These functions bring together various nonlinear characteristics. We have also shown that partially-bent functions are a proper subset of plateaued functions, which settles an open problem related to partially-bent functions. We have further demonstrated methods for constructing plateaued functions that have many cryptographically desirable properties.

Acknowledgement

The second author was supported by a Queen Elizabeth II Fellowship (227 23 1002).

References

1. P. Camion, C. Carlet, P. Charpin, and N. Sendrier. On correlation-immune functions. In *Advances in Cryptology - CRYPTO'91*, volume 576, Lecture Notes in Computer Science, pages 87–100. Springer-Verlag, Berlin, Heidelberg, New York, 1991.

2. Claude Carlet. Partially-bent functions. *Designs, Codes and Cryptography*, 3:135–145, 1993.

3. Friedhelm Erwe. *Differential And Integral Calculus*. Oliver And Boyd Ltd, Edinburgh And London, 1967.

4. Xiao Guo-Zhen and J. L. Massey. A spectral characterization of correlation-immune combining functions. *IEEE Transactions on Information Theory*, 34(3):569–571, 1988.

5. F. J. MacWilliams and N. J. A. Sloane. *The Theory of Error-Correcting Codes.* North-Holland, Amsterdam, New York, Oxford, 1978.

6. W. Meier and O. Staffelbach. Nonlinearity criteria for cryptographic functions. In *Advances in Cryptology - EUROCRYPT'89*, volume 434, Lecture Notes in Computer Science, pages 549–562. Springer-Verlag, Berlin, Heidelberg, New York, 1990.

7. B. Preneel, W. V. Leekwijck, L. V. Linden, R. Govaerts, and J. Vandewalle. Propagation characteristics of boolean functions. In *Advances in Cryptology - EUROCRYPT'90*, volume 437, Lecture Notes in Computer Science, pages 155–165. Springer-Verlag, Berlin, Heidelberg, New York, 1991.

8. O. S. Rothaus. On "bent" functions. *Journal of Combinatorial Theory*, Ser. A, 20:300–305, 1976.

9. J. Seberry, X. M. Zhang, and Y. Zheng. Improving the strict avalanche characteristics of cryptographic functions. *Information Processing Letters*, 50:37–41, 1994.

10. J. Wang. The linear kernel of boolean functions and partially-bent functions. *System Science and Mathematical Science*, 10:6–11, 1997.

11. A. F. Webster and S. E. Tavares. On the design of S-boxes. In *Advances in Cryptology - CRYPTO'85*, volume 219, Lecture Notes in Computer Science, pages 523–534. Springer-Verlag, Berlin, Heidelberg, New York, 1986.

12. Y. Zheng X. M. Zhang and Hideki Imai. Duality of boolean functions and its cryptographic significance. In *Advances in Cryptology - ICICS'97*, volume 1334, Lecture Notes in Computer Science, pages 159–169. Springer-Verlag, Berlin, Heidelberg, New York, 1997.

13. R. Yarlagadda and J. E. Hershey. Analysis and synthesis of bent sequences. *IEE Proceedings (Part E)*, 136:112–123, 1989.

On the Linear Complexity of the
Naor-Reingold Pseudo-Random Function

Frances Griffin and Igor E. Shparlinski

Department of Computing, Macquarie University
Sydney, NSW 2109, Australia
{fgriffin,igor}@ics.mq.edu.au

Abstract. We show that the new pseudo-random function, introduced recently by M. Naor and O. Reingold, possesses one more attractive and useful property. Namely, it is proved that, under a certain not too restrictive condition on the parameters of this function, for almost all seeds it produces a sequence with exponentially large linear complexity.

1 Introduction

Let p and l be primes with $l|p-1$ and let $n \geq 1$ be an integer.

Denote by $\mathbb{F}_p$ the finite field of p elements which we identify with the set $\{0, \ldots, p-1\}$. Select an element $g \in \mathbb{F}_p^*$ of multiplicative order l, that is,

$$g^i \neq 1, \ 1 \leq i \leq l-1, \qquad g^l = 1.$$

Then for each n-dimensional vector $\mathbf{a} = (a_1, \ldots, a_n) \in (\mathbb{F}_l^*)^n$ one can define the function

$$f_\mathbf{a}(X) = g^{a_1^{x_1} \cdots a_n^{x_n}} \in \mathbb{F}_p,$$

where $X = x_1 \ldots x_n$ is the bit representation of an n-bit integer X, $0 \leq X \leq 2^n - 1$, with some extra leading zeros if necessary. Thus, given $\mathbf{a} = (a_1, \ldots, a_n) \in (\mathbb{F}_l^*)^n$, for each $X = 0, \ldots, 2^n - 1$ this function produces a certain element of $\mathbb{F}_p$. After that it can be continued periodically.

For a randomly chosen vector $\mathbf{a} \in (\mathbb{F}_l^*)^n$, M. Naor and O. Reingold [3] have proposed the function $f_\mathbf{a}(x)$ as an efficient pseudo-random function (it is assumed in [3] that n is the bit length of p but similar results hold for much more general settings).

It is shown in [3] that the function $f_\mathbf{a}(x)$ has some very desirable security property, provided that certain standard cryptographic assumptions about the hardness of breaking the Diffie-Hellman cryptosystem holds. It is also shown in [3] that this function can be computed in parallel by threshold circuits of bounded depth and polynomial size.

The distribution properties of this function have been studied in [7] and it has been proved that the statistical distribution of $f_{\mathbf{a}}(X)$ is exponentially close to uniform for almost all $\mathbf{a} \in (\mathbb{F}_l^*)^n$. More precisely, for a set $\mathcal{M} \subseteq \mathbb{F}_p$ we define the *discrepancy* $D(\mathcal{M})$ modulo p as

$$D(\mathcal{M}) = \sup_{\mathcal{I} \subseteq [0,1]} \left| \frac{N(\mathcal{I})}{\# \mathcal{M}} - |\mathcal{I}| \right|,$$

where $N(\mathcal{I})$ is the number of fractional parts $\{m/p\}$ with $m \in \mathcal{M}$ which hit the interval $\mathcal{I} = [\alpha, \beta] \subseteq [0, 1]$ of length $|\mathcal{I}| = \beta - \alpha$.

Throughout the paper $\log z$ denotes the binary logarithm of z.

Let us denote by $D_{l,p,g}(\mathbf{a})$ the discrepancy of the set $\{ f_{\mathbf{a}}(x) \mid x = 0, 1, \ldots, 2^n - 1 \}$. It is shown in [7] that if $n = \lceil \log p \rceil$ is the bit length of p then for all, except possibly $o(l^n)$, vectors $\mathbf{a} \in (F_l)^n$, the bound

$$D_{l,p,g}(\mathbf{a}) \leq \Delta(l, p)$$

holds, where

$$\Delta(l,p) = \begin{cases} p^{(1-\gamma)/2} l^{-1/2} \log^2 p, & \text{if } l \geq p^\gamma; \\ p^{1/2} l^{-1} \log^2 p, & \text{if } p^\gamma > l \geq p^{2/3}; \\ p^{1/4} l^{-5/8} \log^2 p, & \text{if } p^{2/3} > l \geq p^{1/2}; \\ p^{1/8} l^{-3/8} \log^2 p, & \text{if } p^{1/2} > l \geq p^{1/3}; \end{cases}$$

and $\gamma = 2.5 - \log 3 = 0.9150\ldots$. Thus $\Delta(l,p) = o(1)$ if $l \geq p^{1/3+\varepsilon}$ for some fixed $\varepsilon > 0$. We remark that, because of some technical reasons, $O\left(nl^{n-1}\right)$ vectors $\mathbf{a}$ with zero components have been included in the considerations in [7]. In fact similar results hold for a wider range of n, not necessarily $n = \lceil \log p \rceil$.

It has also been shown that for almost all primes p even stronger estimates hold. For the elliptic curve version of this generator similar results have been obtained in [8].

Here we study one more important cryptographic characteristic of this sequence.

We recall that the *linear complexity* of an N-element sequence $W_1, \ldots, W_N$ over a ring $\mathcal{R}$ is the length L of the shortest linear recurrence relation

$$W_{X+L} = A_{L-1} W_{X+L-1} + \ldots + A_0 W_X, \qquad X = 1, \ldots, N - L + 1,$$

with $A_0, \ldots, A_L \in \mathcal{R}$, which is satisfied by this sequence, see [1, 2, 4–6] for some main properties of this characteristic and for its relevance to cryptography.

In this paper we obtain an exponential lower bound on the linear complexity $L_{\mathbf{a}}$ of the sequence $f_{\mathbf{a}}(X)$, $X = 0, \ldots, 2^n - 1$, which holds for almost all $\mathbf{a} \in (\mathbb{F}_l^*)^n$.

2 Preparations

We need some statements about the distribution in $\mathbb{F}_l^*$ of products of the form

$$\mathbf{b}^{\mathbf{z}} = b_1^{z_1} \ldots b_m^{z_m}, \qquad \mathbf{z} = (z_1, \ldots, z_m) \in \{0, 1\}^m,$$

which could be of independent interest.

Denote

$$\mathbf{i} = (1, \ldots, 1) \in \{0, 1\}^m,$$

thus $b^{\mathbf{i}} = b_1 \ldots b_m$.

Lemma 1. *For all but at most $2^m(l-1)^{m-1}$ vectors $\mathbf{b} = (b_1, \ldots, b_m) \in (\mathbb{F}_l^*)^m$*

$$\mathbf{b}^{\mathbf{z}} \neq \mathbf{b}^{\mathbf{i}},$$

for all $\mathbf{z} \in \{0, 1\}^m$ with $\mathbf{z} \neq \mathbf{i}$.

Proof. Select a vector $\mathbf{b} = (b_1, \ldots, b_m) \in (\mathbb{F}_l^*)^m$ uniformly at random and denote by P the probability

$$P = \mathrm{Prob}_{\mathbf{b}} \left\{ \exists \, \mathbf{z} \in \{0, 1\}^m \mid \mathbf{z} \neq \mathbf{i}, \ \mathbf{b}^{\mathbf{z}} = \mathbf{b}^{\mathbf{i}} \right\}.$$

Obviously,

$$P \leq \sum_{\substack{\mathbf{z} \in \{0, 1\}^m \\ \mathbf{z} \neq \mathbf{i}}} \mathrm{Prob}_{\mathbf{b}} \left\{ \mathbf{b}^{\mathbf{z}} = \mathbf{b}^{\mathbf{i}} \right\}.$$

Because $\mathbf{z} \neq \mathbf{i}$ then there exists j such that $z_j = 0$. Then $\mathbf{b}^{\mathbf{z}} = \mathbf{b}^{\mathbf{i}}$ can be written in the form $A = Bb_j$ where A and B do not depend on b_j and both are elements of $\mathbb{F}_l^*$. Thus, for any $\mathbf{z} \in \{0, 1\}^m$ with $\mathbf{z} \neq \mathbf{i}$,

$$\mathrm{Prob}_{\mathbf{b}} \left\{ \mathbf{b}^{\mathbf{z}} = \mathbf{b}^{\mathbf{i}} \right\} \leq \frac{1}{l-1}.$$

Therefore $P < 2^m(l-1)^{-1}$ and the bound follows. $\square$

We see that the bound of Lemma 1 is nontrivial for any $m < \log(l-1)$. In the next statement we show that for larger m the products $\mathbf{b}^{\mathbf{z}}$, $\mathbf{z} \in \{0, 1\}^m$ are quite dense in $\mathbb{F}_l^*$.

Lemma 2. *For any Δ, for all but $2^{-m}\Delta^{-1}(l-1)^{m+2}$ vectors $\mathbf{b} = (b_1, \ldots b_m) \in (\mathbb{F}_l^*)^m$, the 2^m products $\mathbf{b}^{\mathbf{z}}$, $\mathbf{z} \in \{0, 1\}^m$ take at least $l-1-\Delta$ values from $\mathbb{F}_l^*$.*

Proof. Put $r = l - 1$. Considering the discrete logarithms, it is easy to see that the lemma is equivalent to the statement that for all but maybe at most $2^{-m}\Delta^{-1}r^{m+2}$ vectors $\mathbf{c} = (c_1, \ldots, c_m) \in (\mathbb{Z}/r\mathbb{Z})^m$ the congruence

$$c_1y_1 + \ldots c_my_m \equiv u \pmod{r}, \qquad y_1, \ldots, y_m \in \{0, 1\}$$

is solvable for at least $r - \Delta$ values of $u \in \mathbb{Z}/r\mathbb{Z}$.

We put

$$e(a) = \exp(2\pi i a).$$

Then the number of solutions $T_u(\mathbf{c})$ of this congruence can be expressed as

$$T_u(\mathbf{c}) = \frac{1}{r} \sum_{(y_1,\ldots,y_m)\in\{0,1\}^m} \sum_{\lambda=0}^{r-1} \mathbf{e}\left(\lambda\left(\sum_{j=1}^{m} c_j y_j - u\right)/r\right)$$

$$= \frac{2^m}{r} + \frac{1}{r}\sum_{\lambda=1}^{r-1} \mathbf{e}(-\lambda u/r) \sum_{(y_1,\ldots,y_m)\in\{0,1\}^m} \mathbf{e}\left(\lambda\sum_{j=1}^{m} c_j y_j/r\right),$$

where $2^m/r$ is the contribution of the term corresponding to $\lambda = 0$. Denote

$$W(\mathbf{c}) = \sum_{u=0}^{r-1}\left(T_u(\mathbf{c}) - \frac{2^m}{r}\right)^2.$$

Then

$$W(\mathbf{c}) = \frac{1}{r^2}\sum_{u=0}^{r-1}\sum_{\lambda,\eta=1}^{r-1}\mathbf{e}((\eta-\lambda)u/r)$$

$$\times \sum_{(y_1,\ldots,y_m)\in\{0,1\}^m}\sum_{(z_1,\ldots,z_m)\in\{0,1\}^m}\mathbf{e}\left(\sum_{j=1}^{m} c_j\,(\lambda y_j - \eta z_j/r)\right)$$

$$= \frac{1}{r^2}\sum_{\lambda,\eta=1}^{r-1}\sum_{(y_1,\ldots,y_m)\in\{0,1\}^m}\sum_{(z_1,\ldots,z_m)\in\{0,1\}^m}\mathbf{e}\left(\sum_{j=1}^{m} c_j\,(\lambda y_j - \eta z_j)/r\right)$$

$$\times \sum_{u=0}^{r-1}\mathbf{e}((\eta-\lambda)u/r).$$

The inner sum is equal to r if $\lambda \equiv \eta \pmod{r}$ and is equal to zero otherwise. Therefore

$$W(\mathbf{c}) = \frac{1}{r}\sum_{\lambda=1}^{r-1}\sum_{(y_1,\ldots,y_m)\in\{0,1\}^m}\sum_{(z_1,\ldots,z_m)\in\{0,1\}^m}\mathbf{e}\left(\lambda\sum_{j=1}^{m} c_j(y_j - z_j)/r\right)$$

$$= \frac{1}{r}\sum_{\lambda=1}^{r-1}\left|\sum_{(y_1,\ldots,y_m)\in\{0,1\}^m}\mathbf{e}\left(\lambda\sum_{j=1}^{m} c_j y_j/r\right)\right|^2$$

$$= \frac{1}{r}\sum_{\lambda=1}^{r-1}\left|\prod_{j=1}^{m}(1+\mathbf{e}(\lambda c_j/r))\right|^2 = \frac{1}{r}\sum_{\lambda=1}^{r-1}\prod_{j=1}^{m}|1+\mathbf{e}(\lambda c_j/r)|^2$$

$$= \frac{1}{r}\sum_{\lambda=1}^{r-1}\prod_{j=1}^{m}|\mathbf{e}(-\lambda c_j/(2r)) + \mathbf{e}(\lambda c_j/(2r))|^2$$

$$= \frac{2^{2m}}{r}\sum_{\lambda=1}^{l-2}\prod_{j=1}^{m}\cos^2\left(\frac{\lambda c_j\pi}{r}\right).$$

Hence

$$
\sum_{\mathbf{c}\in(\mathbb{Z}/r\mathbb{Z})^m} W(\mathbf{c}) = \frac{2^{2m}}{r} \sum_{\lambda=1}^{l-2} \sum_{\mathbf{c}\in(\mathbb{Z}/r\mathbb{Z})^m} \prod_{j=1}^{m} \cos^2\left(\frac{\lambda c_j \pi}{r}\right)
$$

$$
= \frac{2^{2m}}{r} \sum_{\lambda=1}^{r-1} \prod_{j=1}^{m} \sum_{c_j=0}^{r-1} \cos^2\left(\frac{\lambda c_j \pi}{r}\right)
$$

$$
= \frac{2^{2m}}{r} \sum_{\lambda=1}^{r-1} \left(\sum_{c=0}^{r-1} \cos^2\left(\frac{\lambda c \pi}{r}\right)\right)^m .
$$

Let S_λ be the inner sum. It is easy to show that for any integer $k \geq 2$

$$
\sum_{c=0}^{k-1} \cos^2\left(\frac{c\pi}{k}\right) = \frac{1}{2}k.
$$

Denote $d = \gcd(\lambda, r)$ and put $\mu = \lambda/d$, $k = r/d$. Then $\gcd(\mu, k) = 1$ and we obtain

$$
S_\lambda = \sum_{c=0}^{r-1} \cos^2\left(\frac{\mu c \pi}{k}\right) = d \sum_{c=0}^{k-1} \cos^2\left(\frac{\mu c \pi}{k}\right) = d \sum_{c=0}^{k-1} \cos^2\left(\frac{c\pi}{k}\right) = \frac{1}{2}r.
$$

Therefore,

$$
\sum_{\mathbf{c}\in(\mathbb{Z}/r\mathbb{Z})^m} W(\mathbf{c}) = 2^m r^m . \tag{1}
$$

Denote by Q the number of $\mathbf{c} \in (\mathbb{Z}/r\mathbb{Z})^m$ for which $T_u(\mathbf{c}) = 0$ for at least Δ values of $u \in \mathbb{Z}/r\mathbb{Z}$. For each such $\mathbf{c} \in (\mathbb{Z}/r\mathbb{Z})^m$ we obviously have

$$
W(\mathbf{c}) \geq \Delta \frac{2^{2m}}{r^2} .
$$

From (1) we obtain

$$
Q\Delta \frac{2^{2m}}{r^2} \leq 2^m r^m
$$

and the desired result follows. $\qquad\square$

3 Lower Bound of the Linear Complexity

Now we are prepared to prove our main result.

Theorem 1. *Assume that for some $\gamma > 0$*

$$
n \geq (1+\gamma)\log l.
$$

*Then for any $\delta > 0$ and sufficiently large l, the linear complexity, L_a, of the
sequence $f_a(X)$, $X = 0, \ldots, 2^n - 1$, satisfies*

$$L_a \geq \begin{cases} l^{1-\delta}, & \text{if } \gamma \geq 2; \\ l^{\gamma/2-\delta} & \text{if } \gamma < 2; \end{cases}$$

for all except possibly at most $3(l-1)^{n-\delta}$ vectors $a \in (\mathbb{F}_l^*)^n$

Proof. If $\delta \geq \max\{1, \gamma/2\}$ then the bound is trivial. Otherwise we put

$$t = \lceil (\min\{1, \gamma/2\} - \delta) \log l \rceil, \qquad \Delta = \lceil (l-1)2^{-t} \rceil - 1, \qquad s = n - t.$$

We have

$$2^t \leq 2l^{1-\delta} \quad \text{and} \quad 2^{2t} \leq 4l^{\gamma-2\delta}. \tag{2}$$

Therefore

$$2^{-s} = 2^{t-n} \leq 2^t l^{-1-\gamma}. \tag{3}$$

We also have

$$\Delta \geq (l-1)2^{-t-1} \tag{4}$$

provided that l is large enough.

Let $\mathcal{A}$ be the set of vectors $a \in (\mathbb{F}_l^*)^n$ such that simultaneously

$$a_{s+1}^{k_1} \ldots a_n^{k_t} = a_{s+1} \ldots a_n$$

is possible for $(k_1, \ldots, k_t) \in \{0, 1\}^t$ only if $k_1 = \ldots = k_t = 1$ and

$$\#\{a_1^{y_1} \ldots a_s^{y_s} \mid (y_1, \ldots, y_s) \in \{0, 1\}^s\} \geq l - 1 - \Delta.$$

From Lemmas 1 and 2 and the bounds (2), (3) and (4) we derive

$$\begin{aligned} \#\mathcal{A} &\geq (l-1)^n - 2^t(l-1)^{n-1} - 2^{-s}\Delta^{-1}(l-1)^{n+2} \\ &\geq (l-1)^n - 2^t(l-1)^{n-1} - 2^{2t+1}(l-1)^{n-\gamma} \\ &\geq (l-1)^n - 2(l-1)^{n-\delta} - 8(l-1)^{n-2\delta} \\ &\geq (l-1)^n - 3(l-1)^{n-\delta}, \end{aligned}$$

provided that l is large enough. We show that the bound of the theorem holds
for any $a \in \mathcal{A}$.

Let us fix $a \in \mathcal{A}$. Assume that $L_a \leq 2^t - 1$. Then there exist some coefficients

$$C_0, \ldots, C_{2^t-1} \in \mathbb{F}_p,$$

such that

$$\sum_{K=0}^{2^t-1} C_K f_a(X+K) = 0, \qquad X = 0, \ldots, 2^n - 2^t - 1,$$

and such that $C_{2^t-1} = 1$.

Consider X of the form $X = 2^t Y$, where $Y = y_1 \ldots y_s$ is an s-bit integer. We remark that the bits of K occupy only the rightmost zero bits of X in the sum $X + K$. Then we have

$$\sum_{K=0}^{2^t-1} C_K f_{\mathbf{a}}(2^t Y + K) = \sum_{K=0}^{2^t-1} C_K g^{a_1^{y_1} \ldots a_s^{y_s} e_K} = 0, \qquad Y = 0, \ldots, 2^s - 1,$$

where

$$e_K = a_{s+1}^{k_1} \ldots a_n^{k_t}, \qquad K = 0, \ldots, 2^t - 1,$$

and $K = k_1 \ldots k_t$ is the bit expansion of K. Denote

$$F_{\mathbf{a}}(u) = \sum_{K=0}^{2^t-1} C_K g_K^u,$$

where

$$g_K = g^{e_K}, \qquad K = 0, \ldots, 2^t - 1.$$

Because of the choice of the set $\mathcal{A}$ the value of

$$g_{2^t-1} = g^{a_{s+1} \ldots a_n}$$

is unique. Collecting together terms with equal values of exponents and taking into account that $C_{2^t-1} = 1 \neq 0$, we obtain that $F_{\mathbf{a}}(u)$ can be expressed in the form

$$F_{\mathbf{a}}(u) = \sum_{\nu=1}^{R} D_\nu h_\nu^u,$$

where $1 \leq R \leq 2^t$, $D_\nu \in \mathbb{F}_p^*$ and $h_\nu \in \mathbb{F}_p^*$ are pairwise distinct, $\nu = 1, \ldots, R$.

Recalling that $\mathbf{a} \in \mathcal{A}$, we conclude that $F_{\mathbf{a}}(u) \neq 0$ for at most Δ values of $u = 1, \ldots, l - 1$. On the other hand, from the properties of Vandermonde determinants, it is easy to see that for any $u = 1, \ldots, l - 1$, $F_{\mathbf{a}}(u + v) \neq 0$ for at least one $v = 0, \ldots, R - 1$. Therefore, $F_{\mathbf{a}}(u) \neq 0$ for at least $(l - 1)/R \geq (l - 1)2^{-t}$ values of $u = 1, \ldots, l - 1$, which is not possible because of the choice of Δ. The obtained contradiction implies that $L_{\mathbf{a}} \geq 2^t$ and we obtain the desired result. $\quad\square$

4 Remarks

It is useful to recall that typically the bit length of p and l are of the same order as n. Thus

$$\log p \asymp \log l \asymp n.$$

In the most interesting case n is the bit length of p, that is, $n \sim \log p$. In this case Theorem 1 implies a lower bound on $L_{\mathbf{a}}$ which is exponential in n, if $l \leq p^{1-\varepsilon}$

for some $\varepsilon > 0$. On the other hand the uniformity of distribution results of [7] are nontrivial for $l \geq p^{1/3+\varepsilon}$. Therefore, although there is a nontrivial overlap between these two restrictions, it would be interesting to estimate the linear complexity for all values of $l \leq p$.

It would also be interesting to improve the bound of Theorem 1. In particular, it is known that the expected values of the linear complexity of a truly random N-element sequence over $\mathbb{F}_p$ is $0.5N + O(1)$ (much more precise results are given in [5] as well as a very good account of other related results, see also [4]), thus one can expect that $L_{\mathbf{a}} \geq 2^{n+O(1)}$.

It is also an interesting open question to study the linear complexity of single bits of $f_{\mathbf{a}}(X)$. For example, one can form the sequence $\beta_{\mathbf{a}}(X)$ of the rightmost bits of $f_{\mathbf{a}}(X)$, $X = 0, \ldots, 2^n - 1$, and study its linear complexity (as elements of $\mathbb{F}_2$). Unfortunately we do not see any approaches to this question.

References

1. T. W. Cusick, C. Ding and A. Renvall, *Stream Ciphers and Number Theory*, Elsevier, Amsterdam, 1998.
2. A. J. Menezes, P. C. van Oorschot and S. A. Vanstone, *Handbook of Cryptography*, CRC Press, Boca Raton, FL, 1996.
3. M. Naor and O. Reingold, 'Number-theoretic constructions of efficient pseudo-random functions', *Proc. 38th IEEE Symp. on Foundations of Comp. Sci.*, 1997, 458–467.
4. H. Niederreiter, 'Some computable complexity measures for binary sequences', *Proc. Intern. Conf. on Sequences and Their Applications*, Singapore, 1998, (to appear).
5. H. Niederreiter and M. Vielhaber, 'Linear complexity profiles: Hausdorff dimension for almost perfect profiles and measures for general profiles', *J. Compl.*, **13** (1996), 353–383.
6. R. A. Rueppel, 'Stream ciphers', *Contemporary Cryptology: The Science of Information Integrity*, IEEE Press, NY, 1992, 65–134.
7. I. E. Shparlinski, 'On the uniformity of distribution of the Naor–Reingold pseudo-random function', *Preprint*, 1999, 1-11.
8. I. E. Shparlinski, 'On the Naor–Reingold pseudo-random number generator from elliptic curves', *Preprint*, 1999, 1-9.

On the Channel Capacity of Narrow-Band Subliminal Channels *

Kazukuni Kobara and Hideki Imai

Institute of Industrial Science, The University of Tokyo
Roppongi, Minato-ku, Tokyo 106-8558, Japan
TEL : +81-3-3402-6231 Ext 2327
FAX : +81-3-3402-7365
E-mail: kobara@imailab.iis.u-tokyo.ac.jp

Abstract. Subliminal channels, discovered by Simmons, yield a possibility to transmit covert messages by embedding them in cryptographic digital data, such as (EC)DSA signatures. The embedded messages are used for transmitting important information, or as watermarks or imprints of the data. Anyone can use or abuse these channels very easily because most cryptographic digital data widely used in the Internet is not subliminal-free. For example, DSA signatures are not subliminal-free and they are widely used in various applications, such as PGP, SSH2 and so on. It is very important to evaluate the ability of such channels. In this paper, we evaluate the channel capacity of narrow-band subliminal channels where a transmitter tries only the limited number of input values. Then, we apply this result to a practical model where carriers are transmitted one after another successively. (A carrier denotes data in which a covert message is embedded.) We show, under this model, memories can be used to increase the channel capacity, and then we compare their channel capacities. In addition, we show it is possible to reduce the computational complexity of establishing a narrow-band subliminal channel.

Key words: subliminal channels, narrow-band, channel capacity, computational complexity, information hiding.

1 Introduction

Subliminal channels [10, 12, 11, 13, 16], discovered by Simmons, yield a possibility to transmit covert messages by embedding them in cryptographic digital data, such as (EC)DSA signatures [1, 2]. The embedded messages are used for transmitting important information, or are used as watermarks or imprints of the cryptographic digital data. For example, by embedding IDs in digital signatures in secret, they can be used for distinguishing forgeries from the real ones. On the other hand, these channels might be abused to leak your important data

* This work was performed in part of Research for the Future Program (RFTF) supported by Japan Society for the Promotion of Science (JSPS) under contact no. JSPS-RETF 96P00604.

from your IC cards [17, 4]. Anyone can use or abuse these channels very easily because cryptographic digital data which are not subliminal-free are widely used in the Internet. For example, DSA signatures are not subliminal-free [14] and they are widely used in various applications, such as PGP, SSH2 and so on. It is very important to evaluate the ability of such channels.

According to the relationship between the capacity of the subliminal channel and the computational complexity of making the channel, subliminal channels are mainly divided into the following two categories: broad-band channels and narrow-band ones [13]. (The Newton channel [3] are categorized as a broad-band channel, and then the halting attack [6] is categorized as a narrow-band channel in [15].) Over a broad-band channel, one can transmit as much information as random numbers included in a carrier. (A carrier denotes data in which a covert message is embedded.) Although the amount of is larger than that of a narrow-band channel, broad-band channels have the disadvantage that they cannot be established on any digital data, e.g. broad-band channels on ElGamal public-key cryptography [7] are not known yet [2]. Moreover, in some digital signature schemes, a subliminal transmitter has to reveal his/her signing-key to his/her subliminal receivers to establish a broad-band channel. On the contrary, narrow-band channels can be established on almost any data including fluctuation such as random numbers. Subliminal transmitters do not have to reveal their signing-keys to their subliminal receivers. The only disadvantage is the channel capacity. Only a small amount of information can be transmitted over a narrow-band channel. The reason why "only a small amount" is due that a one-way function is in between a transmitter and a receiver [9].

On narrow-band subliminal channels, it is known that a transmitter has to try 2^x different input values on average to embed x bit-information in a carrier. However this "on average" does not mean trying 2^x different input values is enough to embed x bit-information, and then it is still unknown how much information can be embedded in a carrier by trying n different input values [3]. In this paper, we evaluate it at first in Section 3. Then, we apply this result to a practical model where carriers are transmitted one after another successively in Section 4. Under the practical model, memories play an important role to increase the channel capacity. We show some concrete schemes in Subsection 4.2 and 4.3, then their channel capacities are compared in Section 5. In addition, we propose how to reduce the computational complexity of trying the n input values in Subsection 2.1. This reduces the computational complexity of making a narrow-band subliminal channel.

---

[2] While kleptography [18, 19] can make a trapdoor to random numbers of ElGamal public-key cryptosystems, it cannot transmit arbitrary message symbols by itself.

[3] How small the channel capacity of a subliminal channel can be is considered in [5, 6, 15]. The result is that it can be negligible by applying some additional procedures [15].

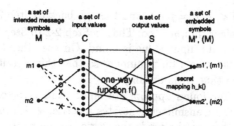

Fig. 1. Conceptual figure for explaining how to make a narrowband subliminal channel.

2 How to make a narrow-band subliminal channel

A lot of cryptosystems, including DSA signatures, Schnorr signatures and El-Gamal public-key cryptosystems (over finite fields or elliptic curves), require a random number c which is put in a one-way function $f()$, e.g. $f(c) = g^c \bmod p$ where g is a generator of a multiplicative subgroup of Z_p^*, or $f() = [c]G$ where $[c]$ is the multiplication by c on an elliptic curve and G is a base point. Then its output $s = f(c)$ is transmitted to receivers. Narrow-band subliminal channels can be established on all the data including such an output s. For example, D-SA signatures are in the following format "(msg, d, sig)" where "msg" denotes a signing message, "sig" denotes an authenticator and $d = (g^c \bmod p) \bmod q$. And then the format includes "s" which is an output value of a one-way function $f()$ whose input value is a random number c because $s = f(c) = g^c \bmod p = Y^{(((g^c \bmod p) \bmod q)/sig)} \cdot g^{(h(msg)/sig)} \bmod p$ where Y is the singer's public-key.

Generalized algorithm for embedding a covert message in such an output value "s" is described bellow. In this algorithm, C and S denote sets of input values and output values of a one-way function $f()$, respectively. Let m denote a covert message symbol a transmitter intends to embed in the carrier, and then let m' denote a symbol which is actually embedded in the carrier. (If a transmitter find an appropriate input value in the following algorithm, $m = m'$. Otherwise $m \neq m'$.) Let $\mathcal{M}$ denote a set of m, and let $\mathcal{M}'$ denote a set of m, respectively, ($\mathcal{M} \equiv \mathcal{M}'$). Let $|\mathcal{M}|$ denote the cardinality of $\mathcal{M}$.

Algorithm

Step 1: Both a transmitter and a receiver share a secret mapping $h_k()$ which transforms S into $\mathcal{M}'$, $(\mathcal{M})$ like Fig.1.

Step 2: The transmitter finds an appropriate input value $c \in C$ which is transformed into an intended message symbol $m \in \mathcal{M}$ satisfying $m' = m = h_k(f(c))$ by the following way.

    Step 2-1: $j := 0$

    Step 2-2: Generate a random number c_j.

    Step 2-3: Calculate $s_j = f(c_j)$.

    Step 2-4: Calculate $m' = h_k(s_j)$.

    Step 2-5: If $m = m'$ go to Step 3.

Step 2-6: If $j < n$, $j := j + 1$, and then go back to Step 2-2.

(In Simmons' algorithm, $n = \infty$. That is, Step 2-2 to Step 2-6 are continued until an appropriate input value is found. On the other hand, we consider the situation that n is a finite number so that this algorithm should finish within a certain time without fail.)

Step 3: By using the found appropriate input value c (or one of the n tried input values), the transmitter generates a carrier which includes "s" where $s = f(c)$. And then he/she sends the carrier to the receiver.

Step 4: The receiver decodes a message symbol m' by applying the secret mapping $h_k()$ to the received "s", i.e. $m' = h_k(s)$.

2.1 How to reduce computational complexity

In this subsection, we propose to reduce the computational complexity of step 2 in the above algorithm. This applies (at least) when $f()$ is in the following forms, $f(c) = g^c \bmod p$ or $f(c) = [c]G$ where $[c]$ is the multiplication by c on an elliptic curve. In the case of $f(c) = g^c \bmod p$, our proposal uses the following Step 2-6' instead of Step 2-6,

Step 2-6': If $j < n$, $j := j + 1$, $c_j := c_{j-1} + 1$, $s_j = s_{j-1} \cdot g$, and then go back to Step 2-4.

The only difference is that s_j is obtained from s_{j-1} with single multiplication operation, i.e. $s_j = s_{j-1} \cdot g$, while the original requires an exponential operation, i.e. $s_j = g^{c_j} \bmod p$. Because single multiplication operation is by far faster than an exponential operation, the computational complexity of making a narrow-band subliminal channel is reduced.

The probability distribution of s_j of our proposal is the same as the original one, even though the joint probability $Pr(s_0, s_j)$ or $Pr(s_{j-1}, s_j)$ of our proposal is different from the original one. The difference cannot be used by others because only one member is transmitted in the generated output values $\{s_0, \cdots, s_{n-1}\}$ and the rest are discarded. Therefore it is no problem that the joint probability is different.

3 Channel capacity

Let $Capa(S; C)$ denote the channel capacity between S and C which are sets of the output values and the input values of a one-way function $f()$, respectively. The capacity $Capa(S; C)$ is defined by the following equation

$$Capa(S; C) = \max_{Pr(C)} I(S; C)$$

$$= \max_{Pr(C)} \sum_{c \in C} \sum_{s \in S} Pr(s|c) Pr(c) \cdot \log_2 \frac{Pr(s|c)}{\sum_{c \in C} Pr(s|c) Pr(c)}. \tag{1}$$

If the output value of $f()$ deterministically depends on c, the probability $Pr(s|c)$ takes on 0 or 1. Thus $Capa(\mathcal{S};\mathcal{C}) = \log_2 |Im(f)|$ where $|Im(f)|$ denotes the number of the unique output values obtained by putting all the available input values in $f()$. For example, when $f()$ is in the following form $s = f(c) = g^c \bmod p$ and $1 \leq c \leq q$ where q is the order of g, $Capa(\mathcal{S};\mathcal{C}) = \log_2 q$ bits/carrier. This means, from an information theoretical point of view, a transmitter should be able to transmit $Capa(\mathcal{S};\mathcal{C})$ bit-information to a receiver by transmitting an output value of the function $f()$ no matter whether $f()$ is a one-way function. However in a real life, it is hard when $f()$ is a one-way function. Because in order to transmit $Capa(\mathcal{S};\mathcal{C})$ bit-information practically, at least $2^{Capa(\mathcal{S};\mathcal{C})}$ symbols have to be mapped onto both $\mathcal{S}$ and $\mathcal{C}$ adequately. This mapping is computationally infeasible when $f()$ is a one-way function. On the contrary, the assumption that the mapping is computationally feasible means that computing an input value from each output value is also feasible. This contradicts to the assumption that $f()$ is a one-way function.

We call $Capa(\mathcal{S};\mathcal{C})$ "**potential channel capacity**" meaning that the channel capacity potentially can reach this amount with unlimited computational power. Accordingly, we call $Capa(\mathcal{M}';\mathcal{M})$ "**practical channel capacity**" (or simply channel capacity) meaning that the channel capacity can practically reach this amount with limited computational power. The practical channel capacity $Capa(\mathcal{M}';\mathcal{M})$ is defined by

$$
\begin{aligned}
&Capa(\mathcal{M}';\mathcal{M}) \\
&= \max_{Pr(\mathcal{M})} \sum_{m \in \mathcal{M}} \sum_{m' \in \mathcal{M}'} Pr(m'|m) Pr(m) \cdot \log_2 \frac{Pr(m'|m)}{\sum_{m \in \mathcal{M}} Pr(m'|m) Pr(m)}
\end{aligned}
\tag{2}
$$

where $Pr(m'|m)$ is the conditional probability of m' being embedded in s after the intended symbol is m. The conditional probability $Pr(m' = m|m)$ that an embedded symbol is the same as the intended symbol is given by

$$
Pr(m' = m|m) = 1 - SER
\tag{3}
$$

where SER is the symbol error rate (the probability that an symbol which is different form m is embedded). The symbol error rate SER is approximated to

$$
SER = \left(1 - \frac{1}{|\mathcal{M}|}\right)^n
\tag{4}
$$

because one random input value c is transformed into the intended message symbol m with probability $\frac{1}{|\mathcal{M}|}$. The conditional probability $Pr(m' \neq m|m)$ of the symbol $m' \neq m$ being embedded in a carrier depends on the rule for choosing a symbol when no appropriate input value is found. Accordingly, we define: (i) random methods where a randomly selected input value is used when no appropriate input value is found, (ii) Hamming distance methods where a transmitter uses an input value being transformed into a message symbol whose Hamming distance is the smallest from the intended message symbol m, and (iii) maximal partition methods which are slightly modified versions of (ii). The

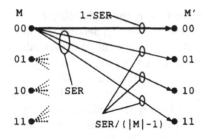

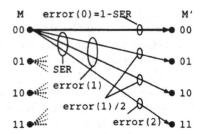

Fig. 2. The channel diagram of a random method for $|\mathcal{M}| = 4$. (The conditional probability $Pr(01|00)$ of a symbol "10" being embedded in a carrier is given by $SER/(|\mathcal{M}| - 1)$ after the intended symbol is "00".)

Fig. 3. The channel diagram of a Hamming distance method where $|\mathcal{M}| = 4$. (The conditional probability $Pr(10|00)$ of a symbol "10" being embedded in a carrier is given by $error(1)/2$ after the intended symbol is "00".)

practical channel capacity of these three methods are evaluated in the following subsections.

3.1 Practical channel capacity for random method

A random method uses a randomly selected input value when no appropriate input value is found within the n trials. In this case,

$$Pr(m' \neq m|m) = \frac{SER}{(|\mathcal{M}| - 1)} \tag{5}$$

because each wrong symbol m' where $m' \neq m$ is embedded (in a carrier) with the same probability, and the number of the wrong symbol is $(|\mathcal{M}| - 1)$. Needless to say, the probability $Pr(m' = m|m)$ is given by (3).

For $|\mathcal{M}| = 4$, the channel diagram is illustrated in Fig.2. The conditional probability $Pr(01|00)$ of a symbol "01" being embedded in a carrier after the intended symbol is "00", is $SER/(|\mathcal{M}| - 1)$. The practical channel capacity $Capa(\mathcal{M}'; \mathcal{M})$ is written in the following form after substituting (5), (3) and $Pr(m) = 1/|\mathcal{M}|$ in (2),

$$
\begin{aligned}
&Capa(\mathcal{M}'; \mathcal{M}) \\
&= \log_2 |\mathcal{M}| + (1 - SER) \cdot \log_2(1 - SER) + SER \cdot \log_2 \frac{SER}{|\mathcal{M}| - 1}.
\end{aligned} \tag{6}
$$

We show the relationship between the practical channel capacity of a random method and *Rate* (the number of embedded bits $\log_2 |\mathcal{M}|$) where $n = 10^8$ in Fig.4(c). The practical channel capacity increases proportionally to *Rate* until around *Rate* = 22 bits/carrier, but beyond this value, it decreases drastically.

3.2 Practical channel capacity for Hamming distance method

A Hamming distance method uses, when no appropriate symbol is found, an input value which is transformed into a message symbol whose Hamming distance is the smallest from his/her intended message symbol.

Under the assumption that $|\mathcal{M}| = 2^x$ and x is a natural number, the probability that the Hamming distance between m and m' becomes greater than or equal to i ($0 \le i \le x$) is given by

$$errover(i) = \left(\frac{|\mathcal{M}| - a(i)}{|\mathcal{M}|} \right)^n \tag{7}$$

where

$$a(i) = \sum_{j=0}^{i-1} b(j), \qquad b(j) = \binom{\log_2 |\mathcal{M}|}{j}.$$

The function $b(j)$ calculates the number of symbols whose Hamming distance form the intended symbol m is j. The function $a(i)$ calculates the number of symbols whose Hamming distance is smaller than i, and then $a(0) = 0$ and $a(1) = 1$. The equation (7) is obtained from the probability that one random input value is transformed into a message symbol m' whose Hamming distance is greater than or equal to i is $(|\mathcal{M}| - a(i))/|\mathcal{M}|$.

The probability that the Hamming distance between m and m' is equal to i is given by

$$error(i) = errover(i) - errover(i+1). \tag{8}$$

Thus, $Pr(m'|m)$ is obtained

$$Pr(m'|m) = Pr(m', Ham(m', m) = i|m) = error(i)/b(i) \tag{9}$$

where $Ham(m', m)$ denotes the Hamming distance between m' and m. For $|\mathcal{M}| = 4$, the channel diagram is illustrated like Fig.3. In this figure, for example,

$$Pr(01|00) = Pr(01, Ham(01, 00) = 1|00) = error(1)/b(1) = error(1)/2,$$
$$Pr(11|00) = Pr(11, Ham(11, 00) = 2|00) = error(2)/b(2) = error(2)/1,$$
$$Pr(10|01) = Pr(10, Ham(10, 01) = 2|01) = error(2)/b(2) = error(2)/1.$$

The practical channel capacity $Capa(\mathcal{M}'; \mathcal{M})$ is written in the following form after substituting (9) and $Pr(m) = 1/|\mathcal{M}|$ in (2)

$$Capa(\mathcal{M}'; \mathcal{M}) = \log_2 |\mathcal{M}| + \sum_{i=0}^{\log_2 |\mathcal{M}|} error(i) \cdot \log_2 \frac{error(i)}{b(i)}. \tag{10}$$

Fig.4(b) illustrates the practical channel capacity of a Hamming distance method versus *Rate* for $n = 10^8$. Note that, the practical channel capacity does not decrease even if *Rate* becomes larger than 22 bits/carrier where the practical channel capacity of the random method begins to decrease.

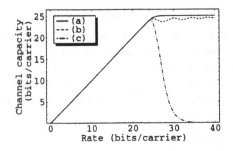

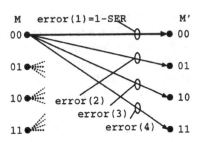

Fig. 4. The practical channel capacity versus *Rate*: (a) maximal partition method, (b) Hamming distance method, (c) random method. (*Rate* = $\log_2 |\mathcal{M}|$, $n = 10^8$.)

Fig. 5. The channel diagram of a maximal partition method for $|\mathcal{M}| = 4$.

3.3 Practical channel capacity for maximal partition method

In the Hamming distance method, the priority of embedding symbols is the same when they have the same Hamming distance from the intended message symbol. For example, when $|\mathcal{M}| = 4$ and the intended symbol is "00", the priority is in the following order $\{00, \{01, 10\}, 11\}$. The priority of "01" and "10" are the same. We call the highest priority as 1 and the second highest priority as 2. That is, the priority of "00" is 1 in the above order, and ones of both "01" and "10" are 2.

In maximal partition methods, each message symbol has a unique priority even if they have the same Hamming distance from the intended message symbol. For example, the priority becomes like this $\{00, 01, 10, 11\}$. The priority of "01" is 2, and the priority of "10" is 3. The symbols "01" and "01" have the different priority.

The probability that m' has the priority greater than or equal to i (where $1 \le i \le |\mathcal{M}|$) is given by

$$errover(i) = \left(\frac{|\mathcal{M}| - (i-1)}{|\mathcal{M}|}\right)^n. \tag{11}$$

This is obtained in the same idea used in 3.2. The probability that the priority of m' is i is given by

$$error(i) = errover(i) - errover(i+1). \tag{12}$$

Thus,

$$Pr(m'|m) = Pr(m', Pri(m', m) = i|m) = error(i) \tag{13}$$

where $Pri(m', m)$ is the priority of m' where m is the intended symbol. For example, suppose that the priority is in the following order $\{00, 01, 10, 11\}$ when "00" is the intended symbol, and then $\{01, 00, 11, 10\}$ when "01" is the intended symbol. In this case, $Pri(01, 00) = 2$, $Pri(01, 01) = 1$, $Pri(10, 01) = 4$, and then

$$Pr(01|00) = Pr(01, Pri(01, 00) = 2|00) = error(2),$$
$$Pr(01|01) = Pr(01, Pri(01, 01) = 1|01) = error(1),$$
$$Pr(10|01) = Pr(10, Pri(10, 01) = 4|01) = error(4).$$

The channel diagram is illustrated like Fig.5. The practical channel capacity $Capa(\mathcal{M}'; \mathcal{M})$ is written in the following form after substituting (13) and $Pr(m) = 1/|\mathcal{M}|$ in (2)

$$Capa(\mathcal{M}'; \mathcal{M}) = \log_2 |\mathcal{M}| + \sum_{i=1}^{|\mathcal{M}|} error(i) \cdot \log_2 error(i). \qquad (14)$$

The relationship between $Capa(\mathcal{M}'; \mathcal{M})$ and $Rate$ is illustrated in Fig.4(a) for $n = 10^8$.

3.4 Achievable channel capacity

We are interested in the relationship between the maximum value of the practical channel capacity (we call this "achievable channel capacity") and the number of tried input values n. For example, in the case of $n = 10^8$, the maximum value of the practical channel capacity is given by the flat place on the curve (a) in Fig.4. Theoretically speaking, such flat place will be found by solving the following equation

$$\frac{\partial Capa(\mathcal{M}'; \mathcal{M})}{\partial |\mathcal{M}|} = 0 \qquad (15)$$

where $Capa(\mathcal{M}'; \mathcal{M})$ is given by (14). However, it is a little bit complicated. Therefore we found such flat places by evaluating (14) on sufficient large $|\mathcal{M}|$ and various n. These evaluated values are plotted as dots in Fig.6. The solid line in Fig.6 denotes their linear fitting. It was given by:

$$\max_{|\mathcal{M}|}(Capa(\mathcal{M}'; \mathcal{M})) \approx -1.44 + \log_2 n. \qquad (16)$$

This yields a necessary condition to embed x bit-information in a carrier. The condition is that at least n_x trials are required to embed x bit-information in a carrier. The number of necessary trials n_x is given by the following equation which can be derived from (16) [4]:

$$n_x = 2^{(x+1.44)} = 2.71 \cdot 2^x. \qquad (17)$$

---

[4] Interestingly, the constant 2.71 in (17) is almost the same as the natural logarithm $e = 2.7183....$

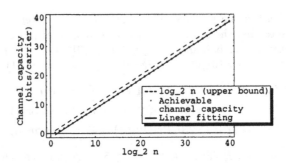

Fig. 6. The relationship between the achievable channel capacity and n.

4 Channel capacities under a practical model

In Section 3, we clarified the relationship between the maximum value of the practical channel capacity (achievable channel capacity) and the number of tried input values n. This result is obtained, provided that only single carrier is transmitted. In this section, we consider a more practical model where a lot of carriers are transmitted one after another successively, and then n' input values are tried until the next carrier is transmitted after current carrier is transmitted.

In this model, the channel capacity depends not only on n' but also the amount of available memory space. Some concrete schemes are proposed, and then their channel capacities are compared.

4.1 Scheme I

In scheme I, a transmitter starts finding an appropriate input value for the next carrier after each current carrier is transmitted. The practical channel capacity is equivalent to the one obtained in Section 3 where n is replaced with n'. Therefore, the achievable channel capacity is about $\log_2 n' - 1.44$ bits/carrier. Our proposals, the following two schemes, can achieve larger capacities than this.

4.2 Scheme II

In scheme II, a transmitter simultaneously finds appropriate input values for the next u carriers between each carrier transmission (see Fig.7(b)). The corresponding algorithm is bellow. This algorithm uses the following u pairs of registers to store u pairs of a covert message symbol and its corresponding input value, $\{(MEMM_0, MEMC_0), \cdots, (MEMM_{u-1}, MEMC_{u-1})\}$.

Step 1: Save u next message symbols m_{t_j} in $MEMM_{(j \bmod u)}$ for $1 \leq j \leq u$, respectively. The symbol m_{t_j} is going to be embedded in the j-th carrier.

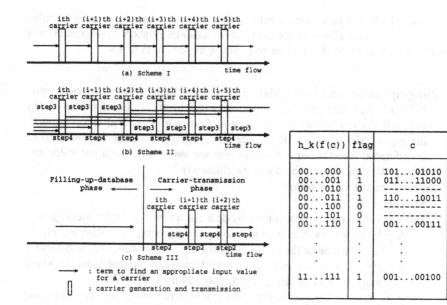

Fig. 7. (a) Scheme I, (b) Scheme II ($u = 4$) and (c) Scheme III.

Fig. 8. An example of a database made in scheme III.

Step 2: $i := 1$

Step 3: Find each appropriate input value c_{t_j} satisfying $MEMM_{(j \bmod u)} = h_k(f(c_{t_j}))$ for $i \le j \le i + u - 1$ by trying n' input values. If found, save it in $MEMC_{(j \bmod u)}$. Otherwise save an input value which has the next highest priority.

Step 4: Generate the i-th carrier using the saved input value in $MEMC_{(i \bmod u)}$, and then transmit it.

Step 5: Clear both $MEMC_{(i \bmod u)}$ and $MEMM_{(i \bmod u)}$, and then save the next message symbol $m_{t_{i+u}}$ in $MEMM_{(i \bmod u)}$.

Step 6: $i := i + 1$

Step 7: Go to Step 3.

A transmitter can try about $n' \cdot u$ input values until he/she transmits a carrier. Therefore, about $(\log_2 n' + \log_2 u) - 1.44$ bit-information can be embedded in each carrier (except the first $u - 1$ carriers).

4.3 Scheme III

In scheme III, a database is made at first, and then it is referred every time a carrier is generated and transmitted (see Fig.7(c)). Suppose that the database can store $|\mathcal{M}|$ sets of an input value c and its corresponding message symbol $m = h_k(f(c))$, and also suppose that covert message symbols appear uniformly due to randomization [8]. The corresponding algorithm is below.

Filling-up-database phase: Make a database of input values c, massage symbols $m = h_k(f(c))$ and flag values like Fig.8 by trying input values one by one. All the flag values are set to 0 at first, and then 1 after corresponding input values are found. If more than one input values are found to be transformed into the same message symbol, leave only one input value and remove the rest to use the database efficiently [5].

Carrier-transmission phase:

Step 1: $i := 1$

Step 2: Whenever a transmitter sends a carrier with a covert message m_{t_i}, he/she generates it using the corresponding entry c_{t_i} satisfying $m_{t_i} = h_k(f(c_{t_i}))$ as long as the flag value of m_{t_i} is 1. Otherwise, he/she generates a carrier having the next prior covert message (among symbols whose flag values are 1) by using the corresponding input value.

Step 3: Remove the used input value from the database, and then make the flag value 0 to prevent that the same input values are used repetitively. Note that when the same input values are used repetitively, indistinguishability becomes weaker. Moreover, in some signature schemes, e.g. (EC)DSA signature schemes, to use the same random numbers more than once is to reveal the signer's secret-key to the receivers.

Step 4: Try n' new input values, and then make the occupation rate of the database larger.

Step 5: $i := i + 1$, Go to Step 2.

The occupation rate v is defined as follows:

$$v = \frac{\text{The number of saved entries of } c \text{ in the database}}{|\mathcal{M}|} \tag{18}$$

where $|\mathcal{M}|$ denotes the number of members of the massage symbol set, and also the storage capacity of the database. The occupation rate v increases after n' new input values are tried between each carrier transmission at step 4, and then decreases every after one used entry is removed from the database at step 3. The occupation rate v is balanced at a certain value v_b where the increments of v is equal to the decrements of v. (Roughly speaking, after $|\mathcal{M}| \cdot \ln \frac{1}{1-v_b}$ input values are tried at the filling-up-database phase, the occupation rate of the database whose storage capacity is $|\mathcal{M}|$ is balanced.) Two values exists as the balanced occupation rate. One is the value after n' input values are tried. The other is one before n' input values are tried (or after each carrier is transmitted). We take

---

[5] If the storage capacity is large enough, it is possible to store some of such input values.

the former as v_b. Then, let v'_b denote the latter one. Both are in the following relationship

$$v_b = v'_b + \frac{1}{|\mathcal{M}|}. \tag{19}$$

This is obtained because only one entry is removed from the database whose size is $|\mathcal{M}|$ after one carrier is transmitted.

The occupation rate is balanced when one entry is found on average by trying n' input values. And then one entry is found by trying $1/(1-v)$ input values on average when the occupation rate is v. This simultaneously means one entry is found by trying n' input values on average when the occupation rate is $v'_b = 1 - 1/n'$. From this and (19), v_b is obtained:

$$v_b = 1 - \frac{1}{n'} + \frac{1}{|\mathcal{M}|}. \tag{20}$$

The symbol error rate SER of this scheme III is given by

$$SER = 1 - v_b = \frac{1}{n'} - \frac{1}{|\mathcal{M}|}. \tag{21}$$

By substituting (21) in (6), the practical channel capacity of this scheme III where the random method is used is obtained. The practical channel capacity of this scheme III (where the maximal partition method is used) is lower-bounded by this value (because the maximal partition methods achieve larger channel capacity that the random methods). Thus,

$$
\begin{aligned}
Capa(\mathcal{M}; \mathcal{M}) &\geq \log_2 |\mathcal{M}| - SER \cdot \log_2(|\mathcal{M}| - 1) - b \\
&> \log_2 |\mathcal{M}| - SER \cdot \log_2 |\mathcal{M}| - b \\
&> a \cdot \log_2 |\mathcal{M}| - b
\end{aligned}
\tag{22}
$$

where $a = 1 - SER$ and $b = (1 - SER) \cdot \log_2(1 - SER) + SER \cdot \log_2 SER$. When $|\mathcal{M}|$ is large enough, the coefficients a and b becomes independent of $|\mathcal{M}|$. Therefore $Capa(\mathcal{M}; \mathcal{M})$ can be proportional to $\log_2 |\mathcal{M}|$.

This result tells us that even if n' is small, the practical channel capacity increases as available memory space increases. Table 1 shows the necessary memory space to make a database with a storage capacity $|\mathcal{M}|$. For example, by using about 20MB memory space, 20 bit-symbols can be embedded in each carrier.

5 Comparison

The practical channel capacities of each scheme are compared in Table 2 . If many message symbols being going to be embedded in the coming carriers are known in advance, the scheme II makes the practical channel capacity the largest. Otherwise the score depends on which can be larger in comparison between n' and available memory space.

Table 1. Memory space required to store $|\mathcal{M}|$ pairs of input values and output values.

$|\mathcal{M}|$ (pairs)	2^{10}	2^{15}	2^{20}	2^{25}	2^{30}	2^{35}
Memory Space MS (Bytes) *1	20.6K	659K	21.1M	675M	21.6G	691G

*1 : Suppose that the size of an input value of a one-way function is 160 bits, and that only input values are stored at the address derived uniquely from the output values. That is, $MS = |\mathcal{M}|(160 + 1)/8$.

Table 2. The practical channel capacity of the considered three schemes.

scheme	practical channel capacity	note
I	$\propto \log_2 n'$	
II	$\propto (\log_2 n' + \log_2 u)$	*1
III	$\propto \log_2 |\mathcal{M}|$	*2

*1 : Message symbols being embedded in the next u carriers have to be known in advance, and then memory space to store u pairs of data is required.

*2 : A database which can store $|\mathcal{M}|$ pairs of data is required.

6 Conclusion

We evaluated how much information can be embedded in a carrier by trying n input values on a narrow-band subliminal channel. The result is $\log_2 n - 1.44$ bits/carrier. This yields that in order to embed x bit-information in a carrier, at least $2^{(x+1.44)} = 2.71 \cdot 2^x$ input values must be tried. Then, we have applied this result to a practical model where the carriers are transmitted one after another successively, and only n' input values are tried between each carrier transmission. Under this model, the practical channel capacity depends not only on the number of trials, but also on the available memory space. We have shown two schemes (scheme II and scheme III) to use memory efficiently. The scheme II is suitable in the case when many message symbols to be embedded in the coming carriers are known in advance. The scheme III can be used even if message symbols are not known in advance and n' is small.

In addition, we proposed how to reduce the computational complexity of trying n input values. This reduces the computational complexity of making a narrow-band subliminal channel.

References

1. "Proposed federal information processing standard for digital signature standard (DSS)". *Federal Register*, 56(169):42980–429892, 1991.
2. "Public key cryptography for the financial services industry: The elliptic curve digital signature algorithm (ECDSA)". *American National Standard X9.62-1998*, 1998.
3. R. Anderson, S. Vaudenay, B. Preneel, and K. Nyberg. "The Newton channel". *1st Int. Workshop on Information Hiding*, pages 151–156, 1996.
4. D. Boneh, G. Durfee, and Y. Frankel. "An attack on RSA given a small fraction of the private key bits". In *Proc. of ASIACRYPT '98, LNCS 1514*, pages 25–34. Springer–Verlag, 1998.
5. M. Burmester, Y. Desmedt, T. Itoh, K. Sakurai, H. Shizuya, and M. Yung. "A progress report on subliminal-free channels". In *Proc. of the Workshop on Information Hiding, LNCS 1174*, pages 157–168. Springer–Verlag, 1996.
6. Y. Desmedt. "Simmons' protocol is not free of subliminal channels". In *Proc. of 9th IEEE Computer Security Foundations Workshop*, pages 170–175, 1996.
7. T. ElGamal. "A public-key cryptosystem and a signature scheme bsed on discrete logarithms". In *Proc. of CRYPTO '84*, pages 10–18, 1985.
8. K. Kobara and H. Imai. "Self-synchronized message randomization methods for subliminal channels". In *Proc. of International Conference on Information and Communications Security (ICICS'97) : LNCS 1334*, pages 325–334. Springer–Verlag, 1997.
9. K. Kobara and H. Imai. "The capacity of a channel with a one-way function". In *Proc. of Japan–Korea Joint Workshop on Information Security and Cryptology (JW-ISC) '97*, pages 173–179, 1997.
10. G. J. Simmons. "The prisoners' problem and the subliminal channel". *Proc. of CRYPTO '83*, pages 51–67, 1984.
11. G. J. Simmons. "The subliminal channel and digital signatures". *Proc. of EUROCRYPT '84, LNCS 209*, pages 364–378, 1985.
12. G. J. Simmons. "A secure subliminal channel (?)". *Proc. of CRYPTO '85*, pages 33–41, 1986.
13. G. J. Simmons. "Subliminal Channels : Past and Present". *European Trans. on Telecommunications*, 4(4):459–473, Jul/Aug 1994.
14. G. J. Simmons. "Subliminal communication is easy using the DSA". In *Proc. of EUROCRYPT '93, LNCS 765*, pages 218–232. Springer–Verlag, 1994.
15. G. J. Simmons. "Results concerning the bandwidth of subliminal channels". *IEEE Journal on Selected Areas in Communication*, 16(4):463–473, May 1998.
16. G. J. Simmons. "The History of Subliminal Channels". *IEEE Journal on Selected Areas in Communication*, 16(4):452–462, May 1998.
17. A. Young and M. Yung. "The dark side of black-box cryptography or: Should we trust capstone?". In *Proc. of CRYPTO '96, LNCS 1109*, pages 89–103. Springer–Verlag, 1996.
18. A. Young and M. Yung. "Kleptography : Using cryptography against cryptography". In *Proc. of EUROCRYPT '97, LNCS 1233*, pages 62–74. Springer–Verlag, 1997.
19. A. Young and M. Yung. "The prevalence of kleptographic attacks on discrete-log based cryptosystems". In *Proc. of CRYPTO '97, LNCS 1294*, pages 264–276. Springer–Verlag, 1997.

AUTHOR INDEX

Lecture Notes in Computer Science

For information about Vols. 1–1652
please contact your bookseller or Springer-Verlag

Vol. 1689: F. Solina, A. Leonardis (Eds.), Computer Analysis of Images and Patterns. Proceedings, 1999. XIV, 650 pages. 1999.

Vol. 1690: Y. Bertot, G. Dowek, A. Hirschowitz, C. Paulin, L. Théry (Eds.), Theorem Proving in Higher Order Logics. Proceedings, 1999. VIII, 359 pages. 1999.

Vol. 1691: J. Eder, I. Rozman, T. Welzer (Eds.), Advances in Databases and Information Systems. Proceedings, 1999. XIII, 383 pages. 1999.

Vol. 1692: V. Matoušek, P. Mautner, J. Ocelíková, P. Sojka (Eds.), Text, Speech and Dialogue. Proceedings, 1999. XI, 396 pages. 1999. (Subseries LNAI).

Vol. 1693: P. Jayanti (Ed.), Distributed Computing. Proceedings, 1999. X, 357 pages. 1999.

Vol. 1694: A. Cortesi, G. Filé (Eds.), Static Analysis. Proceedings, 1999. VIII, 357 pages. 1999.

Vol. 1695: P. Barahona, J.J. Alferes (Eds.), Progress in Artificial Intelligence. Proceedings, 1999. XI, 385 pages. 1999. (Subseries LNAI).

Vol. 1696: S. Abiteboul, A.-M. Vercoustre (Eds.), Research and Advanced Technology for Digital Libraries. Proceedings, 1999. XII, 497 pages. 1999.

Vol. 1697: J. Dongarra, E. Luque, T. Margalef (Eds.), Recent Advances in Parallel Virtual Machine and Message Passing Interface. Proceedings, 1999. XVII, 551 pages. 1999.

Vol. 1698: M. Felici, K. Kanoun, A. Pasquini (Eds.), Computer Safety, Reliability and Security. Proceedings, 1999. XVIII, 482 pages. 1999.

Vol. 1699: S. Albayrak (Ed.), Intelligent Agents for Telecommunication Applications. Proceedings, 1999. IX, 191 pages. 1999. (Subseries LNAI).

Vol. 1700: R. Stadler, B. Stiller (Eds.), Active Technologies for Network and Service Management. Proceedings, 1999. XII, 299 pages. 1999.

Vol. 1701: W. Burgard, T. Christaller, A.B. Cremers (Eds.), KI-99: Advances in Artificial Intelligence. Proceedings, 1999. XI, 311 pages. 1999. (Subseries LNAI).

Vol. 1702: G. Nadathur (Ed.), Principles and Practice of Declarative Programming. Proceedings, 1999. X, 434 pages. 1999.

Vol. 1703: L. Pierre, T. Kropf (Eds.), Correct Hardware Design and Verification Methods. Proceedings, 1999. XI, 366 pages. 1999.

Vol. 1704: Jan M. Żytkow, J. Rauch (Eds.), Principles of Data Mining and Knowledge Discovery. Proceedings, 1999. XIV, 593 pages. 1999. (Subseries LNAI).

Vol. 1705: H. Ganzinger, D. McAllester, A. Voronkov (Eds.), Logic for Programming and Automated Reasoning. Proceedings, 1999. XII, 397 pages. 1999. (Subseries LNAI).

Vol. 1706: J. Hatcliff, T. Æ. Mogensen, P. Thiemann (Eds.), Lectures on Partial Evaluation. Proceedings, 1998. IX, 433 pages. 1999. (Subseries LNAI).

Vol. 1707: H.-W. Gellersen (Ed.), Handheld and Ubiquitous Computing. Proceedings, 1999. XII, 390 pages. 1999.

Vol. 1708: J.M. Wing, J. Woodcock, J. Davies (Eds.), FM'99 - Formal Methods. Proceedings Vol. I, 1999. XVIII, 937 pages. 1999.

Vol. 1709: J.M. Wing, J. Woodcock, J. Davies (Eds.), FM'99 - Formal Methods. Proceedings Vol. II, 1999. XVIII, 937 pages. 1999.

Vol. 1710: E.-R. Olderog, B. Steffen (Eds.), Correct System Design. XIV, 417 pages. 1999.

Vol. 1711: N. Zhong, A. Skowron, S. Ohsuga (Eds.), New Directions in Rough Sets, Data Mining, and Granular-Soft Computing. Proceedings, 1999. XIV, 558 pages. 1999. (Subseries LNAI).

Vol. 1712: H. Boley, A Tight, Practical Integration of Relations and Functions. XI, 169 pages. 1999. (Subseries LNAI).

Vol. 1713: J. Jaffar (Ed.), Principles and Practice of Constraint Programming - CP'99. Proceedings, 1999. XII, 493 pages. 1999.

Vol. 1714: M.T. Pazienza (Eds.), Information Extraction. IX, 165 pages. 1999. (Subseries LNAI).

Vol. 1715: P. Perner, M. Petrou (Eds.), Machine Learning and Data Mining in Pattern Recognition. Proceedings, 1999. VIII, 217 pages. 1999. (Subseries LNAI).

Vol. 1716: K.Y. Lam, E. Okamoto, C. Xing (Eds.), Advances in Cryptology - ASIACRYPT'99. Proceedings, 1999. XI, 414 pages. 1999.

Vol. 1717: Ç. K. Koç, C. Paar (Eds.), Cryptographic Hardware and Embedded Systems. Proceedings, 1999. XI, 353 pages. 1999.

Vol. 1718: M. Diaz, P. Owezarski, P. Sénac (Eds.), Interactive Distributed Multimedia Systems and Telecommunication Services. Proceedings, 1999. XI, 386 pages. 1999.

Vol. 1719: M. Fossorier, H. Imai, S. Lin, A. Poli (Eds.), Applied Algebra, Algebraic Algorithms and Error-Correcting Codes. Proceedings. 1999. XIII, 510 pages. 1999.

Vol. 1721: S. Arikawa, K. Furukawa (Eds.), Discovery Science. Proceedings, 1999. XI, 374 pages. 1999. (Subseries LNAI).

Vol. 1722: A. Middeldorp, T. Sato (Eds.), Functional and Logic Programming. Proceedings, 1999. X, 369 pages. 1999.

Vol. 1723: R. France, B. Rumpe (Eds.), UML'99 - The Unified Modeling Language99. XVII, 724 pages. 1999.

Vol. 1726: V. Varadharajan, Y. Mu (Eds.), Information and Communication Security. Proceedings, 1999. XI, 325 pages. 1999.

Vol. 1727: P.P. Chen, D.W. Embley, J. Kouloumdjian, S.W. Liddle, J.F. Roddick (Eds.), Advances in Conceptual Modeling. Proceedings, 1999. XI, 389 pages. 1999.

Vol. 1728: J. Akoka, M. Bouzeghoub, I. Comyn-Wattiau, E. Métais (Eds.), Conceptual Modeling - ER '99. Proceedings, 1999. XIV, 540 pages. 1999.

Vol. 1729: M. Mambo, Y. Zheng (Eds.), Information Security. Proceedings, 1999. IX, 277 pages. 1999.

Vol. 1734: H. Hellwagner, A. Reinefeld (Eds.), SCI: Scalable Coherent Interface. XXI, 490 pages. 1999.

Vol. 1564: M. Vazirgiannis, Interactive Multimedia Documents. XIII, 161 pages. 1999.

Vol. 1591: D.J. Duke, I. Herman, S. Marshall, PREMO: A Framework for Multimedia Middleware. XII, 254 pages. 1999.